# Lecture Notes in Economics and Mathematical Systems



Managing Editors: M. Beckmann and W. Krelle



274

Willem K. Klein Haneveld

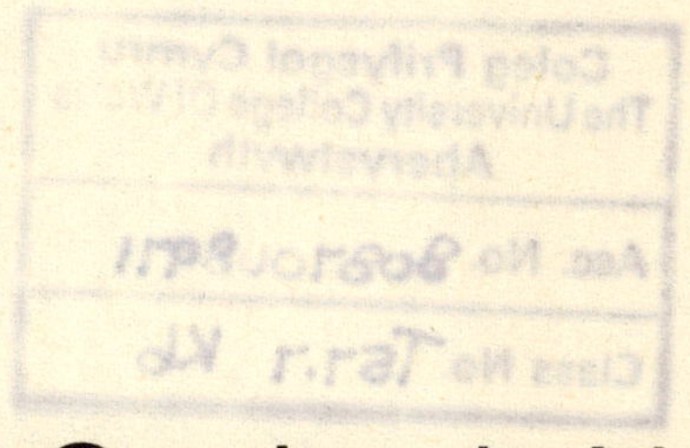

# Duality in Stochastic Linear and Dynamic Programming



Springer-Verlag
Berlin Heidelberg New York London Paris Tokyo

**Managing Editors**

Prof. Dr. M. Beckmann
Brown University
Providence, RI 02912, USA

Prof. Dr. W. Krelle
Institut für Gesellschafts- und Wirtschaftswissenschaften
der Universität Bonn
Adenauerallee 24-42, D-5300 Bonn, FRG

**Author**

Dr. Willem K. Klein Haneveld
Institute of Econometrics, University of Groningen
P.O. Box 800, 9700 AV Groningen, The Netherlands

Sept '87.

ISBN 3-540-16793-5 Springer-Verlag Berlin Heidelberg New York
ISBN 0-387-16793-5 Springer-Verlag New York Berlin Heidelberg

Library of Congress Cataloging-in-Publication Data. Klein Haneveld, Willem K., 1944- Duality in stochastic linear and dynamic programming. (Lecture notes in economics and mathematical systems; 274) Includes bibliographies and index. 1. Stochastic programming. 2. Linear programming. 3. Dynamic programming. 4. Duality theory (Mathematics) I. Title. II. Series.
T57.79.K48 1986 519.7'2 86-20365
ISBN 0-387-16793-5 (U.S.)



Printing and binding: Druckhaus Beltz, Hemsbach/Bergstr.
2142/3140-543210

# ACKNOWLEDGEMENTS

This book has grown out of a number of conference papers and articles. It is a pleasure to record my gratitude to all the colleagues whose comments improved my insight into the subject. In writing the book I benefited greatly from their constructive criticism.

Special thanks are due to Aaf for her continuing moral support. I am indebted to Cor van de Panne, by whose enthousiastic introduction I got fascinated by Operations Research. I want to express my particular gratitude to Roger Wets for his stimulating interest in my work during the last ten years. Thanks to him I could deepen my understanding of stochastic programming during a sojourn at the University of Kentucky at Lexington in 1977/1978. The financial support of the Netherlands Organisation for the Advancement of Pure Research (ZWO), that made this visit possible, is recorded gratefully. Thanks are due to Jaap Ponstein and Jaap Wessels for their careful reading of the manuscript. I wish to thank Tamara Brünner who did an excellent job in typing the manuscript. I thank Evert Jan Bakker and Jan Blaakmeer for their help in running the computer programs and in drawing some of the figures. Finally, I thank Springer-Verlag for making this publication possible.

W.K. Klein Haneveld

# CONTENTS


| | | |
|---|---|---|
| CHAPTER 1 | INTRODUCTION AND SUMMARY | 1 |
| | 1.1 Introduction | 1 |
| | 1.2 Summary of the following chapters | 3 |
| | 1.3 Advices to the reader | 6 |
| CHAPTER 2 | MATHEMATICAL PROGRAMMING AND DUALITY THEORY | 9 |
| | 2.1 Introduction | 9 |
| | 2.2 Mathematical programming problems | 9 |
| | 2.3 Conjugate duality | 11 |
| | References | 19 |
| CHAPTER 3 | STOCHASTIC LINEAR PROGRAMMING MODELS | 21 |
| | 3.1 Introduction | 21 |
| | 3.2 Chance constraints, recourse actions and utility functions | 24 |
| | 3.3 Multistage stochastic linear programming models and their relation to stochastic dynamic programming | 39 |
| | References | 44 |
| CHAPTER 4 | SOME LINEAR PROGRAMS IN PROBABILITIES AND THEIR DUALS | 49 |
| | 4.1 Introduction | 49 |
| | 4.2 Dual linear programs | 51 |
| | 4.3 Linear programming in probabilities | 58 |
| | 4.4 Moment problems | 66 |
| | 4.5 The marginal problem | 80 |
| | 4.6 The stochastic programming problem | 86 |
| | 4.7 Conclusions | 107 |
| | References | 108 |

CHAPTER 5 ON INTEGRATED CHANCE CONSTRAINTS 113
5.1 Introduction 113
5.2 Single integrated chance constraints 115
5.3 Joint integrated chance constraints 134
5.4 Integrated chance constraints and recourse models 135
5.5 Conclusions 137
References 138

CHAPTER 6 ON THE BEHAVIOUR OF THE OPTIMAL VALUE OPERATOR OF DYNAMIC PROGRAMMING 139
6.1 Introduction 139
6.2 The main theorem 143
6.3 Proof of the main theorem 148
6.4 Markovian inventory control models 153
References 163

CHAPTER 7 ROBUSTNESS AGAINST DEPENDENCE IN PERT 165
7.1 Introduction 165
7.2 The marginal problem and its dual 167
7.3 Critical activities in a project network 175
7.4 A minimax approach to project planning 181
7.4.1 Introduction 181
7.4.2 The inner problem, its dual and its reduced dual 185
7.4.3 The dual of the reduced dual problem 194
7.4.4 The optimal solutions of the inner problem 201
7.4.5 The outer problem 208
7.4.6 Worst-case marginals 211
7.4.7 Numerical calculations 212
References 215

CHAPTER 8 A DUAL OF A DYNAMIC INVENTORY CONTROL MODEL: THE DETERMINISTIC AND THE STOCHASTIC CASE 219
8.1 Introduction 219
8.2 Deterministic demands 221
8.2.1 Formulation of the problem 222
8.2.2 The dynamic programming recursion 223
8.2.3 The differentiable form of the dynamic programming recursion 225

8.2.4 The linear program and its dual 226
8.2.5 A dual dynamic programming method 234
8.2.6 A conjugate dual dynamic programming method 237
8.3 Stochastic demands 243
8.3.1 Formulation of the problem 244
8.3.2 Some notations and remarks on conditional expectations 246
8.3.3 The dynamic programming recursion 251
8.3.4 The linear program and its dual 259
8.3.5 The differentiable form of the dynamic programming recursion 266
8.3.6 A dual dynamic programming method 280
8.3.7 A conjugate dual dynamic programming method 282
8.4 Economic interpretations 284
References 289

LIST OF OPTIMIZATION PROBLEMS 291

SUBJECT INDEX 293

# CHAPTER 1

# INTRODUCTION AND SUMMARY

## 1.1 INTRODUCTION

In this book we consider some problems arising in *stochastic linear programming* (SLP) and *stochastic dynamic programming* (SDP). Both branches of mathematical operations research deal with optimization models with random parameters. The probability distribution of these parameters represents the uncertainty of the model builder with respect to their precise values. The backgrounds of SLP models and SDP models are quite distinct. On the one hand, SLP models arise if some of the parameters $(c_j, a_{ij}, b_i)$ in a linear programming problem, e.g.

$$\text{(LP)} \qquad \text{minimize}_{x \in \mathbb{R}^n} \left\{ \Sigma_{j=1}^n c_j x_j : \begin{array}{l} \Sigma_{j=1}^n a_{ij} x_j \geqq b_i, \; i = 1,\ldots,m \\ x_j \geqq 0 \;, \; j = 1,\ldots,n \end{array} \right\},$$

are random variables. If the decision vector $x = (x_1,\ldots,x_n)$ has to be

chosen without the knowledge of the realizations of the random parameters, the problem LP is, essentially, undefined. Therefore, additional specifications have to be made in order to get a well-defined SLP problem. Such specifications can be given in various ways. Which way is appropriate depends, amongst others, on the interpretation and modelling of "risk" and "risk aversion". On the other hand, an SDP model is a description of a controllable dynamic system with costs that is well-defined right from the start. The state of the system, as a function of time, is considered to be a random process with a partly controllable probability distribution. Usually it is assumed that the decision maker observes the current state and that he wants to control the system in such a way that the expected cost is minimum. In this book we only consider discrete-time SDP models with a finite number of stages. In spite of the differences there are close relations between certain SLP models and certain SDP models. In particular, we think of *multistage* SLP models; the characteristic assumption of these models is that the realizations of the random parameters in problem LP come in stages, and that the decision vector x is partitioned according to the stages as well. Multistage SLP models can be formulated as discrete-time SDP models with a finite number of stages.

A third category of problems is formed by *linear programming* problems whose "decision variables" are *probability measures*; such problems are denoted by LPM. Various problems in probability theory can be formulated as LPM problems. A well-known example is the moment problem: find a probability distribution on a given support with prescribed values for certain moments such that the probability of a fixed event is maximum. At first sight it may seem that LPM problems have nothing to do with SLP or SDP problems. Yet, certain LPM models are applicable to SLP and SDP. First of all, it appears that all SDP problems can be dealt with as special LPM problems. Secondly, if for an SLP problem only partial information on the distribution of the random parameters is available, one may be interested in finding a "worst-case" distribution compatible with the available information. In suitable circumstances such a distribution can be found as the solution of an LPM problem. We will come back to both points in detail.

An important approach to the analysis of optimization problems is given by the *duality theory* of mathematical programming. In this theory one defines a so-called dual optimization problem, and the relation between the original problem and the dual problem provides insight in the properties of the optimal solutions for both problems. In this book the duality theory of convex (often even linear) programming is applied, not only to problems in Euclidean spaces, however, but also to problems in "abstract" linear spaces with measures or functions as elements. In particular, we apply sufficient conditions for stability; in that case the original problem and the dual problem have the same optimal value, and optimal dual solutions exist. For stable convex programming problems the well-known Karush-Kuhn-Tucker conditions are necessary and sufficient for optimality. In the linear case these conditions reduce to feasibility and complementary slackness.

In the next section we give a brief overview of the contents of the chapters to come. The final section of the present chapter gives some further information on the interdependence of the various chapters.

## 1.2 SUMMARY OF THE FOLLOWING CHAPTERS

The remainder of this book consists of two parts. The Chapters 2 and 3 form the first part; they are of an introductory nature. In Chapter 2 the terminology of mathematical programming is introduced, and a concise treatment of the duality theory of convex programming problems in abstract spaces is given. Only those elements of the theory are reviewed which are relevant for our study. In particular, a sufficient condition for stability is formulated, and optimal solutions are characterized in the stable case. In Chapter 3 we discuss some SLP models. First we deal with a number of specifications of linear programming models with random constraints and/or a random objective function, including typical concepts such as chance constraints, recourse actions and utility functions. Conceptual issues are emphasized, and comments are made with regard to the relations between the various kinds of models. In the second part of this chapter multistage SLP problems are introduced, and some attention is given to the relation between these models and those of the SDP type. Chapters 2 and 3 do not contain new results; however, some elements of Chapter 3 usually get too little attention in the literature on SLP.

The second part consists of the Chapters 4-8. Those chapters can be read more or less independently of each other. In Chapter 4 a general framework for abstract linear programming duality is given, and it is shown that LPM problems fit nicely in this setting. Special attention is paid to three special LPM problems:

(i) a generalized moment problem, and variants of it,

(ii) a problem with given marginals,

(iii) an SDP problem with a finite number of stages.

In all cases we get as the dual problem an approximation problem in a function space. The complementary slackness conditions, which in LPM problems boil down to conditions on the supports of the probability measures, represent well-known optimality conditions for each of the three problems. For example, the dual problem of SDP can be interpreted as a linear programming formulation of the famous dynamic programming algorithm, in which a sequence of so-called cost-to-go functions is calculated by a backward recursion. It appears, that in all three cases the general duality theory is applicable quite well, but the conditions are stronger than strictly necessary. That is, by exploiting the specific properties of each of the problems it is possible to get similar results under weaker conditions.

In Chapter 5 a new SLP concept is introduced under the name "integrated chance constraint" (ICC). Just as the usual chance constraints ICCs restrict the risk of infeasibility. The distinction is, however, that ICCs also take into account the amounts by which the original constraints are violated. In this respect models with ICCs agree with recourse models. On the other hand, in recourse models costs of infeasibility are introduced, and that is not necessarrily so when ICCs are used. It appears that ICCs not only have attractive conceptual properties but also that they possess mathematical advantages compared to the usual chance constraints. Moreover, by using a duality theorem it is shown, that there is a natural equivalence between certain models with ICCs and certain recourse models.

As mentioned before, it is well-known that SDP problems with a finite number of stages can be solved by means of a backward recursion scheme. However, in the general case it is not trivial that this recursion is well-defined and that optimal policies are characterized by it: complications may arise which have to do with measurability of functions and existence of integrals. After having analyzed the relation between the SDP problem and its recursive algorithm in Chapter 4 we formulate in Chapter 6 conditions not only guaranteeing that the recursion is well-defined but also that optimal policies exist. The proof is based on a measurable selection theorem, and uses the theory of normal integrands. Euclidean state and action spaces are considered, and within this restriction the conditions are weaker than other conditions in the literature. Moreover, it is shown that the conditions are useful for a rather general class of production-inventory control models.

One of most famous SLP problems is concerned with project planning. A project is described as a set of activities together with a set of precedence relations which can be represented by a network. Each activity needs time, and together with the precedence relations these times determine the minimal completion time of the project. Chapter 7 deals with the problem that an agreement has to be made with regard to the completion time of the project, whereas the durations of the activities are random variables. Particular attention is paid to the complication, that in general only little information is available on the joint distribution of these durations. The formulation of the SLP model is of the minimax type. First of all it is assumed that the marginal distributions are known completely, and by applying the duality theory on LPM problems the worst-case joint distributions are characterized completely. It appears that the dual problem can be reduced to a finite-dimensional convex minimization problem, which problem can be dualized again resulting in a flow problem in the project network. Secondly it is shown, that the same analysis can also be applied if only little information with respect to the marginal distributions is available, for instance only the data which are used by the well-known heuristic called PERT. Finally, it is indicated that the minimax formulation has practical value, since computations are relatively simple.

A simple production-inventory model (a single commodity, a finite number of periods, linear production costs, linear holding costs and linear shortage costs) is discussed in Chapter 8. It is assumed that backlogging of demand is possible. Two cases are considered. First it is assumed that all demands are deterministic, and secondly the demands are specified as (possibly mutually dependent) random variables with a known joint probability distribution. In the stochastic case it is assumed that at each moment the previous realized demands are known, whereas the decision on the production has to be taken without knowledge of the realizations of future demands. Usually, production-inventory control models are interpreted as SDP problems, and solved by the backward recursion. In this last chapter we also analyze the linear programming approach. For deterministic demands one gets a simple linear program in a Euclidean space, and for stochastic demands one gets a multistage SLP problem with nonanticipative functions of the realized demands as decision variables. In the linear programming setting one easily derives the dual problem. In the deterministic case the optimal values of the dual variables can be interpreted as the lengths of the shortest (i.e. most inexpensive) paths in the underlying network. In the stochastic case they constitute a "price system" that is related to the condition that all productions must be nonanticipative. Just as the original problem also the dual problem can be solved by means of a backward recursion. In the deterministic as well as in the stochastic case *two* such "dual dynamic programming recursions" are formulated. The dual recursions are closely related to the usual "primal" SDP recursion. The first dual recursion can be identified as a recursion for the derivatives of the primal cost-to-go functions, whereas in the second dual recursion the conjugates of these primal cost-to-go functions occur. By solving the first dual recursion we find explicit expressions for the optimal solution of the dual problem; in the stochastic case this solution is described in terms of probabilities of certain events.

## 1.3 ADVICES TO THE READER

From the foregoing it is clear that the main key words are "duality", "stochastic programming" and "dynamic programming". For readers who have a special interest in one of these subjects in particular we give the following overview.

The main chapters on *duality theory* and its applications are Chapters 2, 4, 7 and 8. Chapter 2 can be skipped by those who have a basic knowledge of the modern approach to duality. Chapter 4 provides a general setting for "abstract" linear programming duality, which is used in Sections 4.5-4.7 of the same chapter, and also in Chapter 7 and 8. Moreover, applications of classical duality of convex (or even linear) programming in Euclidean spaces are contained in Chapters 3, 5 and 7.

The contributions to *stochastic* linear *programming* are contained in Chapters 3, 5, 7 and 8. People who are interested in conceptual issues of stochastic programming are adviced to read Chapters 3 and 5 anyhow. Stochastic programming approaches to particular problems as project planning and inventory control are given in Chapter 7 and 8, respectively.

To readers who have a primary interest in stochastic *dynamic programming* we suggest to take a quick view at Section 3.3, and to go on with Chapter 4 (in particular Section 4.6). For these readers Chapters 6 and 8 will be of interest, too.

In Figure 1.1 we indicate the interdependende between the various chapters. Its arrows describe relations between the chapters. Their significance should not be exaggerated; sometimes they indicate that a formulation presented in a previous chapter is worked out from a different point of view in a later chapter, or that a particular theorem is applied. We repeat that on the whole the chapters can be read independently of each other. For example, people who are interested in project planning might start right away with reading Chapter 7, and only go back to previous chapters when references are made of them.

We conclude this section with a few technical remarks. Within each chapter definitions, lemmas, propositions, theorems and remarks are numbered consecutively. The figures get a separate numbering. The references to the literature are given at the end of each chapter separately. Usually the optimization problems considered are not numbered. To make the tracing easier a list of them is included after Chapter 8. There one finds also a subject index.

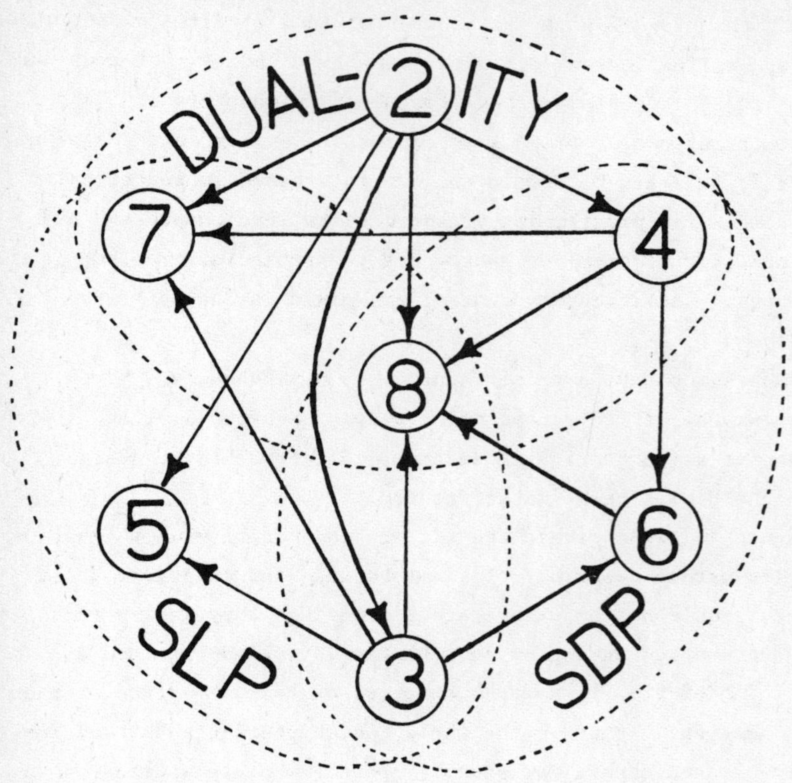


FIGURE 1.1. The interdependence of the chapters. Nodes represent chapters, and arc are from node i to node j indicates that results in Chapter i are relevant for results in Chapter j.

# CHAPTER 2

# MATHEMATICAL PROGRAMMING AND DUALITY THEORY

## 2.1 INTRODUCTION

In this chapter the terminology of mathematical programming and in particular that of the duality theory for convex programming problems is introduced. A few well-known basic duality theorems which will be applied in Chapters 3, 4, 5, 7 and 8 are reviewed. No new results are reported.

## 2.2 MATHEMATICAL PROGRAMMING PROBLEMS

Traditionally, a *mathematical programming* problem is an optimization problem of the following kind:

$$(P_o) \qquad \text{minimize}_{x \in C}\{f_0(x) : f_i(x) \leqq 0,\ i = 1,\ldots,m\}.$$

Here it is understood that C is a given nonempty subset of $\mathbb{R}^n$, that

$f_0 : C \to \mathbb{R}$ is a given *objective function*, and that $f_i : C \to \mathbb{R}$, $i = 1,\dots,m$, are given *constraint functions*. Trivial transformations show, that also constraints of the type $f_i(x) \geqq 0$ or $f_i(x) = 0$ fit in the framework of $P_o$, and that maximization problems are covered as well. The vector $x \in \mathbb{R}^n$ of *decision variables* is called a *feasible* solution of the mathematical programming problem $P_o$ if it satisfies all the *constraints* $x \in C$, $f_i(x) \leqq 0$, $i = 1,\dots,m$; it is called *optimal* if it is feasible and if $f_0(x) = \inf P_o$ where

$$\inf P_o := \inf_{x \in C} \{f_0(x) : f_i(x) \leqq 0,\ i = 1,\dots,m\}$$

is the so-called *optimal value* of $P_o$. Problem $P_o$ is called feasible if it has feasible solutions; if not we set $\inf P_o := +\infty$. It is called *solvable* if it has optimal solutions; in that case one usually denotes the optimal value of $P_o$ by $\min P_o$. If $\inf P_o = -\infty$ we say, by a slight abuse of language, that $P_o$ has *unbounded* solutions. $P_o$ is called a *convex programming* problem if C is a convex set and if each $f_i$, $i = 0,1,\dots,m$, is a convex function. It is called a *linear programming problem*, or a *linear program* if C is a polyhedral set, e.g. $C = \{x \in \mathbb{R}^n : x \geqq 0\}$, if $f_0$ is linear and if each $f_i$, $i = 1,\dots,m$, is affine. It is clear that convexity (and, a fortiori, linearity) of a mathematical programming problem makes the characterization and the calculation of its optimal value and its optimal solutions much easier.

In this book we will consider many convex (often even linear) programming problems which are more "abstract" than problem $P_o$ above, in the sense that C is a subset of an "arbitrary" vector space over the reals instead of $\mathbb{R}^n$; moreover, the restriction that there is only a finite number of constraints in $\mathbb{R}$ will be discarded. In the next section we shall review some results on such *abstract convex programming* problems, established by R.T. Rockafellar in his beautiful monograph [3]. Especially we will deal with those elements of his conjugate duality theory which will be applied in the sequel.

## 2.3 CONJUGATE DUALITY

Consider the abstract convex programming problem

$$\text{(P)} \qquad \text{minimize}_{x \in X} f(x),$$

where X is a linear space and $f: X \to (-\infty,\infty]$ is a convex function. Without much loss of generality we assume that P is feasible, i.e. $\{x \in X: f(x) < \infty\} \neq \emptyset$. P is a generalization of the finite-dimensional mathematical programming problem $P_o$ as described in the previous section (at least if $P_o$ is feasible and convex) as can be seen by the specialization

$$\text{(2.1)} \qquad \begin{aligned} X &:= \mathbb{R}^n, \\ f(x) &:= \begin{cases} f_0(x) & \text{if } x \in C \text{ and } f_i(x) \leqq 0,\ i = 1,\dots,m, \\ +\infty & \text{otherwise} \end{cases} \end{aligned}$$

The central issue of duality theory is the relation between the *primal* problem P and *dual* problems to be formulated below. An important contribution of R.T. Rockafellar to the duality theory is his general framework for generating meaningful dual problems to the given problem P. The basic idea is the embedding of problem P in a parametrized family of optimization problems, where the parameter is interpreted as a *perturbation* of problem P. Analysis of the sensitivity of the problem with respect to the perturbation leads in a natural way to the dual problem. Since the choice of the perturbations is at our disposal, many dual problems for P can be formulated. However, not every parametrization of P is suitable. To get nice results, the problem must be perturbed convexly. Moreover, topology comes in, which is not surprising in a sensitivity analysis. Both side conditions are satisfied in the following definition. For U a local convex topological vector space let $F: X \times U \to (-\infty,\infty]$ be any function which satisfies

$$\text{(2.2)} \qquad \begin{aligned} &F(x,0) = f(x) \text{ for all } x \in X, \\ &F \text{ is convex jointly in } x \text{ and } u, \text{ and} \\ &F(x,.) \text{ is lower semicontinuous for all } x \in X. \end{aligned}$$

Elements of $u \in U$ are interpreted as perturbations, and the problem of minimizing $F(x,u)$ over x is interpreted as a perturbation of P. The corresponding *optimal value function* (or perturbation function, or marginal function) $\varphi: U \to [-\infty,\infty]$ is defined by

$$(2.3)\qquad \varphi(u) := \inf_{x\in X} F(x,u),\ u \in U.$$

The convexity assumption on F implies that $\varphi$ is convex too. Obviously, the unperturbed problem $(u := 0)$ is precisely P, so that

$$(2.4)\qquad \inf P = \varphi(0).$$

The dual problem will be defined in "prices" on perturbations. Formally, the space Y of prices will be the topological dual space $U^*$ of U, that is the linear space of all continuous linear forms on U. The relation between U and Y can be represented by a real-valued bilinear form $\langle .,. \rangle$ on $Y \times U$. In applications the linear space Y and the bilinear form, the *pairing* of the spaces Y and U, are often chosen a priori, and the topology on U is chosen *compatible* with this pairing, that is: $U^* = Y$.

Now we are ready to introduce the *dual problem* of P, relative to its embedding in F:

$$(D)\qquad \text{maximize}_{y\in Y}\, g(y)$$

where the dual function $g: Y \to [-\infty,\infty]$ is a concave function defined by

$$\begin{aligned} g(y) &= \inf_{x\in X} \inf_{u\in U} \{F(x,u) + \langle y,u\rangle\} \\ &= \inf_{u\in U} \inf_{x\in X} \{F(x,u) + \langle y,u\rangle\} \qquad (2.5) \\ &= \inf_{u\in U} \{\langle y,u\rangle + \varphi(u)\} \qquad , y \in Y. \end{aligned}$$

The definition of $g(y)$ expresses the idea that the decision-maker gets the opportunity to perturb problem P at a price y, and the dual problem can be interpreted as the problem of finding a price y which makes perturbations as unprofitable as possible. Since $u = 0$ (no perturbations) is a possible choice, it follows that $g(y) \leqq \varphi(0)$ for all $y \in Y$. Hence *weak duality* holds:

$$\sup D \leqq \inf P.$$

If $\sup D < \inf P$ the difference $\inf P - \sup D$ is called the *duality gap*. Problem P relative to its embedding in F is called *normal* if $\sup D = \inf P$. This terminology expresses the idea that, although abnormal problems can be constructed easily, in practice almost every convex programming problem P is expected to be normal. Finding verifiable sufficient conditions for normality is one of the main issues in duality theory. By the way, abnormality is very likely to occur if the assumptions on F are relaxed in the sense that $\varphi$ is not convex any more.

For normal problems P the optimal solutions $y^o$ of D, which do not have to exist necessarily, have an interpretation as *equilibrium prices*: in that case it follows directly from (2.4) and (2.5) that

$$\varphi(0) = \min_{u \in U}\{<y^o,u> + \varphi(u)\}$$

so that no perturbation $u \in U$ ("small" or "large") of P is profitable, if it has to be paid for at the price of $y^o$ per unit.

The relation between the dual function g and the optimal value function $\varphi$ is given in (2.5). It reveals that $g(y) = -\varphi^*(-y)$, where $\varphi^*$ is the *(convex) conjugate function* of $\varphi$

$$\varphi^*(y) := \sup_{u \in U}\{<y,u> - \varphi(u)\}, \; y \in Y.$$

The geometrical interpretation of the conjugate function of $\varphi$ is, that it describes all non-vertical half-spaces in $U \times \mathbb{R}$ which contain the epigraph of $\varphi$,

$$\text{epi}\,\varphi := \{(u,\alpha) \in U \times \mathbb{R}: \varphi(u) \leqq \alpha\}.$$

That is, for any $y \in Y$ $\varphi^*(y)$ is the smallest number $\lambda$, if any, for which

$$\varphi(u) \geqq <y,u> - \lambda \text{ for all } u \in U.$$

In passing we remark that occasionally we will call the function

$$\hat{\varphi}(y) := \inf_{u \in U}\{<y,u> + \varphi(u)\},\ y \in Y,$$

the *concave conjugate function* of any function $\varphi$. Of course, $\hat{\varphi}(y) = -\varphi^*(-y)$ so that both conjugate functions determine each other completely. Hence g is the concave conjugate function of the optimal value function $\varphi$. In the sequel when using the term "conjugate" we mean always "convex conjugate" unless explicitly stated differently.

For obvious reasons the function $\varphi^{**}: U \to [-\infty,\infty]$

$$\varphi^{**}(u) := \sup_{y \in Y}\{<y,u> - \varphi^*(y)\},\ u \in U,$$

is called the *biconjugate function* of $\varphi$. Since $\varphi^*(y) = -g(-y)$ we have

$$\sup D = \varphi^{**}(0).$$

Hence, P is normal if and only if

$$\varphi^{**}(0) = \varphi(0).$$

This brings the question of gap-free duality into the realm of the analysis of biconjugate functions. It is not difficult to derive [3 Theorems 4 and 5] that for *convex* functions $\varphi$

$$\varphi^{**}(u) = \liminf_{u' \to u} \varphi(u') \text{ if } \varphi(u) < +\infty.$$

Therefore, the convex optimization problem P is normal if it is feasible and if the optimal value function is lower semicontinuous at 0. The convexity of F, hence of $\varphi$, makes it possible to ensure even continuity (rather than only lower semicontinuity) at 0 by boundedness above on a neighbourhood of $0 \in U$; this condition implies even *stability* of P, by which we mean that

$$\inf P = \max D.$$

That is, not only normality holds but also it is certain that the supremum in the dual problem is attained.

THEOREM 2.1. Sufficient conditions for stability of P.
*Consider* P *in its parametrisation given by* F, *and assume that* (2.2) *holds, using a topology on* U *which is compatible with the pairing between* Y *and* U. *If the optimal value function* $\varphi$ *is bounded above on a neighbourhood of* 0, *then it is continuous in* 0, *actually, and* P *is stable. Moreover* $y^* \in Y$ *solves* D *iff*

$$-y^* \in \partial\varphi(0)$$

*where, as usual, the subgradient* $\partial\varphi(u)$ *of* $\varphi$ *in* u *is defined as*

$$\partial\varphi(u) := \{y \in Y\colon \varphi(u') \geqq \varphi(u) + \langle y,u'-u\rangle\ \forall u' \in U\}.$$

PROOF. [3 Theorems 17a and 16], [1 Theorems 3.8.11 and 3.8.16] □

Conjugate duality is closely related to minimax theory. In order to see this, define the *Lagrangean* function L of the perturbed problem P by

$$L(x,y) := \inf_{u\in U}\{F(x,u) + \langle y,u\rangle\},\ x \in X,\ y \in Y.$$

Since F is convex L is convex in x for all y, and concave in y for all x. From the definition of the dual function g it is clear that the dual problem D can be written as

$$\text{maximize}_{y\in Y}\ \inf_{x\in X} L(x,y).$$

On the other hand, the convexity and the lower semicontinuity of F(x,.) assumed in (2.2) imply that the primal problem P can be written as

$$\text{minimize}_{x\in X}\ \sup_{y\in Y} L(x,y),$$

see [3 Theorem 6]. In fact, this result is a corollary of the fact, that the biconjugate of a lower semicontinuous convex function coincides with that function, at least in all points where it is less than $+\infty$. Consequently, normality of P is equivalent to L having a *saddle-value*, and, under normality, solvability of P and D is equivalent to L having a *saddle-point*. This saddle-point interpretation allows for a complete characterization of primal optimality in the stable case.

<u>THEOREM 2.2. Necessary and sufficient conditions for optimality in P under the assumption of stability.</u>

*Consider* P *in its parametrisation given by* F, *and assume that* (2.2) *holds, using a topology on* U *which is compatible with the pairing between* Y *and* U. *Suppose that* P *is stable. Then* $x^o$ *solves* P *iff the Lagrangian* L *has a saddle-point* $(x^o,y^o)$ *for a* $y^o \in Y$. *Such a* $y^o$ *is an optimal solution for* D, *and any optimal solution for* D *will do the job.*

<u>PROOF</u>. [3 Theorem 15] □

In specific applications Theorem 2.2 is the basis of so-called *Karush-Kuhn-Tucker* type theorems, of which we will give an example in the sequel of this section.

The exposition on conjugate duality which we gave above is far from complete. Many more results are known on (necessary and) sufficient conditions for normality or stability, on existence and characterization of primal and dual solutions, on applications on various problems, etc. Nevertheless we restrict the attention to Theorems 2.1 and 2.2, since they contain all duality results which we want to apply. For more details on duality theory, and many examples of applications, we refer to [1, 2, 3] and the references given there. We conclude this section by illustrating the approach to the classical convex programming problem and to linear programming. Although these applications can be seen as "prototypes" for the general approach, it must be emphasized that they give only a faint reflection of the flexibility of duality theory.

Consider the mathematical programming problem $P_o$ as defined in Section 2.2, and assume that it is feasible and convex. The specialization (2.1) shows that is can be considered as a special case of P. It is appropriate to include the constraint $x \in C$ in the objective function; that is

$$f_0(x) := +\infty \text{ for } x \in \mathbb{R}^n \setminus C.$$

Also $f_i(x)$, $i = 1,\ldots,m$, are supposed to be defined and convex all over $\mathbb{R}^n$ instead of only in C. This is no loss of generality; one might take for instance $f_i(x) := +\infty$ for $x \notin C$, but in specific examples other convex extensions might be more appropriate. Perturbation of the right-hand sides of the constraints $f_i(x) \leqq 0$, $i = 1,\dots,m$, leads to the definition

$$F(x,u) := \begin{cases} f_0(x) & \text{if } f_i(x) \leqq u_i,\ i = 1,\dots,m, \\ +\infty & \text{otherwise,} \end{cases}$$

for $x \in X := \mathbb{R}^n$ and $u = (u_1,u_2,\dots,u_m) \in U := \mathbb{R}^m$. On U the Euclidean topology is used, which of course is compatible with the pairing $\langle y,u\rangle := \Sigma_{i=1}^m y_i u_i$ with $y \in Y := \mathbb{R}^m$. The choice of the perturbations gives the Lagrangian function in its well-known form

$$L(x,y) = \begin{cases} +\infty & \text{if } x \notin C, \\ f_0(x) + \Sigma_{i=1}^m y_i f_i(x) & \text{if } x \in C,\ y \geqq 0, \\ -\infty & \text{if } x \in C,\ y \ngeqq 0, \end{cases}$$

as can be verified by simple calculation. Saddle-points of L are characterized by the Karush-Kuhn-Tucker conditions:

THEOREM 2.3. $(x^o,y^o) \in \mathbb{R}^n \times \mathbb{R}^m$ *is a saddle-point of the Lagrangian* L *of the convex program* $P_o$ *if and only if the following conditions hold:*

(KKT)

*a.* $x^o \in C$, $f_i(x^o) \leqq 0$, $i = 1,\dots,m$ ("*primal feasibility*");
*b.* $y^o \geqq 0$ ("*dual feasibility*");
*c.* $\Sigma_{i=1}^m y_i^o f_i(x^o) = 0$ ("*complementary slackness*")
*d.* $0 \in (\partial f_0(x^o) + \Sigma_{i=1}^m y_i^o \partial f_i(x^o))$ ("$x^o$ *minimizes* $L(.,y^o)$").

PROOF. [2 Theorem 28.3] □

Recall that $(x^o,y^o)$ is a saddle-point for L iff $P_o$ is normal, $x^o$ solves $P_o$ and $y^o$ solves the dual problem $D_o$. Hence the KKT conditions are sufficient for optimality for $P_o$ (and $D_o$). They are also necessary for $x^o$ to be optimal for $P_o$ if $P_o$ is stable, see Theorem 2.2. A sufficient condition for stability is *Slater's constraint qualification*

$$\exists \hat{x} \text{ in the (relative) interior of C with } f_i(\hat{x}) < 0,\ i = 1,\dots,m,$$

since it implies that the optimal value function is bounded above in a neighbourhood of 0. Slater's constraint qualification does not help very much if the constraints $f_i(x) \leqq 0$, $i = 1,\dots,m$, contain (reformulations of) equality constraints. In this case the criterion for stability given in Theorem 2.1 is still useful; one may be able, for instance, to come up with a feasible solution $x(u)$ of the perturbed problem which satisfies $f_0(x(u)) < M$ for $\|u\| < \delta$. In any case, if $P_o$ is stable then its optimal solutions are completely characterized by the KKT-conditions.

Let us now specialize $P_o$ further to the *linear programming* problem by assuming that

$$f_0(x) := \begin{cases} \Sigma_{j=1}^n c_j x_j & \text{if } x \in C := \{x \in \mathbb{R}^n : x \geqq 0\}, \\ +\infty & \text{if } x \in \mathbb{R}^n \setminus C, \end{cases}$$

$$f_i(x) := -\Sigma_{j=1}^n a_{ij} x_j + b_i, \; x \in \mathbb{R}^n, \; i = 1,\dots,m,$$

for $c_j, a_{ij}, b_i \in \mathbb{R}$. In this case the dual function can be calculated explicitly. For $y \geqq 0$ we have

$$\begin{aligned} g(y) &= \inf_{x \in \mathbb{R}^n} L(x,y) \\ &= \inf_{x \geqq 0} \{\Sigma_{j=1}^n c_j x_j + \Sigma_{i=1}^m y_i \Sigma_{j=1}^m (-a_{ij} x_j + b_i)\} \\ &= \inf_{x \geqq 0} \Sigma_{j=1}^n (c_j - \Sigma_{i=1}^m y_i a_{ij}) x_j + \Sigma_{i=1}^m y_i b_i \\ &= \Sigma_{i=1}^m y_i b_i + \Sigma_{j=1}^n \inf_{x_j \geqq 0} (c_j - \Sigma_{i=1}^m y_i a_{ij}) x_j \end{aligned}$$

so that

$$g(y) = \begin{cases} \Sigma_{i=1}^m y_i b_i & \text{if } \Sigma_{i=1}^m y_i a_{ij} \leqq c_j, \; j = 1,\dots,n, \\ & \qquad y_i \geqq 0, \; i = 1,\dots,m, \\ -\infty & \text{otherwise.} \end{cases}$$

This exercise shows that the classical duality between the linear programs

(LP) $\quad \text{minimize}_{x \in \mathbb{R}^n} \{cx : Ax \geqq b, \; x \geqq 0\}$

(LD) $\quad \text{maximize}_{y \in \mathbb{R}^m} \{yb : yA \leqq c, \; y \geqq 0\}$

is covered by the conjugate duality theory. In this special case the KKT conditions can be formulated as

$$(\mathrm{KKT}_L)\qquad \begin{array}{ll} \text{(a)} & x^o \geqq 0,\ Ax^o \geqq b \ (\text{"primal feasibility"}); \\ \text{(b)} & y^o \geqq 0,\ y^oA \leqq c \ (\text{"dual feasibility"}); \\ \text{(c)} & y^o(Ax^o-b) = 0,\quad (c - y^oA)x^o = 0 \ (\text{"complementary slackness"}). \end{array}$$

Notice that here condition KKT(d) of Theorem 2.3 has been replaced by $y^oA \leqq c$ and $(c - y^oA)x^o = 0$. Indeed, in the linear case KKT(d) is equivalent to

$$\Sigma_{i=1}^m y_i^o a_{ij} \in \partial f_{0j}(x_j^o),\ j = 1,\dots,n,$$

where $f_{0j}(x_j) := c_jx_j$ if $x_j \in [0,\infty)$ and $f_{0j}(x_j) := +\infty$ if $x_j \in (-\infty,0)$ so that

$$\partial f_{0j}(x_j^0) = \begin{cases} \emptyset & \text{if } x_j^o < 0, \\ (-\infty,c_j] & \text{if } x_j^o = 0, \\ \{c_j\} & \text{if } x_j^o > 0. \end{cases}$$

This shows that KKT in Theorem 2.3 are generalizations of $\mathrm{KKT}_L$ to the convex case.

In Chapter 4 we will discuss "abstract" linear programs. As will be shown, similar explicit dual linear programs can be formulated easily. However, whereas in the finite-dimensional case a simple verification of normality and stability is possible

$$\begin{array}{ll} \inf LP = \sup LD \text{ unless} & \inf LP = +\infty \text{ and } \sup LD = -\infty, \\ \min LP = \max LD \text{ iff} & \inf LP \text{ or } \sup LD \text{ is finite} \end{array}$$

this is not true any more in the abstract case.

REFERENCES

1. J. PONSTEIN (1980). *Approaches to the Theory of Optimization*, Cambridge University Press, Cambridge.

2. R.T. ROCKAFELLAR (1970). *Convex Analysis*, Princeton University Press, Princeton NJ.

3. R.T. ROCKAFELLAR (1974). *Conjugate Duality and Optimization*, SIAM, Philadelphia Pa.

# CHAPTER 3

# STOCHASTIC LINEAR PROGRAMMING MODELS

## 3.1 INTRODUCTION

Linear programming has proven to be a suitable framework for the quantitative analysis of many decision problems. The reasons for its popularity are obvious: many practical problems can be modeled, at least approximately, as linear programs, and powerful software is available. Nevertheless, even if the problem has the necessary linear structure it is not sure that the linear programming approach works. One of the reasons is that the model builder must be able to provide numerical values for each of the coefficients. But in practical situations one often is not sure about the "true" values of all coefficients. Usually the uncertainty is exorcized by taking reasonable guesses or maybe by making careful estimates. In combination with a sensitivity analysis with respect to the most inaccurate coefficients this approach is satisfactory in many cases. However, if it appears that the optimal solution depends heavily on the value of some inaccurate data, it might be sensible to take the uncertainty of the coefficients into consideration in a more fundamental way. Since an evident framework for the quantitative analysis of uncertainty is provided by probability theory it seems only natural to interpret the uncertain coefficient values as realizations of random variables. This approach characterizes *stochastic linear programming*. In this subarea of mathematical programming it is usually assumed that the joint distribution of the random coefficients is known. As will be seen, a wide range of meaningful models can be formulated then. From the point of view of applications, however, this assumption is very demanding, since one is seldom so lucky that the available data are sufficient for a reliable estimation of the joint distribution. Nonetheless it is our opinion that stochastic programming is valuable, also in view of applications. A first reason is that some stochastic programming models do not need a complete knowledge of the distribution. Moreover, by means of a minimax approach sometimes the worst-case distribution compatible with the partial knowledge about the distribution can be determined (see Chapter 7 for an application of this approach). But the main reason is that in our opinion even a subjective specification of the joint distribution may serve the purpose of the decision model, i.e. the analysis of a decision problem under uncertainty, very well. One might object, of course, that subjective probabilities introduce a kind of arbitrariness in the model. We want to stress, however, that in this respect there is no fundamental distinction with applied deterministic linear programming. Also in that case the specification of the coefficients often is arbitrary to a certain extent, and it depends on the judgement of the model builder as well. In fact, there is more subjectivity in applications than sometimes is recognized. In many cases, for instance, there are several conflicting objectives, and in a linear programming framework usually all but one are represented as (goal-)restrictions, whereas the last one is used as the objective function. Hence, a choice has to be made as to the way of representing the goals in the model. In stochastic models this choice is even more intricate, since also the attitude to risk has to be modeled, and this can be done in several ways, as will be seen. But again we claim that with respect to subjectivity there is only a gradual difference between deterministic and stochastic models.

Let us assume, therefore, that a linear programming problem has been completely specified, apart from some coefficients that are random variables with a known joint distribution. Traditionally, two types of models for this situation are considered in stochastic programming, namely the *wait-and-see* and the *here-and-now* models. In the first case the decision maker is assumed to be able to wait for the realization of the random coefficients, whereas in the second case the decisions have to be made before or at least without knowledge of these realizations. In the here-and-now situation a fundamental difficulty comes up: since some coefficients are undetermined the usual meaning of feasibility and/or optimality of the decisions is useless. Therefore additional specification is needed. At stake are the questions of how to measure "risk" and how to incorporate "risk aversion" in the decision model. In this respect, useful concepts are chance constraints, recourse actions and utility functions. For the meaning of these concepts we refer to Section 3.2.

The contents of this chapter can be now summarized as follows. In Section 3.2 some standard modeling concepts for here-and-now problems are introduced. Conceptual issues are emphasized, and it is argued that sometimes alternatives which do not get attention in the literature might be appropriate. This section is concluded with the indication of related literature. In Section 3.3 a short introduction to multistage stochastic linear programs of the recourse type is given. Such models can be interpreted as intermediate between wait-and-see and here-and-now models. It is assumed that the values of the random coefficients become known in stages, and that the decision process also occurs in stages; the decisions at any stage may and will depend on the available information at that stage on the realizations of the random coefficients. Special attention is paid to the relation with stochastic dynamic programming.

This chapter is not meant to be a complete introduction to stochastic programming. No attention is paid to the mathematical theory, to algorithms and to applications. Only linear models are considered. Moreover, only those subjects are covered which are related to the rest of this book.

We conclude this section with an indication of the occurrence of stochastic programming models in the chapters to come. In Chapter 5 some alternative here-and-now formulations, more or less in between chance-constrained and recourse models, are introduced. They can be seen as an elaboration of Remark 3.9 in the present chapter. In Chapter 7 a well-known stochastic programming problem (although not always recognized as such) is considered: the project planning problem. It appears to be possible to overcome the difficulty of incompletely specified distributions, without having to use heuristics as PERT. The main tool is duality theory. In Chapter 8 a simple production-inventory control problem is formulated as a multistage stochastic linear program. Its dual problem is solved. There is, of course, an intimate relation with dynamic programming.

## 3.2 CHANCE CONSTRAINTS, RECOURSE ACTIONS AND UTILITY FUNCTIONS

Consider a decision problem which, apart from the fact that some coefficients are random variables rather than known real numbers, can be modeled as a finite-dimensional linear program. In order to be specific, let us write down formally

$$(3.1) \qquad \text{minimize}_{x \in C} \{\Sigma_{j=1}^{n} c_j(\omega)x_j : \Sigma_{j=1}^{m} a_{ij}(\omega)x_j \geqq b_i(\omega),\ i = 1,\dots,m\}$$

where C is a fixed nonempty polyhedral convex subset of $\mathbb{R}^n$ (determined, for example, by $x \geqq 0$ and other deterministic linear inequalities and equalities), and where $c_j$, $a_{ij}$, $b_i$ are (possibly degenerated) random variables. Without loss of generality we may identify the underlying probability space $\Omega$ with $\mathbb{R}^N$ if N is the number of all random coefficients in (3.1); then $c_j$, $a_{ij}$, $b_i$ are coordinate projections. By definition, the random vector $\omega = (c_j, a_{ij}, b_i) \in \mathbb{R}^N$ has some probability distribution, which, as discussed before, is assumed to be known. As usual in the stochastic programming literature it is also assumed that this distribution does not depend on the decision vector $x \in \mathbb{R}^n$; although a possible dependence upon the decision vector is no serious complication from a conceptual point of view, this would destroy properties which are useful in calculations. Finally, it is assumed that the random coefficients are only known in distribution; that is, their joint distribution is known but no further information on any outcome of any random coefficient will be available before all decisions x have been determined. This here-and-now assumption makes formulation (3.1) useless unless further specifications are made on what is meant by "feasibility" and "optimality" in a random environment. Specifying them is the main subject of this section.

There are several possibilities to give (3.1) a meaning; which one is appropriate does, of course, depend on the problem for which (3.1) is an incomplete formulation. Let us agree that the idea behind (3.1) is that a choice for $x \in C$ is to be made, based on the values of the m+1 *criterion functions*

$$\eta_o(x,\omega) := \Sigma_{j=1}^{n} c_j(\omega)x_j,$$
$$\eta_i(x,\omega) := \Sigma_{j=1}^{n} a_{ij}(\omega)x_j - b_i(\omega),\ i = 1,\dots,m,$$

and that "it is very important to have $\eta_i \geqq 0$ for $i \geqq 1$", and that "one wishes to have a small value for $\eta_o$", if at all possible. The question is how to formulate these wishes for the values of $(\eta_o,\eta_1,\dots,\eta_m)$ in case not the values themselves but their joint probability distribution is a known function of the decision vector x. We shall review the standard answers to this question, and discuss some relations between them. For reasons of exposition we will distinguish between the definition of stochastic feasibility (dealing with $\eta_i(x,\omega)$, $i \geqq 1$) and the definition of stochastic optimality (dealing with $\eta_o(x,\omega)$), but we repeat that in practical situations this distinction is not strict. Moreover, we will interprete $\eta_o$ as monetary cost. This is not essential, but it makes the terminology juicier.

Let us first consider the definition of stochastic feasibility in (3.1). For the sake of convenience we assume in this discussion that $\eta_o$ is deterministic. There are at least three approaches.

a. Reluctantly *infeasibility is accepted, but only with a small probability*. This approach leads to the models with *chance constraints:* $x \in C$ is accepted as a feasible solution if and only if for $\alpha_i \in [0,1]$, to be specified in advance,

(3.2) $$P(\eta_i(x,\omega) < 0) \leqq \alpha_i,\ i = 1,\dots,m,$$

or, if and only if for $\alpha \in [0,1]$, to be specified in advance,

(3.3) $$P(\exists i \in \{1,\dots,m\}: \eta_i(x,\omega) < 0) \leqq \alpha.$$

In (3.2) we have *individual* (or *separate*) *chance constraints*, whereas the condition in (3.3) is called a *joint chance constraint*. In these cases, "risk" is interpreted as "probability of infeasibility" and risk aversion is modelled by means of the specification of maximal acceptable risk levels $\alpha_i$, $i = 1,\dots,m$, or $\alpha$, respectively.

b. Reluctantly *infeasibility is accepted, but only at a certain cost*. In this approach each $x \in C$ is accepted as a feasible solution but the objective function is augmented with the expected value of penalties imposed on the *shortages* $(\eta_i)^- := \max(0,-\eta_i)$. Linear costs of infeasibility lead to the *penalty cost* model

$$(3.4) \qquad \text{minimize}_{x\in C}\{\eta_o(x) + E\, \Sigma_{i=1}^m q_i \cdot (\eta_i(x,\omega))^-\}.$$

Obviously, "risk" is interpreted as "mean shortage", and risk aversion is expressed by the choice of the unit penalties $q_i$, $i = 1,\dots,m$. Only for positive values of $q_i$ the model (3.4) is, of course, in accordance with the idea behind (3.1); their numerical values represent a balance between the m+1 criterion functions. Notice that in this way a possible objective "maximize $\eta_i$" for an $i \geqq 1$ via the goal-restriction $\eta_i \geqq 0$ in (3.1) might end up in the objective function of (3.4).

c. *Infeasibility is not accepted, but corrected afterwards at a certain cost*. That is, the model is extended with a so-called *second stage*; the decision variables of the second stage represent recourse actions, and their values may depend on the realizations of $\eta_i(x,\omega)$. With a fixed linear recourse technology and a fixed linear recourse cost this approach leads to the fixed linear *recourse model*

$$(3.5) \qquad \text{minimize}_{x\in C}\{\eta_o(x) + E\, Q(\eta_1(x,\omega),\, \eta_2(x,\omega),\dots,\eta_m(x,\omega))\}$$

where Q is the (optimal) recourse cost function defined by

$$(3.6) \qquad Q(z) := \inf_{y\in Y}\{qy\colon Wy \geqq -z\},\ z = (z_1,\dots,z_m) \in \mathbb{R}^m .$$

Here Y is a polyhedral convex subset of $\mathbb{R}^K$ with $0 \in Y$, representing the set of feasible recourse actions, q is a $1 \times K$ vector of recourse prices, and W is an $m \times K$ matrix called recourse matrix. The rationale of the recourse model can be seen from its representation in linear programming format

$$\text{(3.7)}\qquad \begin{aligned} \min_x\{ \quad & cx + E \min_y\{qy(\omega) \\ & A(\omega)x + \quad Wy(\omega) \geqq b(\omega),\ \omega \in \Omega, \\ & x \in C \quad\quad y(\omega) \in Y \quad ,\ \omega \in \Omega\}\} \end{aligned}$$

Of course, (3.5) and (3.6) are only in accordance with the idea behind (3.1) if Y, q and W are such that

$$\text{(3.8)}\qquad \begin{aligned} & Q(z) = 0 \text{ if } z \geqq 0 \\ & Q(z) > 0 \text{ otherwise.} \end{aligned}$$

Later on in this section we will give some comments on these three approaches to the specification of feasibility and their relationships.

Let us first consider the definition of stochastic optimality in (3.1). For the sake of convenience we assume in this discussion that $\eta_i$ is deterministic for i = 1,...,m. The question is: which random cost $\eta_o(x,\omega)$, for x in a fixed subset $C_o$, determined by C and possible constraints on $(\eta_1,\eta_2,...,\eta_m)$, is preferred by most? Basically, we are urged to come up with a preference relation on the family of probability distributions of $\{\eta_o(x,\omega),\ x \in C_o\}$, since that is all the available information. We will formulate a number of well-known preference criteria based on the minimization of a real-valued function of the probability distribution of $\eta_o$.

d. *Minimize the expected cost:*

$$\text{(3.9)}\qquad \text{minimize}_{x\in C_o} E\ \eta_o(x,\omega).$$

The motivation is clear: each of the original objective functions $\eta_o(.,\omega)$, $\omega \in \Omega$, is weighted with the probability of its occurrence. This criterion is the most popular, for good reasons, but it has drawbacks. For example, an alternative with a deterministic cost is rejected in favor of an alternative with a large dispersion but a higher mean value, even if the difference of the mean values is small. In many situations this is felt to be too risky. Therefore, the need comes up to find formulations in which risk aversion is modelled explicitly.

e. *Use the variance as measure for risk.* This idea leads to Markowitz's mean-variance analysis, in which two objectives are formulated: minimize the mean value of the cost, and minimize its variance, too. Balancing between both objectives leads to the model

$$\text{(3.10)} \qquad \text{minimize}_{x \in C_o} \{E\, \eta_o(x,\omega) + \lambda.\sigma(\eta_o(x,\omega))\}$$

where $\sigma$ denotes standard deviation, and where $\lambda$ is a positive parameter expressing the degree of risk aversion on the part of the decision maker. Since mean value and variance are relatively simple as far as specification and computability is concerned, model (3.10) is attractive. However, one of its weak points is the inherent symmetry in the measure for risk, whereas it is clear that high costs are much more risky than low costs.

f. *Minimize the probability of bankruptcy*. That is, specify a maximum acceptable cost level $c_o$, and maximize the probability that the cost does not exceed $c_o$:

$$\text{(3.11)} \qquad \text{maximize}_{x \in C_o} P(\eta_o(x,\omega) \leqq c_o).$$

Here the measure for risk is asymmetric, as we want it to be, but on the other hand it should be noted that in (3.11) only the sign of $\eta_o(x,\omega) - c_o$ is important.

g. *Maximize the expected utility*. That is, a so-called *utility function* u: $\mathbb{R} \to \mathbb{R}$ is specified, and the problem in defined as

$$\text{(3.12)} \qquad \text{maximize}_{x \in C_o} E\, u(-\eta_o(x,\omega)).$$

Clearly, (3.12) can only be in accordance with the starting point "minimize $\eta_o$" if u is nondecreasing. Formulation (3.12) is a generalization of (3.9) and (3.11): take $u(z) := z$ for all z, or $u(z) := 0$ for $z < -c_o$, $u(z) := 1$ for $z \geqq -c_o$, respectively. In the theory of expected utility it is shown that risk aversion is modeled by concave (nonlinear) and increasing utility functions.

Without aiming at completeness we now comment on the approaches (a)-(g) and their relationships.

REMARK 3.1. Suppose that in the underlying linear program (3.1) the inequalities are replaced by *equalities*. This expresses the idea that "it is very important to have $\eta_i = 0$ for $i \geqq 1$". A direct application of chance constraints is then useless, since

$$\{x \in C: P(\eta_i(x,\omega) = 0) \geqq 1 - \alpha_i\}$$

will be empty for reasonable values of $\alpha_i$. Instead it is more appropriate to use chance constraints of the type

$$P(-d_i \leqq \eta_i(x,\omega) \leqq e_i) \geqq 1 - \alpha_i\}$$

for positive $d_i$ and $e_i$ to be chosen. In the penalty cost model equalities do not cause difficulties; just replace (3.4) by

$$\text{(3.13)} \qquad \text{minimize}_{x\in C}\{\eta_o(x) + E\, \Sigma_{i=1}^m \{q_i\cdot(\eta_i(x,\omega))^- + q_i'\cdot(\eta_i(x,\omega))^+\}\}$$

where $q_i'$ is the unit penalty for the *surplus* $(\eta_i(x,\omega))^+ := \max(0,\eta_i(x,\omega))$. A similar statement holds for the recourse model: just replace (3.6) by

$$\text{(3.14)} \qquad Q(z) := \inf_{y\in Y}\{qy: Wy = -z\},\ z \in \mathbb{R}^m.$$

REMARK 3.2. Whereas it is sometimes appropriate to represent several constraints in (3.1) as a single chance constraint (3.3) also the reverse might be appropriate: formulate *several chance constraints for one original constraint*. For example, (3.2) might be replaced by

$$\text{(3.15)} \qquad P(\eta_i(x,\omega) < -d_{ih}) \leqq \alpha_{ih},\ h = 1,\ldots,H_i,\ i = 1,\ldots,m,$$

where $0 = d_{i1} < d_{i2} < \ldots$ and $\alpha_{i1} > \alpha_{i2} > \ldots$ . In this way one specifies that large shortages are disliked more than small ones.

REMARK 3.3. Although not expressed in (3.1) it is quite possible that positive values for $\eta_i$, $i \geqq 1$, are appreciated better than just the value zero. This feature can be implemented in the penalty cost model by adding *rewards for surplusses*. That is, use (3.13) with negative $q_i'$. However, $q_i' + q_i$ must be positive; for if not, the strive for surplusses would be seen as more important than the need for avoiding shortages, and that would contradict the idea behind (3.1). Similar provisions can be made in the recourse models.

REMARK 3.4. It is also possible that more *refined penalty cost* functions are to be preferred over the linear ones introduced in (3.4) and (3.13).

For example, it might be felt that large shortages should be penalized heavier than proportional to their sizes. One possibility is to use piecewise linear convex penalty functions; (3.13) is then replaced by

$$\text{(3.16)}\quad \begin{aligned} &\text{minimize}_{x\in C}\{\eta_o(x) + E\,\Sigma_{i=1}^{m}\{\Sigma_{h=1}^{H_i} q_{ih}\cdot(\eta_i(x,\omega) + d_{ih})^- + \\ &\qquad \Sigma_{h=1}^{H_i'} q_{ih}'\cdot(\eta_i(x,\omega) - e_{ih})^+\}\} \end{aligned}$$

for positive $q_{ih}$, $q_{ih}'$, $d_{ih}$ and $e_{ih}$; $0 = d_{i1} < d_{i2} < \ldots$ and $0 = e_{i1} < e_{i2} < \ldots$ .

<u>REMARK 3.5</u>. The use of a penalty cost model such as (3.4) implies that, although reluctantly, one is willing to accept infeasibility in (3.1) if $\omega$ is random. But the choice of model (3.1) implies that one would not do that if the value of $\omega$ would be known. Therefore it is reasonable to ask for sufficiently high values of the penalties $q_i$ in (3.4). How to specify "sufficiently high"? It seems reasonable to require that if the model for risky situations is used under conditions of complete certainty it should be equivalent to the underlying deterministic model. Suppose therefore for a moment that the distribution of the random vector $\omega$ is replaced by the degenerated distribution $P(\omega=\bar\omega) = 1$, where $\bar\omega = (\bar c,\bar A,\bar b)$ is the mean value of $\omega$. Then both (3.1) and (3.4) make sense. The question is: under what conditions do they have the same optimal solutions? For this to be true the penalties $q_i$ in (3.4) should be large enough to enforce $\eta_i(x,\bar\omega) \geqq 0$, $i = 1,\ldots,m$, in the optimum. It is obvious to expect that this property holds if $q_i > \lambda_i^o$, where $\lambda_1^o,\ldots,\lambda_m^o$ are optimal dual variables in (3.1). Indeed, notice that in the deterministic case (3.4) is equivalent to the linear program:

$$\text{(3.17)}\quad \underset{\substack{x\in\mathbb{R}^n\\ y\in\mathbb{R}^m}}{\text{minimize}} \left\{\bar c x + qy:\ \begin{array}{ll} \bar A x + y & \geqq \bar b \\ Tx & \geqq p \\ x \geqq 0 & y \geqq 0 \end{array}\right\}.$$

Here we have represented C by $C = \{x \in \mathbb{R}^n:\ x \geqq 0,\ Tx \geqq p\}$. The dual of (3.17) is, see Section 2.3,

$$(3.18) \qquad \underset{\substack{\lambda\in\mathbb{R}^m\\ \mu\in\mathbb{R}^s}}{\text{maximize}} \left\{ \lambda\bar{b} + \mu p : \begin{array}{ll} \lambda\bar{A} + \mu T & \leqq \bar{c} \\ \lambda & \leqq q \\ \lambda \geqq 0 & \mu \geqq 0 \end{array} \right\}.$$

Remark that (3.17) reduces to (3.1), with $\omega$ replaced by $\bar{\omega}$, if $y = 0$. Therefore, the dual of (3.1) is (3.18) if there the constraints $\lambda \leqq q$ are deleted. Let $(\lambda^o,\mu^o)$ be any solution for the dual of (3.1). If we specify $q$ such that $q \geqq \lambda^o$, then the same solution is optimal for (3.18). Complementary slackness implies then that if $q > \lambda^o$ optimal solutions $(x^o,y^o)$ for (3.17) necessarily satisfy $y^o = 0$.

Similar statements can be made regarding the specification of the penalties in recourse models.

Notice that (3.17), seen as relaxation of (3.1), is a model of *goal programming*.

REMARK 3.6. The penalty cost model (3.13), hence a fortiori the special case (3.4), can be seen as a specialization of the recourse model (3.5), (3.14), at least from a *mathematical* point of view. Indeed, with

$$(3.19) \qquad Y := \{y \in \mathbb{R}^{2m} : y \geqq 0\},\ W := (I,-I),\ q := (q,q')$$

the recourse cost function of (3.14) reduces to

$$Q(z) = \Sigma_{i=1}^{m}\{q_i \cdot z_i^- + q_i' \cdot z_i^+\},$$

provided that $q + q' \geqq 0$, so that (3.5) reduces to (3.13). For this reason, the models (3.4) and (3.13) usually are called *simple recourse* models. When applying recourse models simple recourse models are the most popular, primarily because they can be specified relatively easily and because of computational reasons. One ought to realize, however, that more often than not one does not aim at recourse actions at all. For such cases the indication "penalty cost" is more appropriate than "simple recourse".

Similarly, the refined penalty cost model (3.16) can be reformulated as a "recourse" model à la (3.14). In order to see this one has to specify $\tilde{Y}$, $\tilde{W}$ and $\tilde{q}$ in such a way that the recourse cost function

$$\tilde{Q}(z) := \inf_{y\in\tilde{Y}} \{\tilde{q}y : \tilde{W}y = -z\},\ z \in \mathbb{R}^m,$$

satisfies

$$(3.20)\qquad \tilde{Q}(z) = \Sigma_{i=1}^{m}\{\Sigma_{h=1}^{H_i} q_{ih}\cdot(z_i+d_{ih})^- + \Sigma_{h=1}^{H_i'} q_{ih}'\cdot(z_i-e_{ih})^+\},$$
$$z = (z_1,z_2,\ldots,z_m) \in \mathbb{R}^m.$$

In other words, we are looking for a linear programming representation for the piecewise linear penalty cost function of (3.16). This representation will be described in detail. First introduce "recourse" variables $y_{ih}$, $h = 1,\ldots,H_i$, and $y_{ih}'$, $h = 1,\ldots,H_i'$, both for $i = 1,\ldots,m$ and define $\tilde{Y}$ as the rectangle in the $\Sigma_{i=1}^{m}(H_i+H_i')$-dimensional Euclidean space determined by

$$(3.21)\qquad \begin{array}{l} 0 \leqq y_{ih} \leqq d_{i,h+1} - d_{ih},\ h = 1,\ldots,H_i,\ i = 1,\ldots,m, \\ 0 \leqq y_{ih}' \leqq e_{i,h+1} - e_{ih},\ h = 1,\ldots,H_i',\ i = 1,\ldots,m, \end{array}$$

where $d_{i,H_i+1} := \infty$, $e_{i,H_i'+1} := \infty$. Furthermore, introduce the following *multiple simple "recourse" matrix*

$$(3.22)\qquad \tilde{W} := \begin{bmatrix} +1..+1\ -1..-1 & & & O \\ & +1..+1\ -1..-1 & & \\ & & \ddots & \\ O & & & +1..+1\ -1..-1 \end{bmatrix}.$$

It has m rows, and the elements in the i-th row are zero except in the columns corresponding to $y_{ih}$, $h = 1,\ldots,H_i$, and $y_{ih}'$, $h = 1,\ldots,H_i'$; those elements are +1 and -1, respectively. Finally, define the vector $\tilde{q}$ of corresponding "recourse" prices

$$\tilde{q} := (\tilde{q}_{11},\ldots,\tilde{q}_{1H_1},\tilde{q}_{11}',\ldots,\tilde{q}_{1H_1'}',\ldots,\tilde{q}_{m1},\ldots,\tilde{q}_{mH_m},\tilde{q}_m',\ldots,\tilde{q}_{mH_m'}')$$
$$(3.23)\qquad \text{with } \tilde{q}_{ih} := \Sigma_{k=1}^{h} q_{ik},\ h = 1,\ldots,H_i,\ i = 1,\ldots,m,$$
$$\text{and } \tilde{q}_{ih}' := \Sigma_{k=1}^{h} q_{ik}',\ h = 1,\ldots,H_i',\ i = 1,\ldots,m.$$

Let us verify that with these definitions (3.20) is true. Remark first that the verification can be done for each $i = 1,\ldots,m$ separately, due to the separated structure of $\tilde{Y}$ and $\tilde{W}$. Suppressing the fixed index i, we have to show that for each $z \in \mathbb{R}$ the optimal value of the linear program

$$\underset{\substack{y\in\mathbb{R}^H \\ y'\in\mathbb{R}^{H'}}}{\text{minimize}} \quad \Sigma_{h=1}^{H}(\Sigma_{k=1}^{h}q_k)y_h + \Sigma_{h=1}^{H'}(\Sigma_{k=1}^{h}q_k')y_h'$$

$$(3.24) \qquad \text{subject to} \quad \Sigma_{h=1}^{H}y_h - \Sigma_{h=1}^{H'}y_h' = -z,$$
$$0 \leqq y_h \leqq d_{h+1} - d_h,\ h = 1,\dots,H,$$
$$0 \leqq y_h' \leqq e_{h+1} - e_h,\ h = 1,\dots,H',$$

where $0 = d_1 < d_2 < \dots < d_H < d_{H+1} = \infty$, $0 = e_1 < e_2 < \dots < e_{H'} < e_{H'+1} = \infty$, is given by

$$(3.25) \qquad \Sigma_{h=1}^{H}q_h\cdot(z+d_h)^+ + \Sigma_{h=1}^{H'}q_h'\cdot(z-e_h)^+.$$

But this is a direct consequence of the fact that all $q_k$ and $q_k'$ are positive. In order to see this, remark first that any optimal solution of (3.24) must satisfy

$$\Sigma_{h=1}^{H}y_h = z^-, \quad \Sigma_{h=1}^{H'}y_h' = z^+.$$

If these relations would not hold one could decrease both $\Sigma_{h=1}^{H}y_h$ and $\Sigma_{h=1}^{H'}y_h'$ with the same positive amount; this would decrease the objective function, too, since all $q_k$ and $q_k'$ are positive. Therefore, without loss of generality we may restrict the attention to either the variables $y_h$, $h = 1,\dots,H$, or $y_h'$, $h = 1,\dots,H'$, depending on the sign of z. In order to be specific, assume that z is nonpositive; if z is positive the analysis is completely similar. Define t by $-d_{t+1} < z \leqq -d_t$. Then a feasible solution of (3.24) is given by

$$y_h := d_{h+1} - d_h,\ h = 1,\dots,t-1,$$
$$y_t := -z - d_t$$

where the remaining variables are put to zero. It is an elementary exercise to show that this solution gives (3.25) as the value of the objective function. Its optimality for (3.24) follows from the fact that all coefficients of $y_h$ in the joint constraint are equal to 1 whereas those in the objective function increase with h since all $q_k$ are positive. This property induces a "priority" in the variables: $y_h$ will only be positive in the optimal solution if $y_{h-1}$ is maximal. This completes the verification of (3.20).

Having shown that the penalty cost models can be treated as simple recourse models, let us now take a look from the opposite direction, and interpret the recourse model as a refined penalty cost model. Just as the penalty cost function in (3.16), the recourse cost function (3.14) is piecewisely linear and convex. A basic difference, however, is that for general Y and W the original constraints in (3.1) are no longer penalized separately. In that sense general recourse models are comparable with joint chance constraints, whereas simple recourse models may be compared with separate chance constraints.

REMARK 3.7. Recourse models, and therefore also penalty cost models, go together with the expected utility idea very well. That is, one might interpret the function on $\mathbb{R}^{1+m}$

$$u(z_o,z_1,\dots,z_m) := -z_o - Q(z_1,\dots,z_m)$$

as a *multi-attribute utility* function, and find that in (3.5) expected utility is maximized. Clearly, the utility functions in the penalty cost models (3.4), (3.13) and (3.16) are then additive. On the other hand, the use of chance constraints is sometimes criticized by supporters of the expected utility approach, for various reasons. Apparently, the objection concerns primarily the use of constraints as a means for modeling risk aversion and not so much the minimization of the probability of an undesirable event: we noticed already that (3.11) is a special case of (3.12). In our opinion the fundamental issue is, that the methodology of mathematical programming is on strained terms with a strict application of the often axiomatically founded expected utility theory. As explained in the introduction, many constraints in mathematical programming are *goal constraints* expressing certain *aspirations* rather than logical or technical necessities. Typically chance constraints are such constraints. From the set of possible decisions the model builder rejects first those decisions which do not satisfy them, and after that the most attractive alternative is chosen. Depending on the result he might change the levels of the aspirations, or even change the specification of the goal constraints, and repeat the optimization. This two-stage approach, characterized by first restricting attention to what is accepted as feasible solutions and then optimizing over them, which is typical for mathematical programming, does not fit smoothly in the expected utility framework. There the decision maker is asked to give his multi-attribute utility function which describes once and for all not only his risk attitude but also his balancing of different goals; after that the model analist will calculate the expected utility of each alternative and come up with the definite optimal decision. The main difficulty in this approach is the specification of the utility function. For that reason we do not reject models where constraints are introduced by choice. On the other hand, in practice it seems advisable indeed to use an expected utility approach (with a utility function of a simple type, e.g. linear, or piecewisely linear concave) in the specification of optimality, as soon as feasibility has been specified. This approach might be called *restricted rational*, as a relaxation of the presumptuous terminology of [30], for example. Moreover, in our view it will be very "rational" in practice to consider the "optimal" solution as the start of the mathematical analysis of the decision problem rather than as the end of it: usually, a careful analysis of its properties will often give rise to an adapted definition of feasibility and optimality, so that a sequence of restricted rational models will be the result.

REMARK 3.8. For a long time there has been an interest in the relationship between chance-constrained models and recourse models. Attempts have been made to show equivalence. For example, the elementary production-inventory control models, with a random demand $\omega$,

$$(3.26) \qquad \text{minimize}_{x\in\mathbb{R}} \{x: P(x < \omega) \leqq \alpha\}$$

and

$$(3.27) \qquad \text{minimize}_{x\in\mathbb{R}} \{cx + E\{q.(x-\omega)^- + h.(x-\omega)^+\}\}$$

have the same optimal solution if $-h < c < q$ and $\alpha := (h+c)/(h+q)$, at least if the distribution function of $\omega$ is strictly increasing on its support, and if $E|\omega| < \infty$. Such a *mathematical equivalence* can be shown to hold for a class of problems, but it is not valid in general. A typical difference is, for example, that recourse models always give rise to convex programming problems, whereas the convexity of the feasibility sets determined by chance constraints depends on the probability distribution involved. In our opinion this feature cannot be credited to the chance constraints: people who are willing to use linear programs as a model for decision making should be willing to accept convex combinations of feasible solutions as such, and we do not see why this willingness should be abandoned in the presence of randomness.

Even if mathematical equivalence of a chance-constrained model and a recourse model can be established, this does not say they can simply be exchanged: different types of data are asked for, so that there is no *economic equivalence*, necessarily. It is quite conceivable, that one is willing to specify the maximal acceptable probability of shortage, whereas the evaluation of shortage in terms of costs causes difficulties, or the other way around. It is even possible that both specifications can be made, but that the results do not coincide.

It has been argued, see e.g. [12], that the chance-constrained models are inferior to recourse models in the sense that the former models do not take into consideration the consequences of possible violations of the constraints in (3.1) when $\omega$ gets its realization. We do not agree with this statement. First of all, a chance constraint might represent an important goal as such, but moreover its use does not exclude simultaneous incorporation of a penalty cost for infeasibility, e.g. those in (3.4) [23]. In [12] it has also been indicated that the choice of the risk parameters in the chance-constraints - $\alpha_i$ in (3.2), $\alpha$ in (3.3) - is far from easy in practice. But it must be said that similar difficulties arise in penalty models - except for recourse models with physical recourse actions which can be evaluated in monetary terms, but there have been only a few applications of this type. It has been advocated [38] that the maximal acceptable risk $\alpha$ in (3.3) should be considered as a decision variable, and that a decreasing function $h(\alpha)$ should be added to the objective function. We agree that the choice of $\alpha$ is at the disposal of the model builder, who may and should vary its value during his experiments, but we do not see how the risk cost function $h$ has to be specified in practice.

REMARK 3.9. There is a difference between chance-constrained and recourse models which seems trivial but which nevertheless is often overlooked: they use a different measure for risk. In chance constraints risk is measured *qualitatively* whereas in recourse models this is done *quantitatively*. That is, in chance constraints only the possibility of infeasibility is at stake regardless of the amounts by which the constraints are violated, whereas in recourse models these amounts themselves are important. In practice it might be acceptable to have a constraint violated, perhaps even with a high probability if necessary, if the amount by which it is violated is small enough. In such cases the quantitative approach of infeasibility of recourse models is more appropriate. This does not mean, however, that at the same time one has to adopt the penalty cost structure of recourse models. In certain circumstances it might be more appropriate to fix an upper bound on the risk, as in chance constraints, rather than to penalize it. This is true in particular if it is very difficult to specify the costs of infeasibility. Mixing the ideas of (3.2) and (3.4) in the above sense, we introduce the constraints

$$(3.28) \qquad E(\eta_i(x,\omega))^- \leqq \beta_i, \; i = 1,\dots,m,$$

where $\beta_i$ are positive risk aversion parameters to be specified in advance. We call the constraints in (3.28) *integrated chance constraints*, since the following equality is satisfied, as we shall show later on, see (3.32):

$$(3.29) \qquad E(\eta_i(x,\omega))^- = \int_{-\infty}^{0} P(\eta_i(x,\omega) < t)dt.$$

That is, (3.28) can be seen as the integrated version of (3.15). Constraints of this type, and related ones, are studied in Chapter 5. It appears, that the mathematical properties of integrated chance constraints are nicer than those of the usual chance constraints. This can be explained as follows. Both $P(\eta_i(x,\omega) < 0)$ and $E(\eta_i(x,\omega))^-$ can be written as an *expected loss* $E\, v(\eta_i(x,\omega))$; in the former case the loss function v is defined by

$$(3.30) \qquad v(z) = \begin{cases} 1 \text{ if } z < 0, \\ 0 \text{ if } z \geqq 0, \end{cases}$$

and in the latter case by

$$(3.31) \qquad v(z) = \begin{cases} -z \text{ if } z < 0, \\ 0 \text{ if } z \geqq 0. \end{cases}$$

Comparison of both loss functions illustrates not only the difference between the "qualitative" and the "quantitative" measurement of risk, but it also reveals the different mathematical properties. Although both loss functions are nonincreasing, the former decreases strictly in $z_i = 0$ only, whereas the latter decreases strictly on the entire negative halfline. Moreover, the latter is convex and continuous, whereas the former is not. For more details on integrated chance constraints we refer to Chapter 5.

It is interesting to notice that *expected loss constraints* can be interpreted as *weighted integrated chance constraints*. In order to see this, let $\mu$ be a (not necessarily finite) measure on the Borel sets of $(-\infty,0]$ and define $v(z) := \mu((z,0])$, $z \in \mathbb{R}$. Then, with $\pi$ the probability measure on the Borel sets of $\mathbb{R}$ corresponding to the distribution of $\eta_i(x,.)$, we conclude from Fubini's theorem that

$$\int_{(-\infty,z]}\{\int_{(z,0]}\mu(dt)\}\pi(dz) = \int_{(-\infty,0]}\{\int_{(-\infty,t)}\pi(dz)\}\mu(dt).$$

But this is nothing other than

$$\text{(3.32)} \qquad E\, v(\eta_i(x,\omega)) = \int_{(-\infty,0]}P(\eta_i(x,\omega) < t)\mu(dt).$$

This formula shows that the use of a loss function is equivalent to a weighting of the risks $P(\eta_i(x,\omega) < t)$. In (3.30) all the weight is concentrated in $t = 0$; in (3.31) the weights are uniform on $(-\infty,0]$.

We conclude this digression with the remark that it seems possible to formalize the theory of expected loss constraints along the same lines as is done in the expected utility theory. This would imply that risk aversion should be modeled by a convex decreasing loss function, which is analogous to a concave increasing utility function. We will not work out the details here. Instead we confine ourselves to remarking that uncareful use of chance constraints, which as we have seen correspond to a nonconvex loss function, in certain cases leads to apparently risky decisions.

We close this section with a review of selected references. For general introductions to stochastic programming we refer to [6,38]. The standard mathematical theory is covered in [14]. Numerical results are described in [40]. An up-to-date bibliography on stochastic programming is [41]. Chance constraints were introduced by Charnes and Cooper [3], whereas important theoretical contributions are due to Prékopa [24,23]. By many applications he and his coworkers have shown that chance constraints are useful in practice. Schweigman [31] uses chance constraints and other concepts from stochastic programming in a study on the problems of farmers in developing countries. Recourse models were introduced by Dantzig [4,5], Beale [1] and studied systematically by Wets [32,34,35,36, 37]. The mean-variance analysis is due to Markowitz [17]. Dupaçova [7] is one of those who recognized the possibilities of what we called refined penalty cost models. She also initiated the minimax approach to partly known distributions [39]. Wessels [33] introduced the term discrepancy evaluation function for what we named loss function; actually, he already indicated the idea of expected loss constraints. Garstka [10] invented the term economic equivalence. Ziemba and several of his students [15,16] were very successful in applying simple recourse models to real-life financial problems. For extensive treatments on utility theory we refer to von Neumann and Morgenstern [18], Savage [29], Fishburn [9] and Schneeweiss [30]. Pratt [22] introduces measures for risk aversion. Elementary introductions to the closely related decision analysis have been given by Raiffa [25] and Holloway [13].

## 3.3 MULTISTAGE STOCHASTIC LINEAR PROGRAMMING MODELS AND THEIR RELATION TO STOCHASTIC DYNAMIC PROGRAMMING

As we have seen in the previous section in recourse models two stages are distinguished: the decisions in the first stage correspond to the here-and-now decisions, whereas those in the second stage are of the wait-and-see type. Dantzig [4] already recognized that a more flexible modeling of the process of collecting information is possible. He introduced multistage stochastic linear programs. In each of a finite number of stages the realizations of a subset of the random parameters become known, and each decision variable is assigned to the stage which corresponds to the total information available. Also the constraints are partitioned according the stages: they have a recourse interpretation. Apart from these "coupling" constraints there are also constraints within each stage separately; they will be represented by nonempty polyhedral convex sets $C_t$. The underlying linear program has then the following dynamic structure:

$$
\begin{array}{lllll}
\text{minimize} & c_1x_1 + c_2x_2 + c_3x_3 + \dots + c_Tx_T & & & \\
\text{subject to} & A_{11}x_1 & & & = b_1 \\
& A_{21}x_1 + A_{22}x_2 & & & = b_2 \\
& A_{31}x_1 + A_{32}x_2 + A_{33}x_3 & & & = b_3 \\
& \vdots & & & \\
& A_{T1}x_1 + A_{T2}x_2 + A_{T3}x_3 + A_{TT}x_T & & & = b_T \\
& x_1 \in C_1,\ x_2 \in C_2,\ x_3 \in C_3,\ \dots,\ x_T \in C_T. & & &
\end{array}
\tag{3.33}
$$

The number of stages is T. The vector of decisions at stage t is $x_t$; it has to be chosen in the set $C_t$. The constraints Ax = b have a lower block triangular structure; the t-th block of constraints represents the recourse at stage t, where $A_{tt}$ is the recourse matrix. For T = 2 the structure in (3.33) is that of the recourse model in the previous section, see (3.7). Let us now define the *information structure* associated with (3.33). It is assumed that all coefficients, not only the elements of the vectors $c_t$ and $b_t$ and of the matrices $A_{st}$ but also those required to define $C_t$, say $c_t'$, constitute a random vector ω with a known probability distribution. This vector is partitioned as $\omega = (\omega_1,\omega_2,\dots,\omega_T)$, where $\omega_t$ contains the elements of $c_t$, $b_t$, $A_{st}$ for $s = 1,\dots,t$ and $c_t'$. The characteristic feature of multistage stochastic linear programming is that it is assumed that the decisions $x_t$ are allowed to depend upon the realizations of $(\omega_1,\dots,\omega_t)$ but not upon those of $(\omega_{t+1},\dots,\omega_T)$. (Since the realization of $\omega_1$ is known at each stage, we consider it to be deterministic.) Moreover, the constraints in (3.33) have to be satisfied for all ω, or at least with probability 1. The objective is, to find decisions which minimize the mean value of the objective function in (3.33). This completes the specification of the data of the multistage stochastic linear programming problem.

Basically, there are two different approaches to the precize definition of multistage problems; Olsen [21] calls them the *dynamic* and the *static* version of the incompletely specified problem. For convenience of exposition we will use this terminology throughout this section. The "dynamic" version seems to be the oldest one. The idea is that the model is meant primarily as a way of finding optimal values for the first-stage variables; therefore one tends to eliminate the other stages in order to find a so-called *deterministic equivalent problem* in $x_1$ only. This can be established by applying the backward recursion of dynamic programming. With the notations

$$\omega^t := (\omega_1,\ldots,\omega_t),\ t = 1,\ldots,T,$$
$$x^t := (x_1,\ldots,x_t),\ t = 1,\ldots,T,$$

this recursion defines the functions $Q_t(x^t,\omega^t)$, successively for $t = T,T-1,\ldots,1$, starting with $Q_{T+1} = 0$, by

$$Q_t(x^t,\omega^t) = E^t[\inf_{x_{t+1}\in C_{t+1}}\{c_{t+1}x_{t+1} + Q_{t+1}(x^{t+1},\omega^{t+1}): \quad \Sigma_{j=1}^{t+1}A_{t+1,j}x_j = b_{t+1}\}]. \tag{3.34}$$

Here $E^t$ denotes a (regular) conditional expectation of $\omega^{t+1}$ given $\omega^t$. The "deterministic equivalent problem" is then

$$\text{minimize}_{x_1\in C_1}\{c_1x_1 + Q_1(x_1): A_{11}x_1 = b_1\} \tag{3.35}$$

where $Q_1(x_1) := Q_1(x_1,\omega_1)$. It should be noted that in the "dynamic" version of multistage stochastic programs with the structure of (3.33), just as in the two-stage recourse model of the previous section, the model (3.34) is seen as the *definition* of the optimization problem, see e.g. [36,19,20]. The authors of these papers are aware of the intricate technical problems associated with the rigorous definition of recursion (3.34) and they impose conditions on the data in order to be certain that everything is indeed well-defined. Conditions have been given which not only guarantee that each $Q_t(x^t,\omega^t)$ is well-defined, but also that it is a lower semicontinuous convex extended real function of $x^t$. A special issue is the characterization of the *induced constraints* on $x_1$ (i.e. induced by the condition that in future stages it must be possible to satisfy the constraints with probability 1), which under certain assumptions can be described as $Q_1(x_1) < +\infty$.

Other studies [8,21,35,26,27,28] consider the "static" version of multistage stochastic programming problems. The idea is to interpret (3.33) as a linear program in function spaces. That is, the decisions are interpreted as (Borel measurable) functions of the random vector $\omega$, in such a way that they are *nonanticipative*: $x_t := x_t(\omega^t)$, $t = 1,\ldots,T$. The constraints in (3.33) are then interpreted as equalities in suitably chosen function spaces. Usually, both the spaces of the variables and those of the constraints are chosen to be $L_p$-spaces, $1 \leqq p \leqq \infty$. Omitting the precise choice of the spaces we get the following model as the "static" version of (3.35):

$$\begin{aligned} &\text{minimize} \quad E\, \Sigma_{t=1}^{T} c_t(\omega^t) x_t(\omega^t) \\ &\text{subject to} \quad \Sigma_{t=1}^{s} A_{st}(\omega^t) x_t(\omega^t) = b_s(\omega^s) \text{ a.s., } s = 1,\ldots,T, \\ &\qquad\qquad\quad x_t(\omega^t) \in C_t(\omega^t) \text{ a.s., } t = 1,\ldots,T. \end{aligned}$$

As can be seen in the indicated literature, the "static" version is very suitable for the development of duality theorems for multistage programs. For this purpose it is not essential that the matrix in (3.33) is lower block triangular; Eisner and Olsen [8] consider the full matrix case. The constraints of the s-th stage are then specified as

$$E^s\, \Sigma_{t=1}^{T} A_{st}(\omega) x_t(\omega^t) = b_s(\omega^s) \text{ for almost all } \omega^s.$$

As a result the primal and the dual problem have a similar structure. Rockafellar and Wets [26,27,28] consider convex rather than linear problems.

The "dynamic" and the "static" versions of multistage stochastic linear programs are supposed to model the same problem. Intuitively, one would expect that they are equivalent. Probably the slightly confusing term "deterministic equivalent (to what?) problem" hints in that direction. Equivalence is indeed not difficult to prove if the random vector has a finite discrete distribution. Also it is not difficult to prove equivalence if the recursion (3.34) *behaves properly* (that is, if the functions $Q_t$ are well-defined and Borel measurable, if each infimum is attained and if for all t and all $\omega^t$ minimizing values $x_t^o(\omega^t)$ can be chosen such that $x_t^o$ is Borel measurable), at least if the function spaces in the "static" version are large enough to contain the "dynamic" solution $x_t^o$. But in general equivalence is far from trivial, see e.g. [21]. In fact, the proof of this equivalence is a main subject in the theory of (finite horizon) *stochastic dynamic programming*. We refer to Section 4.6 for an introduction to the model and to [11,2] for a thorough treatment of the theory, and to Chapter 8 for a special case. Here we will restrict ourselves to a short outline of dynamic programming. A first basic feature is the introduction of *states* as part of the problem formulation. The evolution of the state with the time is considered as a controlled random process. At each stage t the current state $s_t$ is observed, and the decision $x_t$ of that stage may and will depend on the realization of $s_t$. The effect of this decision is twofold: the cost at stage t as well as the probability distribution of the next stage may depend on it. A second basic feature is the introduction of *policies*: these are sequences $(x_1(s_1),\dots,x_T(s_T))$ of (Borel measurable, say,) decision rules $x_t(s_t)$. The corresponding optimization problem is then:

find a policy with minimum expected total cost.

Notice that this optimization problem is of the "static" type. On the other hand, the main tool of the analysis of stochastic dynamic programming problems is the famous backward recursion, which gave its name to the whole field. Notice that such a recursion is typically of the "dynamic" type.

It is not difficult to see that recursion (3.34) can be recasted in the framework of stochastic dynamic programming: just define as state at stage t the vector of known realizations of the random parameters and the previous decisions

$$s_t := (x^{t-1},\omega^t),\ t = 1,\dots,T.$$

So it is not surprising that dynamic programming theory is of help to show equivalence between both versions of multistage stochastic programs. This is not always recognized as such. We conjecture, for example, that the proofs of [21] could be simplified if an appropriate use of the results of dynamic programming theory were made. This issue is however not taken up any further in this book.

We conclude this section with a summary of the subjects in the areas of multistage stochastic linear programming and stochastic dynamic programming which are dealt with in this book.

In Section 4.6 we study the general finite horizon stochastic dynamic programming problem by formulating it as a linear programming problem in which the decision variables are probability measures on state and action spaces. The basic tool is the duality theory for abstract linear programs. The dual problem appears to be a representation of the backward recursion.

In Chapter 6 we give rather weak sufficient conditions under which the backward recursion behaves properly. This result can be useful for proving equivalence between the "dynamic" and the "static" versions of multistage stochastic programming models. This is illustrated in Section 6.4 where it is applied to a rather general class of Markovian production-inventory control models. The special form of the state transformation in inventory models will allow us to use rather weak conditions on the data of the problem; these conditions fit in the "dynamic" formulation.

In Chapter 8 again a production-inventory control model is studied. It is more general than the one in Chapter 6 in the sense that it does not have to be Markovian; on the other hand it is more simplified (e.g. one product, linear costs) giving it the structure of (3.33). In particular, the matrix is lower block staircase rather than lower block triangular, and only the right-hand sides are random. The problem is defined as a multistage stochastic linear program in the "static" sense. Its dual problem is easily formulated, and the primal as well as the dual problem is solved explicitly. Optimal productions are characterized by optimal inventory levels, which is not surprising for anyone who has a basic knowledge of elementary inventory control models. Text books on Operations Research consider the "dynamic" version of such problems. The main reason to consider the "static" version is the study of the dual linear program. It is not amazing that also in the "static" formulation of the problem the "dynamic" aspects are dominant, not only in the primal but also in the dual problem. It will appear that the dual problem describes other backward recursions which are equivalent to the original one. It will also appear that the optimal dual solution can be considered as a price system related to the condition that all productions must be nonanticipative. For the sake of comparison also the corresponding deterministic problem is analyzed by means of linear programming duality.

## REFERENCES

1. E.M.L. BEALE (1955). On minimizing a convex function subject to linear inequalities. *J. Roy. Statist. Soc. Ser. B 17*, 173-184.
2. D.P. BERTSEKAS, S.E. SHREVE (1978). *Stochastic Optimal Control: the Discrete-Time Case*, Academic Press, New York-San Francisco-London.

3. A. CHARNES, W.W. COOPER (1959). Chance-constrained programming. *Management Sci. 6*, 73-79.

4. G.B. DANTZIG (1955). Linear programming under uncertainty. *Management Sci. 1*, 197-206.

5. G.B. DANTZIG (1963). *Linear Programming and Extensions*, Princeton University Press, Princeton NJ.

6. M.A.H. DEMPSTER (1980). Introduction to stochastic programming. M.A.H. DEMPSTER (ed.). *Stochastic Programming*, Academic Press, New York-San Francisco-London, 3-59.

7. J. DUPAÇOVA (1980). Water resources system modelling using stochastic programming with recourse. P. KALL, A. PREKOPA (eds.). *Recent Results in Stochastic Programming*. Lecture notes in economics and mathematical systems 179, Springer, Berlin-Heidelberg-New York, 121-133.

8. M.J. EISNER, P. OLSEN (1975). Duality for stochastic programming interpreted as L.P. in $L_p$-space. *SIAM J. Appl. Math. 28*, 779-792.

9. P.C. FISHBURN (1970). *Utility Theory for Decision Making*, Wiley, New York.

10. S.J. GARSTKA (1980). The economic equivalence of several stochastic programming models. M.A.H. DEMPSTER (ed.). *Stochastic Programming*, Academic Press, New York-San Francisco-London, 83-91.

11. K. HINDERER (1970). *Foundations of Nonstationary Dynamic Programming with Discrete Time Parameter*. Lecture notes in operations research and mathematical systems 33, Springer, Berlin-Heidelberg-New York.

12. A.J. HOGAN, J.G. MORRIS, H.E. THOMPSON (1981). Decision problems under risk and chance-constrained programming: Dilemmas in the transition. *Management Sci. 27*, 698-716.

13. C.A. HOLLOWAY (1979). *Decision Making under Uncertainty; Models and Choices*, Prentice-Hall, Englewood Cliffs NJ.

14. P. KALL (1976). *Stochastic Linear Programming*, Springer, Berlin-Heidelberg-New York.

15. J.G. KALLBERG, R.W. WHITE, W.T. ZIEMBA (1982). Short term financial planning under uncertainty. *Management Sci. 28*, 670-682.

16. M.I. KUSY, W.T. ZIEMBA (1983). *A Bank Asset and Liability Management Model*, Collaborative paper CP-83-59, IIASA, Laxenburg.

17. H.M. MARKOWITZ (1959). *Portfolio Selection*, Wiley, Ney York.

18. J. VON NEUMANN, O. MORGENSTERN (1947). *Theory of Games and Economic Behaviour*, Princeton University Press, Princeton NJ.

19. P. OLSEN (1976). Multistage stochastic programming with recourse: the equivalent deterministic problem. *SIAM J. Control 14*, 495-517.

20. P. OLSEN (1976). When is a multistage stochastic programming problem well-defined? *SIAM J. Control 14*, 518-527.

21. P. OLSEN (1976). Multistage stochastic programming problems with recourse as mathematical programming in an $L_p$-space. *SIAM J. Control 14*, 528-537.

22. J.W. PRATT (1964). Risk aversion in the small and in the large. *Econometrica 32*, 122-136.

23. A. PREKOPA (1973). Contributions to the theory of stochastic programming. *Math. Programming 4*, 202-221.

24. A. PREKOPA (1980). Logarithmic concave measures and related topics. M.A.H. DEMPSTER (ed.). *Stochastic Programming*, Academic Press, New York-San Francisco-London, 63-82.

25. H. RAIFFA (1968). *Decision Analysis; Introductory Lectures on Choices under Uncertainty*, Addison-Wesley, Reading Mass.

26. R.T. ROCKAFELLAR, R.J.-B. WETS (1974). Continuous versus measurable recourse in N-stage stochastic programming. *J. Math. Anal. Appl. 48*, 836-859.

27. R.T. ROCKAFELLAR, R.J.-B. WETS (1977). Measures as Lagrange multipliers in multistage stochastic programming. *J. Math. Anal. Appl. 60*, 301-313.

28. R.T. ROCKAFELLAR, R.J.-B. WETS (1978). The optimal recourse problem in discrete time: $L_1$-multipliers for inequality constraints. *SIAM J. Control 16*, 16-36.

29. L.J. SAVAGE (1954). *The Foundations of Statistics*, Wiley, New York.

30. H. SCHNEEWEISS (1967). *Entscheidungskriterien bei Risico*, Springer, Berlin-Heidelberg-New York.

31. C. SCHWEIGMAN (1985). *Operations Research Problems in Agriculture in Developing Countries*, to be published.

32. D. WALKUP, R.J.-B. WETS (1967). Stochastic programs with recourse. *SIAM J. Appl. Math. 15*, 1299-1314.

33. J. WESSELS (1967). Stochastic programming. *Statist. Neerlandica 21*, 39-53.

34. R.J.-B. WETS (1966). Programming under uncertainty: the complete problem. *Z. Wahrsch. Verw. Gebiete 4*, 316-339.

35. R.J.-B. WETS (1970). Problèmes duaux en programmation stochastique. *C.R. Acad. Sci. Ser. A-B 270*, A47-A50.

36. R.J.-B. WETS (1972). Stochastic programs with recourse: a basic theorem for multistage problems. *Z. Wahrsch. Verw. Gebiete 21*, 201-206.

37. R.J.-B. WETS (1974). Stochastic programs with fixed recourse: the equivalent deterministic program. *SIAM Rev. 16*, 309-339.

38. R.J.-B. WETS (1983). Stochastic programming: solution techniques and approximation schemes. A. BACHEM, M. GROETSCHEL, B. KORTE (eds.). *Mathematical Programming: The State of the Art - Bonn 1982*, Springer, Berlin-Heidelberg-New York-Tokyo, 566-603.

39. J. ZACKOVA (1966). On minimax solutions of stochastic linear programming problems. *Časopis Pěst. Math. 91*, 423-429.

40. Numerical methods in stochastic programming, IIASA, Laxenburg; to be published.

41. Bibliography of stochastic programming; to be contained in [40].

CHAPTER 4

# SOME LINEAR PROGRAMS IN PROBABILITIES AND THEIR DUALS

## 4.1. INTRODUCTION

As has been indicated in Chapter 2 the optimality conditions for abstract convex programming problems can be formulated in terms of dual problems. In this chapter we are dealing with the special case of *linear* programming duality. It has its own flavour, since the dual problem of a linear programming problem is easily formulated in an explicit form. The optimality conditions boil down then to primal and dual feasibility and complementary slackness, see Chapter 2.

Many problems in mathematics can be formulated as abstract linear programming problems, and it may be expected that dualization provides insight in the structure of the problem. In what follows we consider a number of linear programs in which the decision variables are *probability measures*. These problems are well-known: they stem from robustness analysis where the probability distribution is partly unknown (e.g. moment problems or marginal problems) or from dynamic programming where the state probabilities are partly controllable. In each case the nonnegativity of the variables is obvious, and also the linearity of objective function and constraint mappings is quite natural, since many operations on probabilities such as expectations (but also passing to a marginal distribution, or composition with a fixed transition probability) are linear. In fact, there is a natural duality between "measures" and "functions" provided by "integration", the integral being interpreted as bilinear form. Unless the underlying spaces are finite, the linear programs in probabilities are, of course, infinite-dimensional. However, since "abstract" duality theory is formulated in the framework of general topological vector spaces it is not unreasonable to expect that this stumble-block can be overcome. This point of view is the basic idea of this chapter. Since the theory of moment problems and that of stochastic dynamic programming are well-established, new results are not to be expected. Instead, the question is: is it possible to derive the well-known results by stressing the linear structure of the problem and by applying duality theory? It will appear that although "the" dual linear programs are easily formulated, the main difficulty is the precize choice of the spaces involved. This choice is, of course, crucial for the applicability of general duality theorems as Theorem 2.1. As a matter of fact, whereas the dual problems have precisely the form which is to be expected, straight application of general duality theory appears to be of restricted value, since the verification of sufficient conditions for normality and stability is only possible under too restrictive assumptions. That is, by exploiting the special properties of each of the problems considered direct proofs of duality results have been given under weaker conditions than probably is possible just as a corollary of abstract linear programming duality.

In Section 4.2 we build up the framework of abstract linear programming duality, and in the next section some basic features of linear programs in probability measures are formulated. Section 4.4 deals with the generalized moment problem and its dual. Special attention is paid to the moment problem with (lattice) bounds on the probabilities. The relation with Chebychev approximation and the Neyman-Pearson lemma is indicated. The marginal problem, a generalization of the classical transportation problem, is the subject of Section 4.5. Just as the moment problem the marginal problem is related to many inequalities existing in probability theory. In Section 4.6 we study the finite horizon stochastic dynamic programming problem. The last section provides the conclusions.

## 4.2 DUAL LINEAR PROGRAMS

In the formulation of linear programming problems inequality conditions are fundamental. An inequality "$\geqq$" in a vector space X over the reals is defined by

$$x_1 \geqq x_2 \text{ iff } x_1 - x_2 \in X^+ \quad (x_1, x_2 \in X)$$

where $X^+$ is any convex cone in X with apex at 0. The vectors of $X^+ = \{x \in X: x \geqq 0\}$ are called positive, and $X^+$ is called the positive cone. Extreme specifications such as $X^+ := X$, in which case the condition $x \geqq 0$ is not restrictive, or $X^+ := \{0\}$, in which case 0 is the only positive vector, are not excluded. The inverse relation is, of course, defined by $x_1 \leqq x_2 \Leftrightarrow x_2 \geqq x_1$, and the vectors of $-X^+ = \{x \in X: x \leqq 0\}$ are called negative. If $X^+ \cap (-X^+) = \{0\}$, $X = (X, \geqq)$ is called an ordered vector space, since the relation $\geqq$ is a partial order in that case.

In a general setting, a linear programming problem is defined as an optimization problem of one of the following types

$$\text{minimize}_{x \in X}\{cx: Lx \geqq b,\ x \geqq 0\}$$

$$\text{maximize}_{x \in X}\{cx: Lx \leqq b,\ x \geqq 0\}$$

where $b \in U$, X and U are real vector spaces with fixed positive cones, $L: X \to U$ is a linear map and $c: X \to \mathbb{R}$ is a linear form on X. Notice that also the case $Lx = b$ is covered (take $U^+ = \{0\}$). Similarly, if one likes to suppress the constraints $x \geqq 0$, $X^+$ can be defined as X. Maximization can, of course, be transformed into minimization, and $\leqq$ inequalities into $\geqq$ inequalities, but we will deal with both standard forms since it is convenient in the treatment of duality.

A pair of real vector spaces (V,X) is called a *duality* if a bilinear form $\langle .,. \rangle: V \times X \to \mathbb{R}$ has been singled out. Then $\langle v,. \rangle$ is a linear form on X, for each $v \in V$, and $\langle .,x \rangle$ is a linear form on V, for each $x \in X$. A shorthand notation for the duality is $\langle V,X \rangle$.

DEFINITION 4.1. A dual pair of linear programming problems.
*The pair of linear programs*

$$(LP_1) \qquad \text{minimize}_{x \in X}\{\langle c,x\rangle : L_1 x \geqq b,\ x \geqq 0\}$$
$$(LP_2) \qquad \text{maximize}_{y \in Y}\{\langle y,b\rangle : L_2 y \leqq c,\ y \geqq 0\}$$

*is called a dual pair if their data satisfy the following conditions:*
*a.* X, U, Y, V *are real vector spaces, each with a fixed positive cone.*
*b.* $\langle V,X\rangle$ *and* $\langle Y,U\rangle$ *are two dualities. The dualities are compatible with the positivities:*

$$(4.1) \qquad \begin{array}{l} x \geqq 0 \text{ and } v \geqq 0 \text{ imply } \langle v,x\rangle \geqq 0, \\ u \geqq 0 \text{ and } y \geqq 0 \text{ imply } \langle y,u\rangle \geqq 0. \end{array}$$

*c.* $L_1: X \to U$ *and* $L_2: Y \to V$ *are adjoint linear maps, i.e.*

$$(4.2) \qquad \langle L_2 y,x\rangle = \langle y,L_1 x\rangle \text{ for all } x \in X,\ y \in Y.$$

*d.* $b \in U$, $c \in V$.

If $(LP_1,LP_2)$ is a dual pair then $LP_i$ is called a dual of $LP_{3-i}$, $i = 1,2$. By definition it is a symmetrical relation.

Before going into more details regarding this definition, we first want to indicate that weak duality,

$$(4.3) \qquad \inf LP_1 \geqq \sup LP_2,$$

can easily be shown for arbitrary dual pairs of linear programs. Remark first that (4.3) holds trivially if at least one of the programs is infeasible. Further, if x and y are feasible for $LP_1$ and $LP_2$, respectively, then

$$(4.4) \qquad \langle c,x\rangle \geqq \langle L_2 y,x\rangle = \langle y,L_1 x\rangle \geqq \langle y,b\rangle$$

where the inequalities follow from (4.1) and the equality is given by (4.2). From (4.4) one derives (4.3) by infimization over x and supremization over y. With respect to optimal solutions of $LP_1$ and $LP_2$, it is clear that optimality of x as well as y is guaranteed if everywhere in (4.4) equality holds, that is, if the *complementary slackness* conditions hold:

$$\text{(4.5)}\qquad \begin{aligned} \langle c - L_2 y, x\rangle &= 0,\\ \langle y, L_1 x - b\rangle &= 0. \end{aligned}$$

This trivial analysis proves the following theorem.

THEOREM 4.2. Elementary Duality Theorem.
*Let* $(LP_1, LP_2)$ *be a dual pair of linear programming problems. Then the following statements are true.*

*a.* $\inf LP_1 \geqq \sup LP_2$.

*b.* $\left.\begin{array}{l} x \textit{ is feasible for } LP_1 \\ y \textit{ is feasible for } LP_2 \\ \textit{complementary slackness holds} \end{array}\right\} \Leftrightarrow \left\{\begin{array}{l} x \textit{ is optimal for } LP_1 \\ y \textit{ is optimal for } LP_2 \\ \inf LP_1 = \sup LP_2. \end{array}\right.$

Notice that for the proof of Theorem 4.2 one does not need very detailed information on the data. In particular, no topological assumptions are made. Theorem 4.2 indicates that optimality is guaranteed by feasibility and complementary slackness. On the other hand, complementary slackness is only necessary for optimality if there is no duality gap ($\inf LP_1 = \sup LP_2$) and if both problems have optimal solutions. The last condition can, of course, be relaxed by looking at sequences: if $x^{(k)}$ and $y^{(k)}$ are feasible for $LP_1$ and $LP_2$, respectively, for all $k = 1,2,\ldots$ then $(x^{(k)})$ and $(y^{(k)})$ are optimal sequences and $\inf LP_1 = \sup LP_2$ iff complementary slackness holds asymptotically:

$$\text{(4.6)}\qquad \begin{aligned} &\lim_{k\to\infty} \langle c - L_2 y^{(k)}, x^{(k)}\rangle = 0,\\ &\lim_{k\to\infty} \langle y^{(k)}, L_1 x^{(k)} - b\rangle = 0. \end{aligned}$$

In spite of its elementary nature Theorem 4.2 is important, since it shows that one might circumvent intricate topological questions if complementary slackness can be proved in a direct way. We will come back to this in the sequel (see e.g. Theorem 4.10).

It is clear that Definition 4.1 is useless unless adjoint maps exist. Of course, any linear map $L: X \to U$ has a unique algebraic adjoint map $L': U' \to X'$ defined by $(L'u')(x) := u'(Lx)$, $x \in X$, $u' \in U'$, where $U'$ and $X'$ are the algebraic dual spaces of U and V, respectively. But it will appear to be important to allow for the possibility that V and Y are not the algebraic duals of X and U. It is not difficult to settle the question of existence and uniqueness of the adjoints in this wider setting. To show this we introduce the following definitions. A duality $\langle V,X\rangle$ is called *separated* if X distinguishes points in V

(4.7) $\qquad \langle v,x\rangle = 0 \ \forall x \in X \ \Rightarrow \ v = 0$

and if V distinguishes points in X

(4.8) $\qquad \langle v,x\rangle = 0 \ \forall v \in V \ \Rightarrow \ x = 0.$

Under (4.7) all $v \in V$ represent different linear forms on X so that V may be identified with a subspace of $X'$. Similar, $X \subset V'$ if (4.8) holds.

PROPOSITION 4.3. *Let* $\langle V,X\rangle$ *and* $\langle Y,U\rangle$ *be separated dualities. Then any linear map* $L_1: X \to U$ *has a unique adjoint map* $L_2: Y \to V$ *provided* V *is large enough:* $V \supset L_1'(Y)$. *Similarly, any linear map* $L_2: Y \to V$ *has a unique adjoint map* $L_1: X \to U$ *provided* $U \supset L_2'(X)$.

PROOF. Given $L_1$, define $L_2$ as the restriction of $L_1'$ to Y. This is possible since U distinguishes points in Y. The range of $L_2$ is a subset of V by assumption, and (4.2) holds. The uniqueness of $L_2$ follows from the assumption that X distinguishes points in V. Similarly, for any linear map $L_2: Y \to V$ there exists a unique $L_1: X \to U$ such that (4.2) holds, as V distinguishes points in X, if $L_2'(X) \subset U$ and as Y distinguishes points in U. □

Each linear program has at least one dual in the sense of Definition 4.1. In order to be specific, assume that $LP_1$ with the data X, $X^+$, U, $U^+$, $L_1$, b and c has been given. Then natural choices for Y, V, $L_2$ are $U'$, $X'$, $L_1'$ together with $Y^+ := \{y \in Y: y(u) \geqq 0 \ \forall u \in U^+\}$, $V^+ := \{v \in V: v(x) \geqq 0 \ \forall x \in X^+\}$. This completes the definition of $LP_2$ and all assumptions of Definition 4.1 hold. Often, however, the natural definition of a dual problem is not suitable in practice, since explicit representations for dual variables are missing. Therefore it is convenient to restrict attention to suitable subspaces of the algebraic duals. Usually the choice of the subspaces is made in topological terms. Starting again with $LP_1$, one may choose topologies on X and U such that c and $L_1$ are continuous, and then define $LP_2$ in terms of the topological dual spaces. Then Y is defined as the linear space $U^*$ of all continuous linear forms on U, and V is defined as the linear space $X^*$ of all continuous linear forms on X. Again, if the dualities are separated and $L_1'(U^*) \subset X^*$, then one arrives at a candidate for $LP_2$ (and $L_1'$ is automatically weakly continuous [33 Theorem IV.2.1]). Whether this second construction of $LP_2$ is appropriate depends on the possibility of a suitable choice of the topologies. In particular, one likes to have "nice" dual spaces together with tractable topologies. In the framework of Definition 4.1 the spaces Y and V are chosen a priori, and in applications supposed to be "nice". Given the dualities $\langle V,X\rangle$ and $\langle Y,U\rangle$ only those topologies on X and U are to be considered which satisfy $X^* = V$, $U^* = Y$. Such topologies are called *compatible* with the duality. The weakest topology on X which is compatible with the duality $\langle V,X\rangle$ is the weak topology $\sigma(X,V)$, and the strongest one is the Mackey topology $\tau(X,V)$ [33 Theorem IV.3.3].

As said before, complementary slackness is only necessary for optimality if there is no duality gap. In fact, in addition to the elementary duality theorem we only need results from advanced duality theory in order to prove *normality*: $\inf LP_1 = \sup LP_2$. In the sequel we shall restrict attention to Rockafellar's conjugate duality theory described in Chapter 2. In order to make a connection between his results and the dual pair $(LP_1,LP_2)$ we will first investigate the *conjugate duals* of $LP_1$ and $LP_2$. As the optimal value function of $LP_1$ we define the function $\varphi_1\colon U \to [-\infty,\infty]$,

$$\varphi_1(u) := \inf_{x \geqq 0}\{\langle c,x\rangle : L_1x \geqq b-u\},\ u \in U.$$

The conjugate dual problem of $LP_1$ is then

$$(LP_1^*) \qquad \text{maximize}_{y\in Y}(\inf_{u\in U}(\varphi_1(u) + \langle y,u\rangle)).$$

Similarly, the optimal value function $\varphi_2$ of $LP_2$ is

$$\varphi_2(v) := \sup_{y \geqq 0}\{\langle y,b\rangle : L_2y \leqq c+v\},\ v \in V,$$

and the conjugate dual problem of $LP_2$ is defined as

$$(LP_2^*) \qquad \text{minimize}_{x \in X}(\sup_{v \in V}(\varphi_2(v) - \langle v,x \rangle)).$$

PROPOSITION 4.4. *Let* $(LP_1, LP_2)$ *be a dual pair of linear programs. If the positive cones* $V^+$ *and* $Y^+$ *are induced by the bilinear forms and the positive cones* $X^+$ *and* $U^+$, *that is, if*

$$(4.9) \qquad \begin{array}{l} v \geqq 0 \Leftarrow \langle v,x \rangle \geqq 0 \ \forall x \geqq 0, \\ y \geqq 0 \Leftarrow \langle y,u \rangle \geqq 0 \ \forall u \geqq 0, \end{array}$$

*then* $LP_1^* = LP_2$. *Similarly, if* $X^+$ *and* $U^+$ *are induced by the bilinear forms and* $V^+$ *and* $Y^+$, *that is, if*

$$(4.10) \qquad \begin{array}{l} x \geqq 0 \Leftarrow \langle v,x \rangle \geqq 0 \ \forall v \geqq 0, \\ u \geqq 0 \Leftarrow \langle y,u \rangle \geqq 0 \ \forall y \geqq 0, \end{array}$$

*then* $LP_2^* = LP_1$.

REMARK. The inverse implications in (4.9) and (4.10) are, of course, true because of (4.1).

PROOF. The objective function $\psi$ of $LP_1^*$ can be written as

$$\psi(y) = \inf_{x,u}\{\langle c,x \rangle + \langle y,u \rangle : x \geqq 0,\ L_1 x \geqq b-u\}$$

or, with $u_0 := L_1 x - b + u$, as

$$\begin{aligned} \psi(y) &= \inf_{x,u_0}\{\langle c,x \rangle + \langle y, b - L_1 x + u_0 \rangle : x \geqq 0,\ u_0 \geqq 0\} \\ &= \langle y,b \rangle + \inf_{x \geqq 0}\langle c - L_2 y, x \rangle + \inf_{u_0 \geqq 0}\langle y,u_0 \rangle \end{aligned}$$

because of (4.2). Hence

$$\psi(y) = \begin{cases} \langle y,b \rangle & \text{if } L_2 y \leqq c \text{ and } y \geqq 0, \\ -\infty & \text{otherwise;} \end{cases}$$

the first statement follows from (4.1) and the second from (4.9). Since by definition the supremum over an empty set is $-\infty$, we showed that $LP_2 = LP_1^*$. The proof of $LP_2^* = LP_1$ under (4.10) is similar. □

In the next theorem we apply the sufficient conditions for *stability* (i.e. no duality gap, and the dual extremum is attained) described in Theorem 2.1 to a dual pair of linear programs.

THEOREM 4.5. Advanced Duality Theorem.
*Let* $(LP_1, LP_2)$ *be a dual pair of linear programs with separated dualities.*

a. *If* $\varphi_1$ *is bounded above, hence continuous, in a neighbourhood of* $0 \in U$ *(in a topology on* $U$ *compatible with the duality* $\langle Y,U \rangle$*) and if* (4.9) *holds, then*

$$(4.11) \qquad -\infty \leqq \inf LP_1 = \sup LP_2 < +\infty,$$

*and the supremum is attained if it is finite.*

b. *If* $\varphi_2$ *is bounded below, hence continuous, in a neighbourhood of* $0 \in V$ *(in a topology on* $V$ *compatible with the duality* $\langle V,X \rangle$*) and if* (4.10) *holds, then*

$$(4.12) \qquad \infty \geqq \sup LP_2 = \inf LP_1 > -\infty,$$

*and the infimum is attained if it is finite.*

PROOF. (a) Because of Proposition 4.4 we conclude from (4.9) that $LP_2 = LP_1^*$; then (4.11) is a direct consequence of Theorem 2.1. Both references in the proof of this theorem consider only separated dualities.
(b) Follows by reversing signs. □

REMARK. The boundedness condition on the optimal value function is not necessary for normality or even stability. Weaker sufficient conditions and even necessary and sufficient conditions have been formulated (e.g. [29,32]). However, in applications one needs conditions which are operational; that is, it must be possible to verify whether or not they are satisfied in the problem at hand. Since we were not able to verify these weaker conditions in the examples to be given, we restrict the attention to the more practical conditions of Theorem 4.5.

## 4.3 LINEAR PROGRAMMING IN PROBABILITIES

In this section we introduce a natural duality between "measures" and "functions" which will appear to be useful for all three linear programs to which we want to apply the duality theory. In the next sections we will give more details on these problems; now we give only a rough sketch.

*Moment problem.*

The classical moment problem can be described as:

> For given subset S of $\mathbb{R}$, find a probability measure on S with prescribed first m moments, which minimizes the expectation of a given criterion function on S.

Often the criterion function is the indicator of a subset $S_o \subset S$ so that, in fact, the probability of $S_o$ is minimized.

*Marginal problem.*

> For given spaces $S_i$, $i = 1,\dots,m$, find a probability measure on $S := S_1 \times S_2 \times \dots \times S_m$ with prescribed marginals on each $S_i$, which minimizes the expectation of a given criterion function on S.

Again, the criterion function may be an indicator function.

*Stochastic dynamic programming problem.*

Consider a discrete-time Markovian decision process. That is, the state of the system is random and evoluates in time in a Markovian way. At each stage of time an action is taken which influences the cost of that stage as well as the transition to the state of the next stage. Decisions on the actions are based on perfect observation of the current state, and are to be taken in such a way that the total expected cost is minimal. In terms of partly controllable probabilities, for a finite horizon the problem can be defined as follows.

> For given state and action spaces $S_0, A_0, S_1, \dots, S_{n-1}, A_{n-1}, S_n$ find a probability measure on $S_0 \times A_0 \times S_1 \times A_1 \dots \times S_n$ with prescribed marginal on $S_0$ and prescribed transition probabilities from $S_{i-1} \times A_{i-1}$ to $S_i$, $i = 1,\dots,n$, which minimizes the expectation of the total cost.

Notice that in each problem the decision variables are probability measures. It is also not difficult to see, that the objective function is linear in these variables since it is the expectation of a fixed "cost" function. So it is not too surprising that it appears to be possible to put the three problems in the linear programming duality framework of Definition 4.1. The introduction of the necessary building stones for this is the purpose of the present section. We will introduce suitable linear spaces (for "measures" as well as for "cost functions") with positive cones, bilinear forms and related linear maps and their adjoints.

To begin with, we will embed the set of probability measures on a space S in a *linear* space. Let S be an arbitrary set (in applications a Borel subset of some Euclidean space will do) and $\mathcal{S}$ a σ-algebra of subsets of S which contains all singletons (e.g. $\mathcal{S}$ the Borel subsets of $S \subset \mathbb{R}^n$). A finite signed measure on the measurable space $(S,\mathcal{S})$ is by definition any real (hence finite) countably additive function on $\mathcal{S}$ with $x(\emptyset) = 0$. Because of the assumption of finiteness the *space* $M_o(S)$ *of finite signed measures on* $(S,\mathcal{S})$ is a vector space, which contains the probability measures on $(S,\mathcal{S})$. In fact the probabilities are a convex subset of the natural positive cone of $M_o(S)$, i.e.

$$M_o^+(S) := \{x \in M_o(S)\colon x(E) \geqq 0 \ \forall E \in \mathcal{S}\}$$

consisting of all finite measures on $(S,\mathcal{S})$. This cone defines a *partial order* on $M_o(S)$ since $M_o^+(S) \cap -M_o^+(S) = \{0\}$. Moreover, this cone is *generating*, i.e. $M_o^+(S) - M_o^+(S) = M_o(S)$, since every signed measure in $M_o(S)$ can be decomposed as a difference of two measures in $M^+(S)$: for any $x \in M_o(S)$ a unique Jordan decomposition $x = x^+ - x^-$, $x^+$ and $x^-$ in $M^+(S)$, exists with the property that $x^-$ vanishes on each measurable subset of some $\hat{S} \in \mathcal{S}$ whereas $x^+$ vanishes on each measurable subset of its complement in S. This minimality property makes the decomposition unique, although $\hat{S}$, the *positive set*, is not unique, generally. In fact, $M_o(S)$ is a *vector lattice* under the partial order induced by the positive cone. This means that each pair $(x_1,x_2) \in M_o(S) \times M_o(S)$ has a unique least (order) upper bound $\sup(x_1,x_2) \in M_o(S)$ and a unique greatest (order) lower bound $\inf(x_1,x_2) \in M_o(S)$. These statements are a direct consequence of the existence of a unique Jordan decomposition: just define $\sup(x_1,x_2) := x_1 + (x_2-x_1)^+$

and $\inf(x_1,x_2) := -\sup(-x_1,-x_2)\ \forall x_1,x_2 \in M_o(S)$. Hence for any $x \in M_o(S)$,

$$\begin{aligned} x &= x^+ - x^- \text{ with } \inf(x^+,x^-) = 0, \\ x^+ &= \sup(0,x) \in M_o^+(S), \\ x^- &= \sup(0,-x) \in M_o^+(S), \\ |x| &:= \sup(x,-x) = x^+ + x^- \in M_o^+(S). \end{aligned} \tag{4.13}$$

The measures $x^+$, $x^-$, $|x|$ are called the *positive part*, the *negative part* and the *total variation* of the signed measure x, respectively. As will be seen in the sequel the lattice structure will be helpful in calculations.

A natural companion of $M_o(S)$ is $F_o(S)$, *the vector space of all S-measurable bounded real functions* v *on* S. $F_o(S)$ is also a vector lattice; its partial order is determined by the positive cone

$$F_o^+(S) := \{v \in F_o(S) : v(s) \geqq 0\ \forall s \in S\},$$

and $\sup(v_1,v_2)(s) := \max(v_1(s),v_2(s))$, $\inf(v_1,v_2)(s) := \min(v_1(s),v_2(s))$, $s \in S$, for any $v_1,v_2 \in F_o(S)$. It is easy to verify that for any $v \in F_o(S)$

$$\begin{aligned} v &= v^+ - v^- \text{ with } \inf(v^+,v^-) = 0, \\ v^+ &= \sup(0,v) \in F_o^+(S), \\ v^- &= \sup(0,-v) \in F_o^+(S), \\ |v| &= \sup(v,-v) = v^+ + v^- \in F_o^+(S);\ |v|(s) = |v(s)|,\ s \in S. \end{aligned} \tag{4.14}$$

The definitions of $M_o(S)$ and $F_o(S)$ are such that the bilinear form

$$\langle v,x\rangle := \int_S v(s)x(ds),\ v \in F_o(S),\ x \in M_o(S),$$

defines a duality $\langle F_o(S),M_o(S)\rangle$. This duality has nice properties, as will be seen later on. However, its drawback is, that, in order to be sure that $\langle v,x\rangle$ is finite for all $v \in F_o(S)$ and all $x \in M_o(S)$, we had to restrict ourselves to bounded functions in the definition of $F_o(S)$. Sometimes this function space is too restricted. A useful generalization is given by the following definitions. Let $e: S \to \mathbb{R}$ be a fixed $S$-measurable function with $e(s) \geqq 1\ \forall s \in S$. Then we define $F(S)$ as *the linear space of all S-measurable real functions* v *on* S *which are bounded by a multiple of* e, that is

(4.15) $\sup_{s \in S} \frac{|v(s)|}{e(s)} < \infty.$

The function e is called a *bounding function*. Furthermore we define, using the same bounding function,

$$M(S) := \{x \in M_o(S): \int_S e(s)|x|(ds) < \infty\}$$

and the positive cones

$$F^+(S) := \{v \in F(S): v(s) \geqq 0 \ \forall s \in S\},$$
$$M^+(S) := \{x \in M(S): x(E) \geqq 0 \ \forall E \in S\}.$$

Since for $v \in F(S)$, $x \in M(S)$

$$|\int_S v(s)x(ds)| \leqq \int_S |v(s)||x|(ds) \leqq M \cdot \int_S e(s)|x|(ds) < \infty$$

it is clear that

(4.16) $\langle v,x\rangle := \int_S v(s)x(ds) \quad v \in F(S),\ x \in M(S)$

defines a duality $\langle F(S),M(S)\rangle$. If $e = 1$ this duality reduces to $\langle F_o(S),M_o(S)\rangle$ but in general we have $F(S) \supset F_o(S)$ and $M(S) \subset M_o(S)$. All statements made on the positive cones $M_o^+(S)$ and $F_o^+(S)$ together with the lattice structure of the corresponding linear spaces are also true for $M^+(S)$ and $F^+(S)$, as can easily be verified.

PROPOSITION 4.6. $\langle F(S),M(S)\rangle$ *is a separated duality. Furthermore, it is compatible with the positive cones* $F^+(S)$ *and* $M^+(S)$, *and these cones induce each other in the sense of* (4.9) *and* (4.10) *by the bilinear form* (4.16). *The same statements are true for* $\langle F_o(S),M_o(S)\rangle$, $F_o^+(S)$ *and* $M_o^+(S)$, *respectively.*

PROOF. If $v \in F(S)$ with $v \neq 0$, then $\exists s_0 \in S$ with $v(s_0) \neq 0$. This implies $\langle v,x_0\rangle \neq 0$ for $x_0$ the unit mass in $s_0$, which is in $M(S)$ since $S$ contains all singletons and $e(s_0)$ is finite. This proves (4.7). In order to prove (4.8), let $x \in M(S)$ with $x \neq 0$. Without loss of generality we may assume that $x(\hat{S}) \neq 0$ where $\hat{S}$ is a positive set of x. Then $\langle v_0,x\rangle \neq 0$ for

$v_0$ the indicator function of $\hat{S}$, which is in $F(S)$ since $e(s) \geqq 1$. Hence $\langle F(S),M(S)\rangle$ is separated. The statements about the positive cones have completely similar proofs. By taking $e = 1$ the results on $F_o(S)$ and $M_o(S)$ follow. □

Complementary slackness conditions, see (4.5), have an obvious interpretation in the $\langle F(S),M(S)\rangle$-setting. Let $x \in M^+(S)$, say a probability measure, and let $v \in F^+(S)$. Then it is well-known that

$$(4.17) \qquad \langle v,x\rangle = 0 \;\Leftrightarrow\; x(\{s \in S\colon v(s) > 0\}) = 0.$$

That is, for a given probability measure x complementary slackness of x with respect to a nonnegative function v means that v must be equal to zero almost surely with respect to x. Reversely, given a nonnegative function v complementary slackness of v with respect to a probability measure x is a condition on the support of x: x is to be restricted to the subset of S where v is equal to zero.

Summarizing, the duality $\langle F(S),M(S)\rangle$ together with the positive cones $F^+(S)$ and $M^+(S)$ is a suitable framework for setting up a dual pair of linear programs, one of which having probability measures on $(S,\mathcal{S})$ as variables. Also the description of the probability measures x in $M(S)$

$$\{x \in M(S)\colon x \geqq 0,\ \langle 1,x\rangle = 1\}$$

fits in this framework. The same is true for objective functions of the kind $\langle c,x\rangle$ with $c \in F(S)$, representing the expected value of the real function c on S with respect to x. It will appear that also the other constraints in the three problems mentioned at the beginning of this section are linear. In the moment problem they are of the type $\langle g_i,x\rangle = b_i$, $i = 1,\ldots,m$, for fixed $g_i \in F(S)$ and $b_i \in \mathbb{R}$. In the marginal problem the linear map "passing to the marginals" occurs. Then the constraint space is a space of measures, too. In the stochastic dynamic programming model this is also the case, but the linear map is of a more general type. In the sequel of this section we will discuss a rather general linear map, determined by a transition measure and its adjoint; this covers all we need for the marginal problem as well as for the stochastic dynamic programming model.

Let $(S,\mathcal{S})$, $e_S$, $\langle F(S),M(S)\rangle$, $F^+(S)$ and $M^+(S)$ be as before, and let

$(T,\mathcal{T})$, $e_T$, $<\mathcal{F}(T),\mathcal{M}(T)>$, $\mathcal{F}^+(T)$ and $\mathcal{M}^+(T)$ be defined similarly, starting from a measurable space $(T,\mathcal{T})$. We will use the same notation for the bilinear forms of both dualities, except if confusion might occur; then we make the distinction $<v,x>_S$ and $<y,u>_T$. A nonnegative real function g on $S \times \mathcal{T}$ will be called a *transition measure* from $(S,\mathcal{S})$ to $(T,\mathcal{T})$ if it is a measure on $\mathcal{T}$ for each fixed $s \in S$ and if it is a $\mathcal{S}$-measurable function on S for each fixed $E \in \mathcal{T}$. We use the notation $g(E|s)$, $s \in S$, $E \in \mathcal{T}$. In applications, g mostly will be a *transition probability*, where $g(.|s)$ is a probability measure in that case. Then g can be interpreted as the conditional probability of a random vector $\eta$ taking values in T given the value s of a random vector $\xi$ taking values in S; this explains the notation. We define $\mathcal{M}^+(T|S)$ as the cone of all transition measures g from $(S,\mathcal{S})$ to $(T,\mathcal{T})$ which satisfy

(4.18) $$M_0(g) := \sup_{s\in S} \frac{1}{e_S(s)} \int_T e_T(t) g(dt|s) < \infty.$$

As a consequence

(4.19) $$g(E|.) \in \mathcal{F}^+(S) \text{ for all } E \in \mathcal{T},$$

(4.20) $$g(.|s) \in \mathcal{M}^+(T) \text{ for all } s \in S.$$

For each transition measure $g \in \mathcal{M}^+(T|S)$ two linear maps are introduced, which will appear to be adjoint:

(4.21) $$\begin{aligned} &G_1: \mathcal{M}(S) \to \mathcal{M}(T); \text{ for } x \in \mathcal{M}(S),\ E \in \mathcal{T}: \\ &(G_1x)(E) \equiv xg(E) := \int_S g(E|s)x(ds) = <g(E|.),x>_S, \end{aligned}$$

and

(4.22) $$\begin{aligned} &G_2: \mathcal{F}(T) \to \mathcal{F}(S); \text{ for } y \in \mathcal{F}(T),\ s \in S: \\ &(G_2y)(s) := \int_T y(t)g(dt|s) = <y,g(.|s)>_T. \end{aligned}$$

Before showing that the claims implicit in the definitions are correct, we will first indicate their direct probabilistic interpretations, consider a special case and give some examples. Let $\xi$ be a random vector taking values in S, and $\eta$ a random vector taking values in T, such that the conditional distribution of $\eta$ given $\xi = s$ is given by the transition probability $g(.|s)$. Then $(G_2y)(s)$ represents a version of the conditional expectation $E[y(\eta)|\xi=s]$ whereas $G_1x$ is the probability distribution xg of $\eta$ induced by the probability distribution x of $\xi$ and the transition probability g. In fact, the joint distribution x*g of $(\xi,\eta)$ is characterized by

$$(x*g)(E_1 \times E_2) := \int_{E_1} g(E_2|s)x(ds), \quad E_1 \in \mathcal{S},\ E_2 \in \mathcal{T}.$$

The notations xg and x*g for the marginal distribution on T and the joint distribution on $S \times T$, respectively, will be used in Section 4.6 and in Chapter 8. Recall, by the way, that in the stochastic dynamic programming problem certain transition probabilities are prescribed; it is not surprising, therefore, that in the formal description in Section 4.6 mappings as $G_2$ occur in the constraints.

A very special but important case occurs when the transition probability is *deterministic*, that is when

$$g(E|s) = \begin{cases} 1 \text{ if } h(s) \in E \\ 0 \text{ if } h(s) \notin E \end{cases}$$

for a measurable *transition function* $h: S \to T$. In this case the maps $G_1$ and $G_2$ reduce to

$$(G_1x)(E) = x(\{s \in S: h(s) \in E\}) = xh^{-1}(E),$$

$$(G_2y)(t) = y(h(t)).$$

EXAMPLES of deterministic transition probabilities include the following.

1. Let $(S,\mathcal{S}) \subset (T,\mathcal{T})$, let $e_S$ be the restriction of $e_T$ to S, and let $h(s) := s$. Then $G_1x$ denotes $x \in M(S)$ as a signed measure on $\mathcal{T}$ rather than on $\mathcal{S}$, and $G_2y$ is the restriction of $y \in F(T)$ to S. In particular, if $(S,\mathcal{S},e_S) := (T,\mathcal{T},e_T)$ then $G_1$ and $G_2$ are the identity maps.
2. Let $(S,\mathcal{S}) = (S_1 \times S_2, \mathcal{S}_1 \otimes \mathcal{S}_2)$, $e_S(s_1,s_2) := e_{S_1}(s_1) + e_{S_2}(s_2) - 1$, $(T,\mathcal{T},e_T) = (S_1,\mathcal{S}_1,e_S)$ and define $h(s_1,s_2) := s_1$. Then $G_1x$ denotes the marginal measure with respect to the first coordinate of $x \in M(S)$:

$(G_1x)(E) = x(E \times S_2)$, $E \in S_1$, and $G_2$ describes the embedding of $y \in F(S_1)$ into $F(S_1 \times S_2)$. For obvious reasons we denote mappings which transform a measure on a product space to a marginal measure as *projections*. From the beginning of this section it will be clear that projections will occur in the formal description of the marginal problem as well as of the stochastic dynamic programming problem.

In the deterministic case by transformation of variables it is easily shown that $G_1$ and $G_2$ are adjoint. But, not surprisingly, this holds in the general case as well.

PROPOSITION 4.7. *Let* $g \in M^+(T|S)$. *Then* $G_1 \equiv xg$ *and* $G_2$ *are well-defined and linear. Moreover, they are adjoint: for each* $x \in M(S)$ *and* $y \in F(T)$ *the identity*

$$(4.23) \qquad \langle y, G_1x\rangle_T = \langle G_2y, x\rangle_S$$

*is true.*

PROOF. Because of (4.20), $G_2y$ is a well-defined real function on S for each $y \in F(T)$. Moreover, since $|y(t)|/e_T(t) \leqq M < \infty \quad \forall t \in T$, it follows from (4.18) that $|(G_2y)(s)|/e_S(s) \leqq M.M_0(g)$, so that $G_2y \in F(S)$ if $G_2y$ is measurable. The proof of measurability is standard, i.e. by extension of the subclass for which it is known to be true. By definition it is true if y is the indicator of a set in $T$, hence also if y is a non-negative measurable simple function. From the monotone convergence theorem it follows then that $G_2y$ is measurable for all nonnegative measurable y. Since $G_2y = G_2y^+ - G_2y^-$ the proof of measurability is complete. Hence $G_2$ is well-defined. Obviously it is linear.

In order to show that $G_1$ is well-defined, we first remark that for each $x \in F(S)$ $G_1x$ is a well-defined function on $T$ because of (4.19). For $x \geqq 0$, $G_1x$ is countably additive because of the monotone convergence theorem, and since $x = x^+ - x^-$ this holds for general $x \in M(S)$ as well. Therefore, $G_1x$ is a finite signed measure on $(T,T)$ for each $x \in M(S)$. In order to show that $G_1x \in M(T)$ we show first for $x \in M(S)$, $y \in F(T)$

$$(4.23') \qquad \int_T y(t)(G_1x)(dt) = \int_S\{\int_T y(t)g(dt|s)\}x(ds).$$

Because of the linearity and the decompositions $y = y^+ - y^-$, $x = x^+ - x^-$ we only have to show (4.23') for $y \geqq 0$ and $x \geqq 0$. The proof is again standard. Take $x \geqq 0$ fixed; by definition (4.23') is true for measurable indicators y, hence for measurable nonnegative simple functions y. For a general nonnegative measurable y, apply again the monotone convergence theorem. With (4.23') it follows from (4.18) that $G_1 x \in \mathcal{M}(T)$ for each $x \in \mathcal{M}(S)$. This completes the proof that $G_1$ is well-defined. Its linearity is trivial. Moreover, by proving (4.23') we also showed (4.23), actually. □

## 4.4 MOMENT PROBLEMS

For a long time extremal values of the probability of a fixed event have been studied under partial specification (or knowledge) of the probability distribution. If only the range and the numerical value of certain moments of a one-dimensional distribution are assumed to be known, the classical moment problem given in the previous section comes up. For an overview of the history of this problem and its generalizations we refer to Karlin and Studden [19]. In this section we will discuss the following generalized moment problem and some of its variants.

Let $(S,\mathcal{S})$ be a measurable space, let $g_i$, $i = 1,\dots,m+1$, be measurable (i.e. $\mathcal{S}$-measurable) real functions on S, and let $b_i$, $i = 1,\dots,m$, be real numbers, and consider the problem:

(MO) find a probability measure on $(S,\mathcal{S})$ which minimizes the expected value of $g_{m+1}$ under the condition that for each $i = 1,\dots,m$ the expected value of $g_i$ is equal to $b_i$.

We assume that the σ-field $\mathcal{S}$ contains all singletons. This technical assumption is not very restrictive. It is true for example if $\mathcal{S}$ contains the Borel subsets of a Borel $S \subset \mathbb{R}^n$, or, more generally, if S is a Hausdorff topological space and $\mathcal{S}$ is the σ-field generated by the closed sets.

A natural approach to the analysis of problem MO is provided by the linear programming duality. In order to formulate the corresponding dual pair of linear programs, we specify as follows.

(i) $X := \mathcal{M}(S)$, $V := \mathcal{F}(S)$ where $\mathcal{M}(S)$ and $\mathcal{F}(S)$ are as in Section 4.3 using the bounding function $e_S := \sup_{i=0,1,\dots,m+1} |g_i|$ with

$g_0 := 1 \in F(S)$. The duality $\langle F(S), M(S)\rangle$ and the positive cones $M^+(S)$ and $F^+(S)$ are also defined as in Section 4.3.

(ii) $Y := \mathbb{R}^{1+m}$, $U := \mathbb{R}^{1+m}$ with positive cones $Y^+ := Y$, $U^+ := \{0\}$ and with the duality $\langle Y,U\rangle$ given by the usual inner product.

(iii) $L_1: X \to U$, $(L_1x)_i := \langle g_i, x\rangle$, $i = 0,1,\ldots,m$, $x \in X$.

$L_2: Y \to V$, $(L_2y)(s) := \Sigma_{i=0}^{m} y_i g_i(s)$, $s \in S$, $y \in Y$.

(iv) $b := (b_0, b_1, \ldots, b_m) \in U$ with $b_0 := 1$; $c := g_{m+1} \in V$.

It is easily seen that all conditions of Definition 4.1 are satisfied, so that the linear programs

$$(\mathrm{MO}_1) \qquad \text{minimize}_{x\in M(S)} \{\langle g_{m+1}, x\rangle : \langle g_i, x\rangle = b_i,\ i = 0,\ldots,m,\ x \geqq 0\}$$

and

$$(\mathrm{MO}_2) \qquad \text{maximize}_{y\in \mathbb{R}^{1+m}} \{\Sigma_{i=0}^{m} y_i b_i : \Sigma_{i=0}^{m} y_i g_i(s) \leqq g_{m+1}(s)\ \forall s \in S\}$$

form a dual pair. Moreover, MO is equivalent to $\mathrm{MO}_1$ since $\{x \in M(S): \int_S g_0(s)x(ds) = b_0,\ x \geqq 0\}$ is precisely the set of all probabilities on the measurable subsets of S.

PROPOSITION 4.8. $(\mathrm{MO}_1, \mathrm{MO}_2)$ *is a dual pair of linear programs.* $\mathrm{MO}_1$ *is equivalent to* MO. *Both dualities involved are separated. For both dualities the positive cones induce each other by the bilinear form.*

PROOF. Only the last two statements have not yet been shown. For $\langle V,X\rangle$ the statements in (4.7), (4.8), (4.9), (4.10) follow as in Proposition 4.6; for $\langle Y,U\rangle$ they follow by direct verification. □

Proposition 4.8 shows that not only the Elementary Duality Theorem 4.2 but also both variants of the Advanced Duality Theorem 4.5 are applicable. Of course, part a of Theorem 4.5 is the most attractive here, since U is finite-dimensional but V is not.

THEOREM 4.9. *Suppose that* $\mathrm{MO}_1$ *is feasible, and that the feasibility is maintained under small perturbations of the right-hand sides. Then if* $\hat{y} \in \mathbb{R}^{1+m}$ *and* $\tilde{y} \in \mathbb{R}^{1+m}$ *exist such that*

(4.24) $\Sigma_{i=0}^{m}\hat{y}_i g_i(s) \leqq g_{m+1}(s) \leqq \Sigma_{i=0}^{m}\tilde{y}_i g_i(s)$, $s \in S$,

*then* $\inf MO_1 = \max MO_2$ *and this common value is finite. Moreover,* $x^*$ *is optimal for* $MO_1$ *iff it is feasible and satisfies the complementary slackness condition*

(4.25) $x^*(\{s \in S: \Sigma_{i=0}^{m} y_i^* g_i(s) < g_{m+1}(s)\}) = 0$

*where* $y^*$ *is any optimal solution of* $MO_2$.

PROOF. The feasibility condition and the second inequality in (4.24) imply that the optimal value function of $MO_1$ is bounded above in the neighbourhood of $0 \in U$ in the usual topology of $\mathbb{R}^{1+m}$. The first inequality in (4.24) shows that $MO_2$ is feasible so that $\inf MO_1 > -\infty$. All remaining statements in the theorem follow now from Theorems 4.2 and 4.5 part a. □

Of course, several variations of the dual pair $(MO_1, MO_2)$ exist for which the same results hold. For example, Theorem 4.9 holds under unchanged conditions if minimization and maximization are reversed, together with reversing the inequality in $MO_2$. The same is true if also inequalities are considered in $MO_1$: just change the positive cones $U^+$ and $Y^+$ appropriately and get the same results.

Theorem 4.9 coincides with a well-known result in moment theory [19 Theorem XII.2.1; 30; 18; 7 Theorem 10.7; 22 Theorem 3]. In fact, many well-known inequalities in probability theory are derived in [19 Chapters XII and XIII] as so-called generalized Chebyshev inequalities, and the procedure which is proposed to derive them might be interpreted as a direct application of Theorem 4.2. That is, instead of trying to verify the conditions in Theorem 4.9, try to solve $MO_2$. This involves a lower approximation of $g_{m+1}$ by a linear combination of $g_0,\ldots,g_m$. Often this approximation is exact (no slack) only in a finite number of points. The complementary slackness condition (4.25) then indicates that one has to restrict the attention in $MO_1$ to probabilities with this finite support. The many examples in [19] show that often it is possible to find optimal solutions for $MO_2$ and $MO_1$ by just trying to satisfy primal and dual feasibility and complementary slackness.

We introduced $MO_2$ as dual problem for the generalized moment problem $MO_1$. But it has independent value, too, as an approximation problem for functions. Notice that it can be rewritten as a "one-sided Chebyshev approximation": by eliminating $y_0$ and the constraints, it follows that $MO_2$ is equivalent to

$$\text{maximize}_{y\in\mathbb{R}^m} -\|\Sigma_{i=1}^m y_i(g_i - b_i g_0) - g_{m+1}\|^+$$

where $\|g\|^+ := \sup_{s\in S} g(s)$. Similarly, if in $MO_1$ the constraints $x \geqq 0$, $\langle g_0,x\rangle = 1$ are replaced by $\langle g_0,|x|\rangle \leqq 1$ then for $b_i = 0$, $i = 1,\ldots,m$, $MO_2$ can be rewritten as the well-known "symmetrical" Chebyshev approximation of $g_{m+1}$ by linear combinations of $g_1,\ldots,g_m$:

$$\text{maximize}_{y\in\mathbb{R}^m} -\|\Sigma_{i=1}^m y_i g_i - g_{m+1}\|$$

where $\|g\| := \sup_{s\in S}|g(s)|$. Hence also Chebyshev approximation fits in the framework of linear programming duality. One might object that the constraint $\langle g_0,|x|\rangle \leqq 1$ is not linear. However, it is equivalent to $x = x_1 - x_2$, $\langle g_0,x_1+x_2\rangle \leqq 1$, $x_1 \geqq 0$, $x_2 \geqq 0$ and $\inf(x_1,x_2) = 0$. The last condition is the only nonlinear one, but it is not difficult to show that it can be dropped without changing the problem: if $x_1$ and $x_2$ are feasible, then the same is true for $x_1 - \inf(x_1,x_2)$ and $x_2 - \inf(x_1,x_2)$, and the objective function does not change by this transformation. We will not work out the details of the Chebyshev approximation. The interested reader is referred to [19,7], for example.

The function approximation problem $MO_2$ is the main subject in the so-called *semi-infinite linear programming* theory [7,5,8]. Of course, the duality theory plays an important role there, and the relation with moment theory is recognized. It is interesting to notice, that in the semi-infinite linear programming theory usually *not* $MO_1$, but a restriction of it is considered as the dual problem of $MO_2$. The restriction is, that in $MO_1$ only measures x with a *finite support* are considered. At first sight this may look too limited, but due to the finite number of constraints it is not: it can be shown [30; 22 Lemma 1] that for any feasible solution for $MO_1$ there exists another feasible solution with the same value of the objective function, and with a finite support: applying Caratheodory's theorem, one may even assume that the support contains at most m+1 points. This restriction to finite supports makes it possible to solve semi-infinite linear programs by means of an extension of the Simplex method [7 Chapter VII]. Characteristic assumptions are: $S \subset \mathbb{R}^k$, S compact, $g_i$ continuous, $i = 0,1,\dots,m+1$. Under such conditions it is also not difficult to get a duality result which is well-known in semi-infinite linear programming [7 Theorem 10.12; 22 Theorem 4] by using *part b* of Theorem 4.5. Indeed, assume that

(4.26) S is a *compact metric* space (whose σ-field $\mathcal{S}$ is generated by the closed subsets); and $g_1, g_2, \dots, g_{m+1}$ are continuous.

Not surprisingly, we restrict the attention in $MO_2$ to continuous functions. That is, $F(S)$ is replaced by its subspace $F_c(S)$ of all continuous functions. Since under (4.26) the bounding function $e_S := \sup (g_0, |g_1|, \dots, |g_{m+1}|)$ itself is bounded and continuous, it may simply be replaced by $g_0$. Therefore, $F_c(S)$ can be identified with the normed space $C(S)$ of all bounded continuous functions on S, with the supremum norm. The dual of $C(S)$ is precisely the space of all finite signed measures $M_o(S)$ which coincides here with $M(S)$.

THEOREM 4.9'. *Under the assumptions above, if* $MO_1$ *is feasible then*

$$-\infty < \sup MO_2 = \min MO_1 < +\infty.$$

PROOF. Define $\hat{y}_0 = \inf_{s \in S} g_{m+1}(s) - 1$; it is a finite number since $g_{m+1}$ is continuous and S is compact. Then $\hat{y} := (\hat{y}_0, 0, \dots, 0) \in \mathbb{R}^{1+m}$ is feasible for $MO_2$ for all small perturbations of the right-hand sides of the constraint. Hence, the optimal value function of $MO_2$ is bounded from below by $\hat{y}_0$. Application of Theorem 4.5 part b gives the desired result. Notice that the feasibility of $MO_1$ implies that $\inf MO_1 < \infty$. □

Theorems 4.9 and 4.9' are not the most general duality results known for moment problems. Kemperman [22], for instance, considers the following pair of problems

$\widehat{MO}_1$ minimize$_{x \in X^+}\{\int_S g(s)x(ds): \int_S g_i(s)x(ds) \geqq b_i,\ i \in I\}$

$\widehat{MO}_2$ maximize$_{y \in Y^+}\{\Sigma_{i\in I} y_i b_i: \Sigma_{i\in I} y_i g_i(s) \leqq g(s),\ s \in S\}$

where $S$ is a completely regular topological space,

$X^+$ is the set of all finite inner regular measures on $S$,

$I$ is an *arbitrary* index set,

$g$, $g_i$, $i \in I$ are upper semicontinuous functions on $S$ which are bounded from above,

$b_i$, $i \in I$, are real numbers,

$Y^+$ is the set of all $(y_i,\ i \in I)$ which satisfy $y_i \geqq 0$ for all $i \in I$, and $y_i = 0$ for all but a finite number of them.

Theorem 6 in [22] is a generalization of Theorem 4.9' in the sense that

$$-\infty < \sup \widehat{MO}_2 = \min \widehat{MO}_1 < +\infty$$

is derived under the following conditions:

(i) $\widehat{MO}_1$ is feasible,

(ii) a "Slater-type" of condition for $\widehat{MO}_2$,

(iii) a "tightness" condition on $\{(g_i,b_i),\ i \in I\}$.

We have not been able to derive results of the same generality from Theorem 4.5 part b.

In the remainder of this section we consider a *moment problem with bounds*, as introduced by Gaivoronski [4]. Already in [24] we studied this problem from a linear programming point of view. The corresponding linear program is

$$(\mathrm{MOB}_1) \quad \text{minimize}_{x\in M(S)}\left\{\langle g_{m+1},x\rangle : \begin{array}{l}\langle g_i,x\rangle = b_i,\ i = 0,\dots,m\\ x \geqq \underline{x},\ -x \geqq -\bar{x}\end{array}\right\}$$

where the bounds $\underline{x} \in M^+(S)$, $\bar{x} \in M^+(S)$ are fixed and satisfy $\underline{x} \leqq \bar{x}$. By the way, $\mathrm{MOB}_1$ is *not* a generalization of $\mathrm{MO}_1$ since $\bar{x}$ is a *finite* measure. We assume that

$$g_i(s) \geqq 0 \;\forall s \in S,\; b_i \geqq 0,\; i = 1,\dots,m;$$

(4.27)

$$\langle g_i,\underline{x}\rangle < b_i < \langle g_i,\bar{x}\rangle,\; i = 0,1,\dots,m.$$

The first condition is equivalent to $g_i$ being bounded from below: one might add a function $\gamma_i \cdot g_i$ to $g_i$ and replace $b_i$ by $b_i + \gamma b_0$ without changing $MOB_1$. (Recall that $g_0(s) = 1$, $b_0 = 1$ as before.) The second assumption in (4.27) is not very restrictive if one is not interested in infeasible problems. Indeed, if $\langle g_i,\underline{x}\rangle > b_i$ then $MOB_1$ does not have feasible solutions since $g_i \geqq 0$ and $x \geqq \underline{x}$ imply $\langle g_i,x\rangle > b_i$. If $\langle g_i,\underline{x}\rangle = b_i$, then the constraints $\langle g_i,x\rangle = b_i$, $x \geqq \underline{x}$ and $g_i \geqq 0$ imply $x(E) = \underline{x}(E)$ for all Borel sets $E \subset \{s \in S\colon g_i(s) > 0\} =: E_i$, so that in that case one might reduce $MOB_1$ to a problem on $S \setminus E_i$ and delete the i-th constraint. Similar remarks hold for the last inequality in (4.27).

The dual of $MOB_1$ in the sense of Definition 4.1 is $MOB_2$:

$$\text{maximize}_{(y,\underline{y},\bar{y})}\{\textstyle\sum_{i=0}^{m} y_i b_i + \langle\underline{y},\underline{x}\rangle - \langle\bar{y},\bar{x}\rangle\}:$$

($MOB_2$)

$$\textstyle\sum_{i=0}^{m} y_i g_i(s) + \underline{y}(s) - \bar{y}(s) = g_{m+1}(s),\; s \in S,$$

$$y \in \mathbb{R}^{1+m},\; \underline{y} \in F^+(S),\; \bar{y} \in F^+(S)\}.$$

Unlike $MO_2$, $MOB_2$ has infinite-dimensional variables. It can, however, be reduced to a finite-dimensional concave (not necessarily linear) maximization problem. In order to show this, notice that for any feasible $(y,\underline{y},\bar{y})$ also $(y,\, \underline{y} - \inf(\underline{y},\bar{y}),\, \bar{y} - \inf(\underline{y},\bar{y}))$ is feasible. Moreover, since $\underline{x} \leqq \bar{x}$ and $\inf(\underline{y},\bar{y}) \geqq 0$ the objective function cannot decrease by this transformation, so that the constraint $\inf(\underline{y},\bar{y}) = 0$ may be added without loss. This gives us the opportunity to eliminate $\underline{y}$ and $\bar{y}$. Therefore, $MOB_2$ is equivalent to

$$\text{maximize}_{y \in \mathbb{R}^{1+m}} -h(y)$$

where $h(y)$ is the finite convex function

$$h(y) := -\Sigma_{i=0}^{m} y_i b_i - \langle(\Sigma_{i=0}^{m} y_i g_i - g_{m+1})^-, \underline{x}\rangle + \langle(\Sigma_{i=0}^{m} y_i g_i - g_{m+1})^+, \bar{x}\rangle$$

$$(4.28) \qquad = -\langle g_{m+1}, \underline{x}\rangle - \Sigma_{i=0}^{m} y_i (b_i - \langle g_i, \underline{x}\rangle) + \langle(\Sigma_{i=0}^{m} y_i g_i - g_{m+1})^+, \bar{x}-\underline{x}\rangle$$

$$= -\langle g_{m+1}, \bar{x}\rangle + \Sigma_{i=0}^{m} y_i (\langle g_i, \bar{x}\rangle - b_i) + \langle(\Sigma_{i=0}^{m} y_i g_i - g_{m+1})^-, \bar{x}-\underline{x}\rangle.$$

The reformulations are a direct consequence of $g = g^+ - g^-$ for $g \in F(S)$.

It is easily verified that Proposition 4.8 is still true if $(MO_1, MO_2)$ is replaced by $(MOB_1, MOB_2)$. Therefore, as in Theorem 4.9 and 4.9', it is interesting to investigate the usefulness of the Duality Theorems 4.2 and 4.5 for finding optimal solutions. The complementary slackness conditions are

$$(4.29) \qquad \begin{aligned} &\langle(\Sigma_{i=0}^{m} y_i g_i - g_{m+1})^+, \bar{x}-x\rangle = 0, \\ &\langle(\Sigma_{i=0}^{m} y_i g_i - g_{m+1})^-, x-\underline{x}\rangle = 0. \end{aligned}$$

With the notations

$$\begin{aligned} \bar{Z}(y) &:= \{s \in S: \Sigma_{i=0}^{m} y_i g_i(s) > g_{m+1}(s)\}, \\ Z_0(y) &:= \{s \in S: \Sigma_{i=0}^{m} y_i g_i(s) = g_{m+1}(s)\}, \\ \underline{Z}(y) &:= \{s \in S: \Sigma_{i=0}^{m} y_i g_i(s) < g_{m+1}(s)\}, \end{aligned}$$

we get the following expressions for primal and dual feasibility together with complementary slackness:

$$(4.30) \qquad \begin{aligned} &\langle g_i, x\rangle = b_i, \ i = 0,1,\dots,m, \\ &x(E) = \bar{x}(E) \text{ for } E \in \mathcal{S},\ E \subset \bar{Z}(y), \\ &\underline{x}(E) \leqq x(E) \leqq \bar{x}(E) \text{ for } E \in \mathcal{S},\ E \subset Z_0(y), \\ &\underline{x}(E) = x(E) \text{ for } E \in \mathcal{S},\ E \subset \underline{Z}(y). \end{aligned}$$

Let us first consider the simple case that there are *no moment constraints* (m=0). Then it is possible to solve $MOB_1$ and $MOB_2$ by means of (4.30) by guessing the solution and verifying its optimality by means of Theorem 4.2. In order to see this, let F be the distribution function of $g_{m+1}$ with respect to the measure $\bar{x}-\underline{x}$:

$$F(y_o) := \int_{\{g_{m+1} < y_0\}} (\bar{x}-\underline{x})(ds) = (\bar{x}-\underline{x})(\bar{Z}(y_0)),\ y_0 \in \mathbb{R}.$$

F is nondecreasing and continuous to the left. The limit from the right is

$$F(y_0+0) = \int_{\{g_{m+1} \leq y_0\}} (\bar{x}-\underline{x})(ds) = (\bar{x}-\underline{x})(\bar{Z}(y_0) \cup Z_0(y_0)),\ y_0 \in \mathbb{R}.$$

F increases from $0 = F(-\infty)$ to $F(\infty) = \langle g_0, \bar{x}-\underline{x}\rangle$. Because of the assumption $\langle g_0, \underline{x}\rangle < 1 < \langle g_0, \bar{x}\rangle$, see (4.27), there exists at least one $y_0^* \in \mathbb{R}$ with $F(y_0^*) \leqq 1 - \langle g_0, \underline{x}\rangle \leqq F(y_0^*+0)$. Take any such $y_0^*$, and choose $\alpha \in [0,1]$ such that $(1-\alpha)F(y_0^*) + \alpha F(y_0^*+0) = 1 - \langle g_0, \underline{x}\rangle$. Of course, if $\bar{x}(Z_0(y_0^*)) = \underline{x}(Z_0(y_0^*))$ any choice for $\alpha$ will do, but if $\bar{x}(Z_0(y_0^*)) > \underline{x}(Z_0(y_0^*))$ then

$$\alpha = \frac{1 - \langle g_0, \underline{x}\rangle - F(y_0^*)}{F(y_0^*+0) - F(y_0^*)}.$$

Finally, define a measure $x^*$ as follows:

$$\begin{aligned} &\text{for } E \in \mathcal{S},\ E \subset \bar{Z}(y_0^*) : x^*(E) := \bar{x}(E), \\ &\text{for } E \in \mathcal{S},\ E \subset Z_0(y_0^*) : x^*(E) := (1-\alpha)\underline{x}(E) + \alpha\bar{x}(E), \\ &\text{for } E \in \mathcal{S},\ E \subset \underline{Z}(y_0^*) : x^*(E) := \underline{x}(E). \end{aligned}$$

THEOREM 4.10. *If* $m = 0$ *then* $x^*$ *solves* $MOB_1$ *and* $y_0^*$ *solves* $MOB_2$.

PROOF. Directly from Theorem 4.2 and (4.30). The value of $\alpha$ guarantees that $\langle g_0, x^*\rangle = 1$, and the value of $y_0^*$ implies $\alpha \in [0,1]$. □

The solution of $MOB_1$ with $m = 0$ has some resemblance with the famous Neyman-Pearson Lemma. The reason is that the problem of finding a test with prescribed level of significance and maximal power also can be interpreted as a linear program, the complementary slackness conditions of which have a form similar to (4.29) and (4.30). In order to see this, let $x_0$, $x_1$ be different but fixed probability measures on $(S,\mathcal{S})$, and $\alpha$ a fixed number in $(0,1)$. Then the problem is: find $S_0 \in \mathcal{S}$ such that $x_0(S_0) \leqq \alpha$ and such that $x_1(S_0)$ is maximal. By allowing for randomized tests and asking for a level of significance of exactly $\alpha$ one arrives at the following program, with again $g_0(s) = 1$, $\forall s \in S$,

$$(\mathrm{NP}_1) \qquad \max_{\varphi\in F_o(S)} \{\langle\varphi,x_1\rangle : \langle\varphi,x_0\rangle = \alpha,\ 0 \leqq \varphi(s) \leqq g_0(s),\ s \in S\}.$$

The distinction with moment problems is that now the variables are *functions* (with bounds) instead of measures (with bounds). Using the $\langle F_o(S),M_o(S)\rangle$ duality, we get as the dual of $NP_1$ in the sense of Definition 4.1

$$(\mathrm{NP}_2) \qquad \operatorname{minimize}_{\substack{y_0\in\mathbb{R}\\ y_1,y_2\in M_o(S)}} \left\{\alpha y_0 + \langle g_0,y_1\rangle : \begin{array}{l} y_1 - y_2 + y_0x_0 = x_1 \\ y_1 \geqq 0,\ y_2 \geqq 0 \end{array}\right\}$$

and the complementary slackness conditions are $\langle\varphi,y_2\rangle = \langle g_0 - \varphi,y_1\rangle = 0$. Since $g_0 > 0$ one may eliminate $y_1$ and $y_2$ from the formulation of $NP_2$ by

$$y_1 = (x_1 - y_0x_0)^+,\quad y_2 = (x_1 - y_0x_0)^-$$

so that $NP_2$ is equivalent to the one-dimensional convex minimization problem

$$(\mathrm{NP}_2') \qquad \operatorname{minimize}_{y_0\in\mathbb{R}} k(y_0) := \alpha y_0 + \langle g_0,(x_1 - y_0x_0)^+\rangle,$$

and the complementary slackness conditions then become:

$$(x_1 - y_0x_0)^+(\{s \in S: \varphi(s) < 1\}) = 0,$$
$$(x_1 - y_0x_0)^-(\{s \in S: \varphi(s) > 0\}) = 0.$$

These results can be made more explicit by using densities. Let $\bar{x}$ be any $\sigma$-finite measure on $(S,\mathcal{S})$ such that $x_0$ and $x_1$ have densities $f_0$ and $f_1$ w.r.t. $\bar{x}$. One might take $\bar{x} := x_0 + x_1$, or $\bar{x}$ := the Lebesque measure in appropriate cases. Then $(x_1 - y_0x_0)^+$ has the density $(f_1 - y_0f_0)^+$, and $(x_1 - y_0x_0)^-$ has the density $(f_1 - y_0f_0)^-$. Therefore, complementary slackness can be rewritten in the wellknown form

$$\varphi(s) = 1 \text{ for } s \in S^+_{y_0} := \{s \in S: f_1(s) > y_0f_0(s)\} \text{ a.e.}[\bar{x}],$$
$$0 \leqq \varphi(s) \leqq 1 \text{ for } s \in S^0_{y_0} := \{s \in S: f_1(s) = y_0f_0(s)\} \text{ a.e.}[\bar{x}],$$

$$0 = \varphi(s) \text{ for } s \in S^-_{y_0} := \{s \in S\colon f_1(s) < y_0 f_0(s)\} \text{ a.e.}[\bar{x}].$$

The choice of $y_0$ must be such that $y_0$ solves $NP_2'$: $0 \in \partial k(y_0)$. This condition can easily be worked out as the well-known condition

$$\bar{x}(S^+_{y_0}) \leqq \alpha \leqq \bar{x}(S^+_{y_0} \cup S^0_{y_0}).$$

Finally, feasibility of $\varphi$ for $NP_1$ is guaranteed by $\varphi(s) := \gamma$, $s \in S^0_{y_0}$, with $\gamma \in [0,1]$ such that $\bar{x}(S^+_{y_0}) + \gamma.\bar{x}(S^0_{y_0}) = \alpha$. Hence we showed that the Neyman-Pearson Lemma may be derived as a consequence of the Elementary Duality Theorem 4.2, just as was the case for problem $MOB_1$ with $m = 0$.

Let us now return to the generalized moment problem $MOB_1$ and its dual $MOB_2$. The presence of *moment constraints* ($m \geqq 1$) complicates the application of Theorem 4.2 a little bit. The complications are twofold. First of all, unlike the case $m=0$ the assumptions (4.27) imply in general neither the existence of an optimal solution $y^*$ of $MOB_2$, nor the feasibility of $MOB_1$. Secondly, if $y^*$ has been found there does not seem to exist an easy guess for $x^*$ on $Z_0(y^*)$ which guarantees that all the constraints $\langle g_i, x\rangle = b_i$, $i = 0,1,\ldots,m$, are satisfied. Nevertheless, if in any particular situation one is able to find a feasible $x^*$ and a feasible $y^*$ such that the complementary slackness conditions (4.30) hold, then again Theorem 4.2 guarantees optimality. Under somewhat restrictive assumptions we have the following result.

<u>THEOREM 4.11</u>. *Consider the dual pair* $(MOB_1, MOB_2)$ *with* $m \geqq 1$. *If* $MOB_2$ *has an optimal solution* $y^*$ *which satisfies*

$$\bar{x}(Z_0(y^*)) = \underline{x}(Z_0(y^*)) \tag{4.31}$$

*then* $MOB_1$ *has the unique optimal solution* $x^*$

$$\begin{aligned} x^*(E) &:= \bar{x}(E) \text{ for } E \in \mathcal{S},\ E \subset \bar{Z}(y^*) \cup Z_0(y^*), \\ x^*(E) &:= \underline{x}(E) \text{ for } E \in \mathcal{S},\ E \subset \underline{Z}(y^*) \cup Z_0(y^*). \end{aligned} \tag{4.32}$$

PROOF. Optimal solutions $y^*$ for $MOB_2$ are characterized by $0 \in \partial h(y^*)$. Hence also $0 \in \partial_{y_i} h(y)$, $i = 0,1,\ldots,m$. These partial subgradients can be calculated easily. Using $g_i \geqq 0$ we get

$$\partial_{y_i} h(y) = -b_i + \langle g_i, \underline{x}\rangle + \int_{\bar{Z}(y)} g_i d(\bar{x}-\underline{x}) + [0, \int_{Z_0(y)} g_i d(\bar{x}-\underline{x})].$$

Because of (4.31) it even follows that $\frac{\partial h}{\partial y_i}(y^*)$ exists and vanishes so that for $i = 0,1,\ldots,m$ we have

$$\int_{\bar{Z}(y^*)} g_i d(\bar{x}-\underline{x}) = \int_{\bar{Z}(y^*) \cup Z_0(y^*)} g_i d(\bar{x}-\underline{x}) = b_i - \langle g_i, \underline{x}\rangle.$$

Using this relation one easily verifies that indeed $x^*$ and $y^*$ satisfy (4.30), and Theorem 4.2 implies optimality of $x^*$ (and $y^*$). □

The question arises how restrictive the conditions in Theorem 4.11 are. Assumption (4.27) implies, that $h(y) \to +\infty$ if $|y_i| \to \infty$, $y_j$ fixed for $j \neq i$, so that the coordinate directions are no directions of recession of h. Under the somewhat stronger condition that h does not recede at all, h has bounded level sets [31 Theorem 27.1.d] and the existence of an optimal $y^*$ is guaranteed. As far as condition (4.31) is concerned, it holds for example if a measure $\mu$ on $(S,\mathcal{S})$ exists such that $\mu(Z_0(y^*)) = 0$ and such that $\bar{x}-\underline{x}$ has a density with respect to $\mu$ (e.g. if $S \subset \mathbb{R}^n$, $\mu$ might be the Lebesque measure).

Let us examine how the Advanced Duality Theorem 4.5 can be used to improve Theorem 4.11. Now topology comes in. Application of part a of Theorem 4.5 does not look very attractive, since there does not seem to exist a handy topology on $M(S)$ which is compatible with the duality $\langle F(S),M(S)\rangle$. The norm topology, $\|x\| := |x|(S)$, is too strong since it allows for nonmanageable continuous linear functionals. Part b of Theorem 4.5 provides a better setting, at least if we assume that (4.26) holds, and if the corresponding adjustments after (4.26) are implemented.

THEOREM 4.12. *Assume that* (4.26) *holds, and replace in* $(MOB_1, MOB_2)$ $F(S)$ *by* $F_c(S)$. *Then* $MOB_1$ *has optimal solutions if it has feasible solutions. Moreover, if* $x^*$ *is any feasible solution of* $MOB_1$, *it is* *optimal iff*

$$\lim_{k\to\infty} \langle (\Sigma_{i=0}^{m} y_i^{(k)} g_i - g_{m+1})^+, \bar{x}-x^* \rangle = 0$$

$$\lim_{k\to\infty} \langle (\Sigma_{i=0}^{m} y_i^{(k)} g_i - g_{m+1})^-, x^*-\underline{x} \rangle = 0$$

*where* $(y^{(k)})$ *is any minimizing sequence for* h.

PROOF. The proof is an application of part b of Theorem 4.5. Clearly, the dualities involved are separated, and (4.10) holds. For the optimal value function $\varphi_2$ of $MOB_2$ we have

$$\varphi_2(v) \geqq - \langle |g_{m+1}|, \bar{x} \rangle - \langle |v|, \bar{x} \rangle, \; v \in C(S) = F_c(S). \tag{4.33}$$

This follows easily by replacing $g_{m+1}$ by $g_{m+1} + v$ in $MOB_2$, and setting each $y_i$ equal to zero. Therefore $\varphi_2$ is bounded below in a neighbourhood of 0 in the norm topology of $C(S)$, since $\langle |v|, \bar{x} \rangle \leqq \|v\| . \|\bar{x}\|$. From part b of Theorem 4.5 we conclude therefore that there is no duality gap, and that $MOB_1$ has optimal solutions (since both problems are feasible the common optimal value is finite). By Theorem 4.2 with (4.6) and (4.29) the proof of Theorem 4.12 is complete. □

Under the assumptions of Theorem 4.12 it is not guaranteed that $MOB_2$ has an optimal solution, or equivalently, that the function h attains its minimum. The function h is convex and bounded below since $MOB_1$ is feasible, and it tends to infinity in the coordinate directions because of (4.27), but even so it is possible that its infimum is not attained.

The compact metrizability of S is assumed in order to be certain that $M(S) = M_o(S)$ is the normed dual of $C(S)$. If S is just a normal topological space, then two complications arise. First, continuous functions are not necessarily bounded, so that $C(S)$ does not coincide with $F_c(S)$ any more. But even in the case that it is justified to consider only bounded continuous functions, another complication arises: the norm topology on $C(S)$ is not compatible any more with the $\langle C(S), M_o(S) \rangle$ duality. If S is not compact, then $M_o(S)$ is only a subspace of the normed dual of $C(S)$: the latter contains, apart from the signed measures of $M_o(S)$ which by definition are *countably* additive set functions, also purely finitely additive set functions [3 Theorem IV.6.2].

We prefer to avoid finite additive set functions, not only because they can not easily be handled, but also because the problem $MOB_1$ would loose its probabilistic meaning. Therefore, we want to use a topology weaker than the norm topology, which is compatible with the $<C(S),M_o(S)>$ duality. On the other hand, this topology should be strong enough to make $v \to <|v|,\bar{x}>$ continuous: this guarantees that $\varphi_2$ is bounded below in the neighbourhood of $0 \in C(S)$ in this topology, see (4.33). Such a topology exists as we will show. Moreover, the first complication can be dealt with simultaneously.

THEOREM 4.13. *Theorem 4.12 remains true if the compactness assumption on* S *in* (4.26) *is dropped.*

PROOF. Define the following family of semi-norms on $F_c(S)$: $\{p_x : x \in M^+(S)\}$ with $p_x(v) := <|v|,x>$. This family generates a locally convex topology $T$ on $F_c(S)$. Since each such a semi-norm is $T$-continuous, it follows from (4.33) that $\varphi_2$ is bounded below in a $T$-neighbourhood of 0. We now show, that $T$ is compatible with the duality $<F_c(S),M(S)>$. First of all, recall that a linear functional T on $F_c(S)$ is $T$-continuous iff it is bounded by a semi-norm, that is

$$\exists x \in M^+(S) \text{ s.t. } |T(v)| \leqq <|v|,x> \; \forall v \in F_c(S).$$

Since for any $x \in M(S)$ and $y \in F_c(S)$ we have $|<v,x>| \leqq <|v|,|x|>$, it follows that for any $x \in M(S)$ the linear functional $v \to <v,x>$ is $T$-continuous. Hence $M(S) \subset F_c(S)^*$. In order to prove the reverse inclusion, let T be any $T$-continuous linear functional on $F_c(S)$. Then an $x_0 \in M^+(S)$ exists with $|T(v)| \leqq <|v|,x_0> \; \forall v \in F_c(S)$. Since we would like to change from the $<F_c(S),M(S)>$ duality to $<C(S),M_o(S)>$ we introduce the following linear maps:

$$G_1 : M(S) \to M_o(S) : (G_1 x)(E) := \int_E e_S(s)x(ds), \; E \in S, \; x \in M(S),$$
$$G_2 : C(S) \to F_c(S) : (G_2 \hat{v})(s) := e_S(s)\hat{v}(s), \; s \in S, \; \hat{v} \in C(S).$$

Since

$$\int \hat{v}(s)(G_1 x)(ds) = \int \hat{v}(s) e_S(s) x(ds), \; \hat{v} \in C(S), \; x \in M(S),$$

it follows that $\langle\hat{v},G_1x\rangle = \langle G_2\hat{v},x\rangle$ $\forall\hat{v} \in C(S)$ $\forall x \in M(S)$ so that $G_1$ and $G_2$ are adjoint. Now define the linear functional $\hat{T}$ on $C(S)$ by $\hat{T}(\hat{v}) := T(G_2\hat{v})$, $\hat{v} \in C(S)$. Then

$$|\hat{T}(\hat{v})| = |T(G_2\hat{v})| \leqq \langle|G_2\hat{v}|,x_0\rangle = \langle G_2|\hat{v}|,x_0\rangle = \langle|\hat{v}|,G_1x_0\rangle.$$

This means, that $\hat{T}$ can be interpreted as a continuous linear functional on the subspace of all bounded continuous functions in $L_1(\hat{x}_0)$, with $\hat{x}_0 := G_1x_0 \in M_o^+(S)$. This subspace is dense, so that $\hat{T}$ can be extended to a continuous linear functional on $L_1(\hat{x}_0)$. As a consequence, it has the representation

$$\hat{T}(\hat{v}) = \int_S \hat{v}(s)\rho(s)\hat{x}_0(ds), \quad \hat{v} \in L_1(\hat{x}_0) \supset C(S),$$

for some $\rho \in L_\infty(\hat{x}_0)$. Define $\hat{x}_1(E) := \int_E \rho(s)\hat{x}_0(ds)$, $E \in S$; then $\hat{x}_1 \in M_o(S)$ and $\hat{T}(\hat{v}) = \langle\hat{v},\hat{x}_1\rangle$. Moreover, with $x_1(E) := \int_E (e_S(s))^{-1}\hat{x}_1(ds)$ we have that

$$\int_S e_S(s)|x_1|(ds) = \int_S|\hat{x}_1|(ds) \leqq \|\rho\|.\int_S|\hat{x}_0|(ds) < \infty$$

so that $x_1 \in M(S)$ and $G_1x_1 = \hat{x}_1$. Since $G_2$ is invertible, we return to T via $T(v) = \hat{T}(G_2^{-1}v)$. Putting things together we find

$$T(v) = \hat{T}(G_2^{-1}v) = \langle G_2^{-1}v,\hat{x}_1\rangle = \langle G_2^{-1}v,G_1x_1\rangle$$

$$= \langle G_2G_2^{-1}v,x_1\rangle = \langle v,x_1\rangle, \quad v \in F_c(S).$$

Since $x_1 \in M(S)$ the proof is complete. □

## 4.5 THE MARGINAL PROBLEM

Let $(S_i,S_i)$ be measurable spaces such that all singletons are contained in $S_i$, and let $p_i$ be probability measures on $(S_i,S_i)$, $i = 1,\ldots,m$. Define $S := S_1 \times S_2 \times \ldots \times S_m$ and let $S$ be the σ-field on S generated by $S_1 \times S_2 \times \ldots \times S_m$, and let c be an $S$-measurable real function on S. These data determine the marginal problem defined in Section 4.3.:

$$\text{(MA)} \qquad \underset{x \in M_o^+(S)}{\text{minimize}} \; \{\int_S c(s)x(ds) : x(S) = 1, \; \mathrm{proj}_i x = p_i, \; i = 1,\dots,m\}.$$

Here, as before, $M_o^+(S)$ is the cone of finite measures on $(S,\mathcal{S})$ and the condition $x(S) = 1$ guarantees that only probabilities are considered. The linear map $\mathrm{proj}_i : M_o(S) \to M_o(S_i)$ transforms each finite signed measure x on $(S,\mathcal{S})$ into its marginal on $(S_i,\mathcal{S}_i)$:

$$(\mathrm{proj}_i x)(E) := x(B_1 \times B_2 \times \dots \times B_m), \; E \in \mathcal{S}_i,$$
$$\text{with } B_i := E \text{ and } B_j := S_j \text{ for } j \neq i.$$

Of course, the condition $x(S) = 1$ is redundant, since $\mathrm{proj}_i x = p_i$ implies $x(S) = (\mathrm{proj}_i x)(S_i) = p_i(S_i) = 1$ since each $p_i$ is a probability. In order that MA is a well-posed problem it is necessary to assure that its objective exists for all feasible x, for example by allowing for infinite values and defining $\infty - \infty := \infty$. This destroys the linearity. Therefore we restrict ourselves to marginal problems for which $\int c(s)x(ds)$ is finite for all feasible x, via assumptions on c and $p_i$ in terms of bounding functions. Let, for $i = 1,\dots,m$, $e_i$ be a bounding function on $S_i$ and use this bounding function in the definition of the duality $\langle F(S_i), M(S_i)\rangle$ as given in Section 4.3. Let $\langle F(S), M(S)\rangle$ be the duality determined by the bounding function e, defined by $e(s_1,\dots,s_m) := \Sigma_{i=1}^m e_i(s_i)$. If we assume that

$$(4.34) \qquad p_i \in M(S_i), \; i = 1,\dots,m,$$

then each x which is feasible for MA is in $M(S)$ since

$$\int_S e(s)|x|(ds) = \Sigma_{i=1}^m \int_S e_i dx = \Sigma_{i=1}^m \int_{S_i} e_i d(\mathrm{proj}_i x) =$$
$$= \Sigma_{i=1}^m \langle e_i, p_i\rangle < \infty$$

so that in MA one might replace $M_o(S)$ by $M(S)$. Furthermore, if it is assumed that

$$(4.35) \qquad c \in F(S)$$

it is clear that $\int_S c(s)x(ds) = \langle c,x\rangle$ is finite (even bounded) on the feasible region of MA.

Assumptions (4.34) and (4.35) make it possible to put MA into the linear programming duality framework of Section 4.2. Define the following pair of linear programs

$$(\mathrm{MA}_1) \qquad \text{minimize}_{x \in M(S)} \left\{ \langle c,x\rangle : \begin{array}{l} \mathrm{proj}_i x = p_i \quad i = 1,\dots,m \\ x \geqq 0 \end{array} \right\},$$

$$(\mathrm{MA}_2) \qquad \text{maximize}_{y_i \in F(S_i)} \left\{ \Sigma_{i=1}^m \langle y_i, p_i\rangle : \begin{array}{l} \Sigma_{i=1}^m y_i(s_i) \leqq c(s_1,\dots,s_m) \\ \text{for all } (s_1,\dots,s_m) \in S \end{array} \right\}.$$

PROPOSITION 4.14. *Assume* (4.34) *and* (4.35). *Then* $\mathrm{MA}_1$ *is equivalent to* MA. *Moreover,* $(\mathrm{MA}_1,\mathrm{MA}_2)$ *is a dual pair of linear programs. Both dualities involved are separated. For both dualities the positive cones induce each other by the bilinear form.*

PROOF. The equivalence of MA and $\mathrm{MA}_1$ has been shown already. That $(\mathrm{MA}_1,\mathrm{MA}_2)$ is a dual pair follows from the specifications:

(i) $\langle V,X\rangle := \langle F(S),M(S)\rangle$; $V^+ := F^+(S)$, $X^+ := M^+(S)$.

(ii) $\langle Y,U\rangle := \langle F(S_1) \times F(S_2) \times \dots \times F(S_m),\ M(S_1) \times M(S_2) \times \dots \times M(S_m)\rangle$ with $\langle y,u\rangle := \Sigma_{i=1}^m \langle y_i,u_i\rangle$, $y_i \in F(S_i)$, $u_i \in M(S_i)$;

$$U^+ := \{(0,0,\dots,0)\},\ Y^+ := Y.$$

(iii) $L_1 : X \to U$, $(L_1x)_i := \mathrm{proj}_i x$, $x \in F(S)$, $i = 1,\dots,m$.

$L_2 : Y \to V$, $(L_2 y)(s_1,\dots,s_m) := \Sigma_{i=1}^m y_i(s_i)$, $(s_1,\dots,s_m) \in S$.

It is easily verified that indeed $L_1(X) \subset U$ since for $x \in M(S)$ $\mathrm{proj}_i x \in M(S_i)$:

$$\int e_i \, d|\mathrm{proj}_i x| \leqq \int e_i \, d(\mathrm{proj}_i |x|) = \int e_i \, d|x| \leqq \langle e,|x|\rangle < \infty.$$

The dualities are compatible with the positivities (by Proposition 4.6, and by a direct proof, respectively). $L_1$ and $L_2$ are adjoint, as can be verified directly. Also the remaining statements follow from Proposition 4.6, or directly. □

How restrictive are assumptions (4.34) and (4.35)? The question is: for a given measurable function c and for given probabilities $p_i$, do there exist bounding functions $e_i$ such that (4.34) and (4.35) hold? In general the answer is negative. If this is the case because MA is not well-posed we do not mind. But also well-defined marginal problems exist for which the answer is negative. A typical example arises if $c \geq 0$ and $\int c\,d(p_1 \times p_2 \times \ldots \times p_m) = +\infty$. Therefore the $(MA_1, MA_2)$ framework is not suitable for all marginal problems. Nevertheless, in many situations it can be applied. Of course, if c is bounded (which occurs for example if one wants to minimize the probability of a fixed event; or if every $S_i$ is compact and c is continuous) $e_i = 1$ will do. If c is unbounded but "not growing too fast" then non-trivial bounding functions may be relevant. A typical example occurs, if $S_i = \mathbb{R}$, if $\int |s_i|^k dp_i(s_i) < \infty$ and if $|c(s_1,\ldots,s_m)| \leq M.\Sigma_{i=1}^m (1 + |s_i|^k)$ for an integer k and a finite M. We conclude that the $(MA_1, MA_2)$ framework is suitable for many but not for all marginal problems. In fact, the assumptions (4.34) and (4.35) can be interpreted as regularity conditions: they are sufficient to guarantee that MA, $MA_1$ and $MA_2$ are well-defined, and that inf $MA_1$ and sup $MA_2$ are finite. Indeed, they imply that the objective functions of MA, $MA_1$ and $MA_2$ are well-defined and finite for all feasible solutions, and that $MA_2$ has feasible solutions. The feasibility of $MA_1$ is trivial: take $x_0 := p_1 \times p_2 \times \ldots \times p_m$, the probability measure on S determined by the marginals under the assumption of independence.

Similar as for the moment problem, the dual $MA_2$ of the marginal problem $MA_1$ can be interpreted as an approximation problem in a function space: $MA_2$ is the problem of finding a *separable* lower bound y for c with maximum expected value with respect to any probability measure x which is feasible for $MA_1$, e.g. the independent measure $x_0$. In this context complementary slackness of a pair of feasible solutions (x,y) for $(MA_1, MA_2)$ means, that the measure x must be restricted to the subset of S on which the approximation is exact, i.e.

$$(4.36) \qquad x(\{s = (s_1,\ldots,s_m) \in S: \Sigma_{i=1}^m y_i(s_i) < c(s_1,\ldots,s_m)\}) = 0.$$

This explains for example, that the independent solution $x_0$ can only be optimal if c itself is separable a.s. $[x_0]$. In this trivial case every feasible solution for $MA_1$ is in fact optimal. If the criterion function c is not separable one may expect strong dependence between the marginals.

It is interesting to notice that the linear program $MA_1$ has a special structure: it is a generalization of the well-known transportation problem. Indeed, for $m = 2$ and $S_1$ and $S_2$ finite the problem $MA_1$ reduces to the transportation problem, and its usual linear programming dual is precisely the corresponding reduction of $MA_2$. Some results about the problem $MA_1$ can be seen as generalizations of results regarding the transportation problem; for example those concerning the optimality of the NW Corner Rule Solution, see Section 7.2.

In probability theory marginal problems have been studied extensively in different degrees of generality. We refer to [21] and Section 7.2 for references. Often the basic approach can be described in terms of linear programming duality, although this is not always formulated explicitly, or even recognized. In the $(MA_1,MA_2)$ framework, this approach is merely an application of the Elementary Duality Theorem 4.2: try to find feasible solutions (x,y) for $(MA_1,MA_2)$ which satisfy complementary slackness; if one succeeds then optimality is guaranteed. This approach works also in the analysis of robustness against dependence in project planning to be discussed in Section 7.3: almost every result about the worst-case joint distribution can be derived from the Elementary Duality Theorem 4.2 only (instead of the Duality Theorem 7.1): in that case $S_i \subset \mathbb{R}$ and the bounding functions $e_i(s_i) = 1 + |s_i|$ do the job.

Let us now analyze the use of the Advanced Duality Theorem 4.5 for the dual pair $(MA_1,MA_2)$. Proposition 4.14 shows, that all conditions of part a as well as part b of Theorem 4.5 are satisfied if we are able to find a compatible topology in which the optimal value function is bounded above/below in a neighbourhood of 0. As for moment problems with bounds, application of part a of Theorem 4.5 does not seem very atractive, neither with the norm topology nor with weak topologies. Moreover, it is clear that the optimal value function $\varphi_1(u) = \infty$ for all perturbations u which do not satisfy $u_1(S_1) = u_2(S_2) = \ldots = u_m(S_m)$. Let us therefore consider part b of Theorem 4.5. It is not difficult to prove, that the optimal value function $\varphi_2$ is bounded below in the norm topology of $V = F(S)$, defined by

$$\|v\| := \sup_{(s_1,\ldots,s_m)\in S} \frac{|v(s_1,\ldots,s_m)|}{\Sigma_{i=1}^m e_i(s_i)} .$$

Indeed, since $y_i := -\|c\|e_i$, $i = 1,\dots,m$, is feasible for MA we know that $\sup MA_2 \geqq -\|c\|\Sigma_{i=1}^m \langle e_i,p_i\rangle$, so that the optimal value function $\varphi_2$ of $MA_2$ satisfies

$$\varphi_2(v) \geqq -\|c+v\|.\Sigma_{i=1}^m \langle e_i,p_i\rangle \geqq -M_1 - M_2\|v\|,\ v \in V.$$

On the other hand, without additional assumptions it is not true that the norm topology is compatible with the $\langle F(S),M(S)\rangle$ duality. Since $C(S)^*$ is isometrically isomorfic to the space of all regular countable additive set functions on the Borel sets of S if S is a compact Hausdorff space [3 p.265] and since a countable additive set function is regular if S is metric [27 p.27; 1 Proposition 7.17] the following assumptions are appropriate, just as in the previous section:

(4.37)
$S_i$ is compact metric for $i = 1,\dots,m$;
$c$ is continuous in the product topology.

As a consequence S is compact metric and c is bounded, so that we can just as well take $e_i := 1\ \forall i$. Therefore (4.37) implies (4.34) and (4.35).

THEOREM 4.15. *Assume that* (4.37) *holds, and replace in*$(MA_1,MA_2)$ $F(S_i)$ *by* $C(S_i)$, $F(S)$ *by* $C(S)$ *and* $e_i$ *by* 1. *Then* $MA_1$ *and* $MA_2$ *have optimal solutions, and they are completely characterized by the feasibility and complementary slackness conditions* (4.36).

PROOF. Everything follows from applications of Theorems 4.5 (part b) and 4.2 and the discussion above, except the solvability of $MA_2$. But this is a direct consequence of the fact that a continuous function attains its maximum on a nonempty closed subset of a compact set. □

Application of the Advanced Duality Theorem 4.5 to the marginal problem has thus led us to Theorem 4.15. It is closely related to the results in [6]. We have not been able to apply Theorem 4.5 under weaker conditions than (4.37), using weak topologies for example. This does not mean that Theorem 4.15 gives the weakest conditions under which $\min MA_1 = \max MA_2$ can be proved. To the contrary: in recent papers [20,21] Kellerer gives very weak conditions for normality ($\inf MA_1$ =

sup $MA_2$) as well as for the solvability of each of the problems. His approach is completely different from ours. He does not prove any result for general dual linear programs, but exploits the special structure of the marginal problem and its dual extensively. He shows that for fixed tight probability measures $p_i$ on Hausdorff spaces $S_i$ the optimal values inf $MA_1$ (using the lower integral) and sup $MA_1$ are continuous functions of c, in several topologies. Using these continuity properties, normality is shown for semi-continuous functions c, and this is extended by means of a functional version of Choquet's capacitability theorem to normality for a wide class of functions. Some of Kellerer's results, which will be applied in Chapter 7, are:

inf $MA_1$ = sup $MA_2$ if c has a separable integrable upper bound, and if -c is a Suslin function;

min $MA_1$ = sup $MA_2$ if c is lower semi-continuous (without any boundedness assumption);

sup $MA_2$ = max $MA_2$ if c has a separable integrable upper bound (without any measurability assumption), provided sup $MA_2 > -\infty$.

For the intricate details we refer to [20,21]. It is not likely that similar strong duality results can be proved for general linear programming problems in probabilities. A similar remark can be made on Levin's related results [25].

## 4.6 THE STOCHASTIC DYNAMIC PROGRAMMING PROBLEM

In the beginning of Section 4.3 we gave a rough sketch of a discrete-time Markovian decision process. We shall now give the details of this model. Since in the linear programming duality framework the choice of the spaces is crucial, we consider rather general state and action spaces. On the other hand we restrict ourselves to a finite horizon. The infinite horizon case allows for a similar analysis with similar results.

The *data* of the Markovian decision process are the following.

Stages : $j = 0,1,\dots,n$.

State spaces : measurable spaces $(S_j,\mathcal{S}_j)$, $\mathcal{S}_j$ containing all singletons, $j = 0,\dots,n$.

Action spaces : measurable spaces $(A_j,\mathcal{A}_j)$, $\mathcal{A}_j$ containing all singletons, $j = 0,\dots,n-1$.

Constraint sets: $D_j(s_j) \in \mathcal{A}_j$, $s_j \in S_j$, $j = 0,1,\ldots,n-1$, describing the admissible actions at stage j in state $s_j$. $D_j := \{(s_j,a_j): a_j \in D_j(s_j)\} \in \mathcal{S}_j \otimes \mathcal{A}_j$, $j = 0,1,\ldots,n-1$; its σ-algebra $\mathcal{D}_j$ of measurable subsets is the restriction of $\mathcal{S}_j \otimes \mathcal{A}_j$. It is assumed that each $D_j$ contains the graph of a measurable mapping from $S_j$ to $A_j$. For $j = n$ we define $(D_n,\mathcal{D}_n) := (S_n,\mathcal{S}_n)$.

Probabilities : (initial) probability measure $p_0$ on $(S_0,\mathcal{S}_0)$; transition probabilities (see Section 4.3 for the definition) $p_j$ from $(D_{j-1},\mathcal{D}_{j-1})$ to $(S_j,\mathcal{S}_j)$, $j = 1,\ldots,n$.

Cost functions : Borel measurable functions $c_j: D_j \to \mathbb{R}$, $j = 0,1,\ldots,n$.

The fundamental concept of the optimization problem which will be formulated in terms of the data of the Markovian decision process is that of *policy*. By definition, a (randomized Markovian) policy δ is a sequence $(\delta_0,\delta_1,\ldots,\delta_{n-1})$ of admissible transition probabilities $\delta_j$ from $(S_j,\mathcal{S}_j)$ to $(A_j,\mathcal{A}_j)$; $\delta_j$ is called admissible if $\delta_j(D_j(s_j)|s_j) = 1$ $\forall s_j \in S_j$. A policy δ is called *deterministic* or nonrandomized if each $\delta_j$ is deterministic (see Section 4.3 for the definition): in that case $\delta_j$ can be identified with a measurable mapping (by a slight abuse of notation also denoted by) $\delta_j: S_j \to A_j$ with $\delta_j(s_j) \in D_j(s_j)$ $\forall s_j \in S_j$; such $\delta_j$ is called an (admissible) *decision rule* at stage j. The assumptions on the constraint sets imply that deterministic policies exist. A fortiori, the set of all policies, denoted by Δ, is nonempty.

It is well-known that an initial probability measure and suitable transition probabilities define a unique joint probability on the product space. Therefore, together with the given (transition) probabilities $p_j$, any policy δ determines a probability measure $q_\delta$ on $S_0 \times A_0 \times \ldots \times S_{n-1} \times A_{n-1} \times S_n$, characterized by

$$q_\delta(\bar{S}_0 \times \bar{A}_0 \times \ldots \times \bar{S}_{n-1} \times \bar{A}_{n-1} \times \bar{S}_n) = \int_{\bar{S}_0} \{\int_{\bar{A}_0} \{\ldots \{\int_{\bar{S}_{n-1}} \{\int_{\bar{A}_{n-1}} p_n(\bar{S}_n|s_{n-1},a_{n-1})\delta_{n-1}(da_{n-1}|s_{n-1})\}p_{n-1}(ds_{n-1}|s_{n-2},a_{n-2})\}\ldots\}\delta_0(da_0|s_0)\}p_0(ds_0)$$

where $\bar{S}_j$ is an arbitrary set in $\mathcal{S}_j$ and $\bar{A}_j$ is an arbitrary set in $\mathcal{A}_j$. The probability measure $q_\delta$ will sometimes be denoted by

$p_0 * \delta_0 * \dots * p_{n-1} * \delta_{n-1} * p_n$, and mathematical expectation with respect to $q_\delta$ is denoted by $E_\delta$.

We are now ready to formulate the stochastic dynamic programming problem:

(SDP) find a policy $\delta \in \Delta$ which minimizes $c_\delta := E_\delta \Sigma_{j=0}^{n} c_j$.

Of course, this problem is only well-defined if the mathematical expectation of $\Sigma_{j=0}^{n} c_j$ exists: apart from measurability conditions which are taken care of in the specification of the data also finiteness assumptions have to be made in order to avoid meaningless expressions as $\infty - \infty$. For this reason it is necessary to assume

$$\text{for each } \delta \in \Delta,\ E_\delta(\Sigma_{j=0}^{n} c_j)^- < \infty \text{ and/or } E_\delta(\Sigma_{j=0}^{n} c_j)^+ < \infty .$$

Usually stronger assumptions are appropriate, such as

$$\text{(4.38)} \quad \begin{aligned} &\text{a. } E_\delta(c_j)^- < \infty,\ j = 0,1,\dots,n, \text{ for all } \delta \in \Delta, \text{ and/or} \\ &\text{b. } E_\delta(c_j)^+ < \infty,\ j = 0,1,\dots,n, \text{ for all } \delta \in \Delta. \end{aligned}$$

One of the advantages of (4.38) is, that $\Sigma_{j=0}^{n} E_\delta c_j$ is a well-defined extended real number which is equal to $c_\delta$, $\forall \delta \in \Delta$. As a consequence, $c_\delta$ can be calculated by recursion (see e.g. [1 Lemma 8.1]):

$$\text{(4.39)} \quad \begin{aligned} &v_n^\delta := c_n, \text{ and for } j = n-1,\dots,0: \\ &v_j^\delta := \int_{A_j} \{c_j(.,a) + \int_{S_{j+1}} v_{j+1}^\delta(s) p_{j+1}(ds|.,a)\} \delta_j(da|.); \\ &c_\delta := \int_{S_0} v_0^\delta dp_0 . \end{aligned}$$

Although under condition (4.38) expressions as $\infty - \infty$ are effectively avoided infinite values for integrals remain possible. Since our duality scheme does not allow for infinite-valued bilinear forms, we will be more restrictive than (4.38), see (4.41c).

The linear programming formulation of SDP starts with the idea to consider certain marginal distributions of $q_\delta$ as decision variables rather than the policy $\delta$ itself. Thus, for any $\delta \in \Delta$ we define the marginal probability measures

$$x_j^\delta := \mathrm{proj}_{S_j \times A_j} q_\delta,\ j = 0,1,\ldots,n-1,$$
$$x_n^\delta := \mathrm{proj}_{S_n} q_\delta,$$

where the same projection notation is used as in Section 4.5. Since $\delta$ is admissible, $x_j^\delta$ is a probability on $D_j$. The probabilities $x_j^\delta$ satisfy certain relations: $x_0^\delta, x_1^\delta, \ldots, x_n^\delta$ must agree with each other, and with the given initial and transition probabilities $p_0, p_1, \ldots, p_n$, as far as the distributions on $S_0, S_1, \ldots, S_n$ are concerned. These relations are linear:

$$\mathrm{proj}_{S_0} x_0^\delta = p_0,$$
$$\mathrm{proj}_{S_i} x_i^\delta = x_{i-1}^\delta p_i,\ i = 1,\ldots,n.$$

Here, as in (4.21), the composition $x_{i-1}^\delta p_i$ is defined by

$$(x_{i-1}^\delta p_i)(E) := \int_{D_{i-1}} p_i(E|s_{i-1},a_{i-1})\,dx_{i-1}^\delta(s_{i-1},a_{i-1}),\ E \in \mathcal{S}_i.$$

Since $c_\delta$ can be written in terms of $x_j^\delta$, we are led to the reformulation of SDP as a linear program in $x_0^\delta, \ldots, x_n^\delta$. In order to define suitable spaces for the $x_j^\delta$, and in order to settle the finiteness of the expected cost, we assume:

(4.40)
- for each $S_j$ a bounding function $e_{S_j}$ is given;
- for each $A_j$ a bounding function $e_{A_j}$ is given;
- for each $D_j$, $j < n$, the corresponding bounding function $e_{D_j}$ is defined by $e_{D_j}(s_j,a_j) := e_{S_j}(s_j) + e_{A_j}(a_j) - 1$.

Using these bounding functions, as in Section 4.3 we define the spaces of signed measures $\mathcal{M}(S_j)$ and $\mathcal{M}(D_j)$, each supplied with the natural positive cones $\mathcal{M}^+(S_j)$ and $\mathcal{M}^+(D_j)$, as well as the function spaces $\mathcal{F}(S_j)$ and $\mathcal{F}(D_j)$ with the natural positive cones $\mathcal{F}^+(S_j)$ and $\mathcal{F}^+(D_j)$. Again, by $\mathcal{M}^+(S_j|D_{j-1})$ and $\mathcal{M}^+(A_j|S_j)$ we denote the cones of transition measures which satisfy (4.18) with the bounding functions of (4.40). The following assumptions are appropriate.

(4.41)
a. $p_0 \in M(S_0)$, $p_j \in M^+(S_j|D_{j-1})$, $j = 1,\dots,n$.
b. $\delta_j \in M^+(A_j|S_j)$, $j = 0,1,\dots,n$.
c. $c_j \in F(D_j)$, $j = 0,1,\dots,n$.

Of course, since $p_0$ is a probability, (4.41a) implies $p_0 \in M^+(S_0)$. If in (4.40) one lets all bounding functions be equal to 1, then (4.41a,b) is not restrictive at all, but (4.41c) forces the cost functions to be bounded. Other choices for the bounding functions may enlarge the class of acceptable cost functions, but they are more restrictive for the transition probabilities, either for the given $p_j$ (if $e_{S_j} \not\equiv 1$) of for the $\delta_j$ (if $e_{A_j} \not\equiv 1$) to be determined, or both. The next proposition shows that under (4.40) and (4.41a,b) we have that $x_j^\delta \in M^+(D_j)$. Together with (4.41c) this implies that $E_\delta|c_j| = \langle|c_j|, x_j^\delta\rangle$ is finite, so that both (4.38a) and (4.38b) are true under (4.40) and (4.41), and so that $c_\delta = \Sigma_{j=0}^n \langle c_j, x_j^\delta\rangle$ is finite.

PROPOSITION 4.16. *Assume* (4.40), (4.41a,b). *Then* $x_j^\delta \in M^+(D_j)$, $j = 0,1,\dots,n$.

PROOF. Since each $x_j^\delta$ is a (probability) measure, we only have to show that $x_j^\delta \in M(D_j)$, $j = 0,1,\dots,n$. The proof is by induction to j. For $j = 0$ we have $x_0^\delta = p_0 * \delta_0$ (see Section 4.3 for the definition of the composition *) and therefore $x_0^\delta \in M(D_0)$ since

$$\begin{aligned}\int_{D_0} e_{D_0}\, d|x_0^\delta| &= \int_{D_0} e_{D_0}\, dx_0^\delta \\ &= \int_{S_0} \{\int_{A_0} (e_{S_0}(s) + e_{A_0}(a) - 1)\delta_0(da|s)\} p_0(ds) \\ &\leqq (1+M)M_0 - 1\end{aligned}$$

where

$$M := \sup_{s\in S_0} (e_{S_0}(s))^{-1} \int_{A_0} e_{A_0}(a)\delta_0(da|s)$$

and $M_0 := \int_{S_0} e_{S_0}\, dp_0$ are finite since $\delta_0 \in M^+(A_0|S_0)$ and $p_0 \in M(S_0)$, respectively. Now assume that $x_{j-1}^\delta \in M(D_{j-1})$ for some $j \in \{1,\dots,n\}$.

Then $x_j^\delta \in M(D_j)$. To see this, notice first of all that $x_j^\delta = (x_{j-1}^\delta p_j)*\delta_j$. Since $p_j \in M^+(S_j|D_{j-1})$ we know that $x_{j-1}^\delta p_j \in M(S_j)$ (Proposition 4.7). As for the case where $j = 0$ it follows from $\delta_j \in M^+(A_j|S_j)$ that $x_j^\delta \in M(D_j)$. □

We now define a pair of linear programming problems, the first one of which is inspired by the discussion above.

$$(SDP_1)\quad \begin{array}{ll} \text{minimize}_{x\in\Pi_{j=0}^n M(D_j)} & \Sigma_{j=0}^n <c_j,x_j> \\ \text{subject to} & \text{proj}_{S_0} x_0 = p_0, \\ & -x_{i-1}p_i + \text{proj}_{S_i} x_i = 0\ ,\ i = 1,\dots,n, \\ & x_j \geqq 0\ ,\ j = 0,1,\dots,n. \end{array}$$

$$(SDP_2)\quad \begin{array}{ll} \text{maximize}_{y\in\Pi_{i=0}^n F(S_i)} & <y_0,p_0> \\ \text{subject to} & y_j(s_j) - <y_{j+1},p_{j+1}(.|s_j,a_j)> \leqq c_j(s_j,a_j), \\ & \forall(s_j,a_j) \in D_j,\ j = 0,1,\dots,n-1, \\ & y_n(s_n) \leqq c_n(s_n),\ \forall s_n \in S_n. \end{array}$$

PROPOSITION 4.17. *Assume* (4.40) *and* (4.41a,c). *Then* $(SDP_1,SDP_2)$ *is a dual pair of linear programs. Both dualities involved are separated. For both dualities the positive cones induce each other by the bilinear form.*

PROOF. We show that all conditions of Definition 4.1 are satisfied. Let
a. $X := \Pi_{j=0}^n M(D_j)$, $V := \Pi_{j=0}^n F(D_j)$, both supplied with the natural positive cones $X^+ := \Pi_{j=0}^n M^+(D_j)$, $V^+ := \Pi_{j=0}^n F^+(D_j)$ and $Y := \Pi_{i=0}^n F(S_i)$, $U := \Pi_{i=0}^n M(S_i)$, with positive cones $Y^+ := Y$, $U^+ := \{0\}$. Then
b. $<V,X> := \Sigma_{j=0}^n <F(D_j),M(D_j)>$, $<Y,U> := \Sigma_{i=0}^n <F(S_i),M(S_i)>$ are indeed dualities which are compatible with the positivities (Proposition 4.6). Further let
d. $\tilde{p}_0 := (p_0,0,0,\dots,0) \in U$ and $c := (c_0,\dots,c_n) \in V$ because of (4.41a,c), and let
c. $L_1$: $X \to U$ and $L_2$: $Y \to V$ be defined by

$$(L_1x)_i := \Sigma_{j=0}^n (L_1)_{ij}x_j, \quad (L_2y)_j := \Sigma_{i=0}^n (L_2)_{ji}y_i$$

with $x = (x_0,\dots,x_n)$, $x_j \in M(D_j)$, and $y = (y_0,\dots,y_n)$, $y_i \in F(S_i)$, and where the maps $(L_1)_{ij}: M(D_j) \to M(S_i)$, $(L_2)_{ji}: F(S_i) \to F(D_j)$ are defined as follows:

$$\text{if } i-j = 0: \quad (L_1)_{ii}x_i := \text{proj}_{S_i} x_i,$$
$$((L_2)_{jj}y_j)(s_j,a_j) := y_j(s_j),$$
$$\text{if } i-j = 1: \quad (L_1)_{i,i-1}x_{i-1} := -x_{i-1}p_i,$$
$$((L_2)_{j,j+1}y_{j+1})(s_j,a_j) := -\langle y_{j+1}, p_{j+1}(.|s_j,a_j)\rangle,$$
$$\text{if } i-j \neq 0,1: \quad (L_1)_{ij} := 0, \quad (L_2)_{ji} := 0.$$

The definitions of $L_1$ and $L_2$ are, of course, such that the constraints in $SDP_1$ and $SDP_2$ are equivalent to $L_1x \geqq \tilde{p}_0$ and $L_2y \leqq c$, respectively. We have to verify that $L_1(X) \subset U$ and $L_2(Y) \subset V$, that $L_1$ and $L_2$ are linear, and that they are adjoint. These verifications can be done for each pair $((L_1)_{ij},(L_2)_{ji})$ separately. If $i-j \neq 0,1$ these statements are trivial. If $i-j = 1$, they are a direct consequence of Proposition 4.7, since $p_i \in M^+(S_i|D_{i-1})$ for $i \geqq 1$. Finally, if $i-j = 0$ they also follow from Proposition 4.7 (see Example 2 in Section 4.3).

Finally, because of Proposition 4.6 $\langle V,X\rangle$ is a separated duality, and $V^+$ and $X^+$ induce each other by the bilinear form. For the same reason $\langle Y,V\rangle$ is separated. $Y^+$ and $V^+$ induce each other by the bilinear form (direct proof). □

Proposition 4.17 guarantees that the Elementary Duality Theorem 4.2 as well as the Advanced Duality Theorem 4.5 are applicable to the dual pair of linear programming problems $(SDP_1,SDP_2)$. Before considering such applications, we first analyze the relation between the original stochastic dynamic programming problem SDP and $(SDP_1,SDP_2)$.

The relation between SDP and $SDP_1$ is complicated for two reasons. First, while the construction of $x_j^\delta$ from $\delta \in \Delta$ is direct, we now are faced with the reverse problem: given any probability measure $x_i$ on $D_j \subset S_j \times A_j$, we must decompose it into a probability on $S_j$ and a transition probability from $S_j$ to $A_j$. This decomposition is possible if the spaces $S_j$ and $A_j$ are *Borel spaces* [1 Proposition 7.27]. For our exposition it is sufficient to identify Borel spaces as complete separable metric spaces (also called Polish spaces); for a slightly more general definition and for an exposition of the properties of Borel spaces we refer to [1,14]. Hence we introduce the assumption:

(4.42) $S_j$ and $A_j$ are Borel spaces, and $\mathcal{S}_j$ and $\mathcal{A}_j$ are the σ-fields of the Borel sets, for all j.

Under (4.42) also $(D_j,\mathcal{D}_j)$ is a Borel space for each j. The second complication is assumption (4.41b) which has no counterpart in SDP. We introduce therefore the *restricted* stochastic dynamic programming problem

(SDP') find $\delta \in \Delta' := \{\delta \in \Delta\text{: (4.41b) holds}\}$
which minimizes $c_\delta := E_\delta\Sigma_{j=0}^n c_j$.

PROPOSITION 4.18. *Assume* (4.38), (4.40), (4.41a,c), (4.42). *The linear program* $SDP_1$ *is closely related to* SDP *and* SDP', *in the following sense:*
*a. For each* $\delta \in \Delta'$ *there is an* x, *feasible for* $SDP_1$, *such that* $\langle c,x\rangle = c_\delta < \infty$.
*b. For each* x, *feasible for* $SDP_1$, *there is a* $\delta \in \Delta$ *such that* $c_\delta = \langle c,x\rangle$.
*c. The optimal values satisfy*

$$\text{(4.43)} \qquad \inf \text{SDP} \leqq \inf \text{SDP}_1 \leqq \inf \text{SDP}'.$$

PROOF. (a) Take $x := x^\delta$ as defined earlier. Then Proposition 4.16 and the discussion before it show what has to be proved.
(b) Let $(x_0,x_1,\dots,x_n)$ be feasible for $SDP_1$. Then each $x_j$ is a probability distribution, since it is positive and has total mass 1. This last statement follows by induction with respect to j, using constraints of $SDP_1$ and the fact that all $p_j$ are probabilities. Because of (4.42) a transition probability $\delta_j$ from $S_j$ to $A_j$ exists for each $j = 0,1,\dots,n-1$ such that $x_j = (\text{proj}_{S_j} x_j)\delta_j$. Hence, $\delta := (\delta_0,\dots,\delta_{n-1}) \in \Delta$. A direct verification shows that $x_j^\delta = x_j$, $j = 0,1,\dots,n$.
(c) This is a direct consequence of the parts a and b. □

One is tempted to extend the proof of part b of Proposition 4.18 by showing that the constructed $\delta_j$ is contained in $M^+(A_j|S_j)$, or at least can be approximated by a $\delta_j' \in M^+(A_j|S_j)$, $j = 0,1,\ldots,n-1$. Then the second inequality in (4.43) would become an equality, actually. However, we are not able to do so without making additional assumptions.

Unfortunately, it is not to be expected that in general the equality inf SDP = inf SDP' holds. For example, it is possible that $\Delta' = \emptyset$ whereas $\Delta \neq \emptyset$ by assumption. But there is a simple condition which guarantees $\Delta' = \Delta$ implying inf SDP = inf SDP'.

<u>PROPOSITION 4.19</u>. *Assumption* (4.40), (4.41a,c). *Then* $\Delta' = \Delta$ *if and only if a finite* M *exists with*

$$(4.44)\qquad \sup_{a_j \in D_j(s_j)} e_{A_j}(a_j) \leqq M.e_{S_j}(s_j),\ \forall s_j \in S_j,\ j = 0,1,\ldots,n-1.$$

<u>PROOF</u>. (If) Because of (4.44) we have for each $\delta = (\delta_0,\ldots,\delta_{n-1}) \in \Delta$

$$\sup_{s_j \in S_j} (e_{S_j}(s_j))^{-1} \int_{A_j} e_{A_j}(a_j)\delta_j(da_j|s_j) \leqq M$$

so that $\delta_j \in M^+(A_j|S_j)$, $j = 0,1,\ldots,n-1$. Hence $\delta \in \Delta'$.
(Only if) Suppose that (4.44) is not true. Then for a certain j a sequence $((s_j^{(k)}, a_j^{(k)}),\ k = 1,2,\ldots) \subset D_j$ exists with

$$\lim_{k\to\infty} e_{A_j}(a_j^{(k)})/e_{S_j}(s_j^{(k)}) = +\infty.$$

Define a deterministic decision rule $\hat{\delta}_j$ by $\hat{\delta}_j(s_j^{(k)}) := a_j^{(k)}$, $k = 1,2,\ldots$ and arbitrary elsewhere. Then $\hat{\delta}_j \notin M(A_j|S_j)$, so that $\Delta' \neq \Delta$. □

<u>PROPOSITION 4.20</u>. *Assume* (4.40), (4.41a,c), (4.42) *and* (4.44). *Then both inequalities in* (4.43) *are equalities, so that* SDP *and* $\mathrm{SDP}_1$ *are equivalent, and* inf SDP = inf $\mathrm{SDP}_1 < +\infty$.

<u>PROOF</u>. A direct consequence of Propositions 4.18 and 4.19. Notice that (4.38) is satisfied in this case, since not only (4.40) and (4.41a,c) are true but also (4.41b) for all $\delta \in \Delta = \Delta'$. □

Although sufficient, $\Delta' = \Delta$ is not necessary to have equalities in (4.43). In [23] we provide a production-inventory control model for which $\Delta' \neq \Delta$ but inf SDP' = inf SDP. On the other hand $\Delta' = \Delta$, that is (4.44), is true if each $A_j$ is finite (or compact, with $e_{A_j}$ continuous) or if $e_{A_j} \equiv 1$. In particular, by taking $e_{S_j} \equiv 1$ as well, we get

COROLLARY 4.21. *Assume* (4.42). *If each* $c_j$ *is bounded, then the conclusions of Proposition 4.20 hold.*

Let us now describe the relation between the stochastic dynamic programming problem SDP and the second linear program, $SDP_2$. To start with, notice that the constraints in $SDP_1$ and $SDP_2$ have a special feature: those in $SDP_1$ are "lower staircase" while those in $SDP_2$ are "upper staircase". The former property reflects the fact that the state-action probabilities $x_j^\delta$ for any $\delta \in \Delta$ in a natural way arise by *forward* recursion:

$$x_0^\delta = p_0 * \delta_0; \text{ for } i = 1,\dots,n-1: \; x_i^\delta := (x_{i-1}^\delta p_i) * \delta_i; \quad x_n^\delta := x_{n-1}^\delta p_n.$$

Similarly, the latter property asks for a *backward* recursion of "cost functions". This reveals the intimate relationship between $SDP_2$ and the *dynamic programming algorithm*. This algorithm, together with the associated optimality conditions, is the main tool of analysis (probably even the raison d'être) of dynamic programming theory. It defines by backward recursion the sequence of so-called *cost-to-go functions* $f_j: S_j \to [-\infty,\infty]$, $j = n,n-1,\dots,0$ by

$$f_n(s) := c_n(s), \; s \in S,$$

and for $j = n-1,\dots,0$:

(DPA) $$f_j(s) := \inf_{a \in D_j(s)} \{c_j(s,a) + \int_{S_{j+1}} f_{j+1}(t) p_{j+1}(dt|s,a)\}, \; s \in S_j,$$

and finally:

$$\inf \text{DPA} := \int_{S_0} f_0(s) p_0(ds).$$

In these definitions the integral needs additional explanation, since $f_{j+1}$ is not necessarily measurable and sufficiently bounded. However, under (4.42) it is universally measurable [1 Corollary 8.2.1; 14 Theorem 14.4], so that the integral has a unique meaning, if one also adopts the convention $-\infty+\infty = +\infty-\infty = \infty$. (For the properties of integrals under this convention we refer to [1 Lemma 7.11]; for our treatise the most important property is that the extended definition of an integral reduces to the classical one if the latter exists, under conditions such as (4.38), that is.)

The importance of the algorithm is the fact that it provides optimality conditions for SDP.

PROPOSITION 4.22. *Assume* (4.38). *Then*

(4.45) $$\inf \mathrm{DPA} \leqq \inf \mathrm{SDP}.$$

*For any* $\delta \in \Delta$, *define* $q_\delta$ *and* $v_j^\delta$ *as in the beginning of this section. Then*

$$\left.\begin{array}{l} v_j = f_j \text{ a.s. } [q_\delta] \\ j = 0,1,\dots,n \end{array}\right\} \Leftrightarrow \left\{\begin{array}{l} \delta \textit{ is optimal for } \mathrm{SDP} \\ \inf \mathrm{DPA} = \inf \mathrm{SDP} \end{array}\right.$$

PROOF. Let $\delta \in \Delta$. By induction based on (4.39) and DPA it follows that $v_j^\delta \geqq f_j$, $j = n,n-1,\dots,0$. Therefore, $c_\delta = \int v_0^\delta dp_0 \geqq \int f_0 dp_0 = \inf \mathrm{DPA}$. Hence, $\inf_{\delta\in\Delta} c_\delta = \inf \mathrm{SDP} \geqq \inf \mathrm{DPA}$. If $v_0^\delta = f_0$ a.s. $[q_\delta]$ then $c_\delta = \inf \mathrm{SDP} = \inf \mathrm{DPA}$. Conversely, if $c_\delta = \inf \mathrm{SDP} = \inf \mathrm{DPA}$, then an inspection of the inequalities in the first part of the proof shows that $v_j^\delta = f_j$ a.s. $[q_\delta]$ for all j. □

The conditions $v_j^\delta = f_j$ a.s. $[q_\delta]$, $j = 0,1,\dots,n$, are equivalent to

(4.46) $$\delta_j(D_j^*(s_j)|s_j) = 1, \; j = 0,1,\dots,n-1, \text{ a.s. } [q_\delta],$$

where, for $j = 0,1,\dots,n-1$ and $s \in S_j$

$$D_j^*(s) := \arg\min_{a\in D_j(s)} \{c_j(s,a) + \int_{S_{j+1}} f_{j+1}(t)p_{j+1}(dt|s,a)\}.$$

Suppose for a moment that each $D_j^*(s_j)$ is nonempty, and that each $D_j^*$ admits a measurable selection. Then there exists a $\delta^* \in \Delta$, even a deterministic one, such that $\delta_j^*(D_j^*(s_j)|s_j) = 1$ for all j and $s_j$. We say that such a policy is *generated by DPA*. Obviously, every policy $\delta^*$ generated by DPA is optimal for SDP. In fact, it can be shown to be "super" optimal: it is optimal uniformly for all initial distributions

$p_0$, and moreover, the restriction of $\delta^*$ to any tail problem is optimal for that tail problem, again uniformly for all initial distributions. Unfortunately, it is not true in general that DPA generates a policy at all.

Notice that Proposition 4.22 implies that (4.46) is only necessary for $\delta \in \Delta$ to be optimal for SDP if

$$\inf \text{DPA} = \inf \text{SDP}$$

is true. We shall show that this equality is closely related to the normality of the dual pair of linear programs ($\text{SDP}_1$,$\text{SDP}_2$). Let us first relate $\text{SDP}_2$ with DPA.

<u>PROPOSITION 4.23</u>. *Assume* (4.38), (4.40), (4.41a,c). *Then*

$$\sup \text{SDP}_2 \leqq \inf \text{DPA}. \tag{4.47}$$

*Moreover, if the cost-to-go functions* $f_j$ *satisfy* $f_j \in F(S_j)$, $j = 0,\ldots,n$, *then*

$$\max \text{SDP}_2 = \inf \text{DPA}.$$

*If* (4.44) *holds,* $f_j \in F(S_j)$ *iff* $f_j$ *is* $S_j$*-measurable,* $j = 0,1,\ldots,n$.

<u>PROOF</u>. Without loss of generality we may assume that a feasible solution $y = (y_0,y_1,\ldots,y_n)$ for $\text{SDP}_2$ exists. By backward recursion, using the constraints of $\text{SDP}_2$, one shows that $y_j(s_j) \leqq f_j(s_j)$ for all $s_j \in S_j$, $j = n,n-1,\ldots,0$. Consequently, $\langle y_0,p_0\rangle \leqq \int f_0 dp_0 = \inf \text{DPA}$. Since y is arbitrary, $\sup \text{SDP}_2 \leqq \inf \text{DPA}$. Clearly equality holds if $f_j \in F(S_j)$ for all j, because then $(f_0,f_1,\ldots,f_n)$ is feasible for $\text{SDP}_2$.

Finally we have to show that under (4.44)

$$\gamma_j := \sup_{s\in S_j} (e_{S_j}(s))^{-1} |f_j(s)|, \; j = 0,1,\ldots,n,$$

are finite numbers. This can be done by recursive estimation based on DPA. Let M be the number defined in (4.44), and define

$$M_1 := \max_j \sup_{(s,a)\in D_j} (e_{D_j}(s,a))^{-1} \langle e_{S_{j+1}}, p_{j+1}(\cdot|s,a)\rangle,$$

$$M_2 := \max_j \sup_{d\in D_j} (e_{D_j}(d))^{-1} |c_j(d)|.$$

$M_1$ and $M_2$ are finite positive numbers because of (4.41a,c). From the definition of $f_n$ in DPA we conclude that $\gamma_n \leqq M_2$. Since $|\inf E| \leqq \sup |E|$ for any $E \subset [-\infty,\infty]$ we conclude from the definition of $f_j$, $j < n$, that

$$\begin{aligned}\gamma_j &\leqq \sup_{(s,a)\in D_j} (e_{S_j}(s))^{-1}(|c_j(s,a)| + \int |f_{j+1}(t)| p_{j+1}(dt|s,a)) \\ &\leqq \sup_{(s,a)\in D_j} (e_{S_j}(s))^{-1}(M_2 + \gamma_{j+1}M_1)e_{D_j}(s,a) \\ &\leqq (M_2 + \gamma_{j+1}M_1)(1+M).\end{aligned}$$

Backward induction shows that each $\gamma_j < \infty$. Since $\gamma_j \geqq 0$, too, the proof is complete. □

The last result in Proposition 4.23 indicates, that under suitable finiteness assumptions the equivalence between $SDP_2$ and DPA is "only" a matter of measurability. Since $f_j$ is universally measurable under (4.42), one can make the equivalence complete by taking for $\mathcal{S}_j$ the $\sigma$-algebra of all uniformly measurable functions on the Borel space $S_j$. We come back to this idea later, see Proposition 4.30.

Propositions 4.18 and 4.20 show the relation between $SDP_1$ and SDP, whereas Proposition 4.22 establishes the relations between $SDP_2$ and DPA. Moreover, the combination of (4.43), (4.45) and (4.47) reveals, that under suitable conditions

$$\sup SDP_2 \leqq \inf DPA \leqq \inf SDP \leqq \inf SDP_1. \tag{4.48}$$

Therefore, the dynamic programming result inf DPA = inf SDP can be derived by showing that $(SDP_1, SDP_2)$ has no duality gap. Then the *optimality conditions* of Proposition 4.22 can be viewed as *consequences of complementary slackness*. In order to see this, suppose that $\sup SDP_2 = \inf SDP_1$, and let $(x^*, y^*)$ be feasible for $(SDP_1, SDP_2)$. The complementary slackness conditions reduce to, see (4.17),:

$$(4.49)\quad \begin{aligned} &x_n^*(\{s_n \in S_n: y_n^*(s_n) = c_n(s_n)\}) = 1, \text{ and for } j = 0,1,\dots,n-1: \\ &x_j^*(\{(s_j,a_j) \in D_j: y_j^*(s_j) = c_j(s_j,a_j) + \langle y_{j+1}^*, p_{j+1}(.\,|s_j,a_j)\rangle\}) = 1. \end{aligned}$$

Because of (4.42) there exists a $\delta^* \in \Delta$ such that $x_j^* = x_j^{\delta^*}$, $j = 0,1,\dots,n$, see the proof of Proposition 4.18 part b. Hence, the probabilities $x_j^*$ are the marginals of $q_{\delta^*} = p_0 * \delta_0^* * \dots * \delta_{n-1}^* * p_n$. Therefore, (4.49) is equivalent to $y_j^* = f_j$, $j = 0,1,\dots,n$, a.s. $[q_{\delta^*}]$, and this implies that $\delta^*$ satisfies (4.46).

The previous paragraph reveals the complete analogy between Proposition 4.22 and Theorem 4.2. Both provide sufficient optimality conditions, which are necessary, too, if there is no gap. Both have a trivial proof. In both cases, it is much more difficult to prove that there is no gap under not too restrictive assumptions. Relation (4.48) shows that the dynamic programming problem is gap-free if the pair of linear programs is gap-free. Hence it is only natural to derive results for the stochastic dynamic programming problem by applying advanced duality theory to the pair of linear programs. Moreover, Proposition 4.17 shows, that all conditions of part a as well as part b of Theorem 4.5 are satisfied, if we are able to find a compatible topology in which the optimal value function is bounded above/below in a neighbourhood of 0. Just as in Sections 4.4 and 4.5, application of Theorem 4.5 part a does not seem very attractive, neither with the norm topology nor with weak topologies. Moreover, it is clear that "small" perturbations (e.g. $u_0 \in \mathcal{M}(S_0)$ with $u_0(E) < 0$ for an $E \in \mathcal{S}_0$ with $p_0(E) = 0$) may give $\varphi_1(u) = +\infty$. Let us therefore consider Theorem 4.5 part b, and analyze the optimal value function $\varphi_2$. We shall show that $\varphi_2$ is Lipschitz continuous, hence bounded below in the neighbourhood of 0, in a norm topology to be defined.

Under the assumptions (4.40), (4.41a,c) the optimal value function $\varphi_2$ of $SDP_2$ is well-defined, and it can be written as

$$\varphi_2(v) = \sup\{\langle y,\tilde{p}_0\rangle: y \in W(c+v)\},\ v \in V,$$

where

$$W(v) := \{y \in Y: L_2 y \leqq v\},\ v \in V.$$

Here the same definitions for $Y$, $V$, $V^+$, $L_2$, $c$ and $\tilde{p}_0$ are used as in the proof of Proposition 4.17. Properties of $\varphi_2$ can be derived from the study of $W(v)$. Some elementary properties of $W(v)$ are

$$W(\lambda v) = \lambda W(v);\ \lambda > 0,\ v \in V,$$
$$W(v_1) + W(v_2) \subset W(v_1 + v_2);\ v_1 \in V,\ v_2 \in V,$$
$$v_1 \leqq v_2 \Rightarrow W(v_1) \subset W(v_2);\ v_1 \in V,\ v_2 \in V.$$

In the following lemma we exhibit properties of $W(v)$ which are consequences of special structure of $L_2$. We use the following norms on $F(D_j)$, $j = 0,1,\dots,n$ and $V = \Pi_{j=0}^n F(D_j)$:

$$\|v_j\| := \sup_{(s_j,a_j)\in D_j} (e_{D_j}(s_j,a_j))^{-1} |v_j(s_j,a_j)|,\ j = 0,\dots,n-1,$$

$$\|v_n\| := \sup_{s_n\in S_n} (e_{S_n}(s_n))^{-1} |v_n(s_n)|,$$

$$\|v\| := \max_{j=0,1,\dots,n} \|v_j\| \text{ for } v = (v_0,v_1,\dots,v_n) \in V.$$

Similarly a norm on $Y = \Pi_{i=0}^n F(S_i)$ can be defined, but we do not need it. Finally, let $v^e \in V$, $y^e \in Y$ be given by

$$v^e := (e_{D_0},e_{D_1},\dots,e_{D_n}),\ y^e := (e_{S_0},e_{S_1},\dots,e_{S_n}).$$

Obviously, $v^e \in V^+$, $\|v^e\| = 1$. In fact $v^e$ is the lattice supremum of all $v \in V$ with $\|v\| \leqq 1$. A similar statement characterizes $y^e$, at least if instead of $Y^+$ (=Y) the natural positive cone is used, consisting of vectors of nonnegative functions. This natural positive cone will be denoted by $\tilde{Y}^+$ and will be used in the following lemma.

<u>LEMMA 4.24</u>. *a. If* $y^{(k)} \in W(v)$, $k = 1,2$, *then* $z := \sup\,(y^{(1)},y^{(2)}) \in W(v)$, *where* sup *denotes the lattice supremum with respect to* $\tilde{Y}^+$.
*b. Assume that* (4.44) *holds. Then there exist positive numbers* $\eta_0,\eta_1,\dots,\eta_n$ *such that* $y^\eta := (\eta_0 y_0^e,\dots,\eta_n y_n^e)$ *satisfies*

$$-y^\eta \in W(-v^e) \text{ and } W(v^e) \subset y^\eta - \tilde{Y}^+.$$

*Furthermore,* $-\|v\|.y^\eta \in W(v) \subset \|v\|.y^\eta - \tilde{Y}^+$.

PROOF. (a) Of course, $z \in Y$ since $|z| \leqq |y^{(1)}| + |y^{(2)}|$. In order to show that $L_2 z \leqq v$, notice that $y_n^{(k)} \leqq v_n$ implies $z_n \leqq v_n$. Analogously, for $j < n$, $(s,a) \in D_j$ and $k = 1,2$

$$y_j^{(k)}(s) \leqq v_j(s,a) + \langle y_{j+1}^{(k)}, p_{j+1}(.|s,a)\rangle$$
$$\leqq v_j(s,a) + \langle z_{j+1}, p_{j+1}(.|s,a)\rangle.$$

The first inequality follows from $L_2 y^{(k)} \leqq v$, and the second one from $p_{j+1} \geqq 0$. Maximization over k gives the desired result.
(b) Because of (4.41a) and (4.18)

$$M_1 := \max_j \sup_{(s,a)\in D_j} (e_{D_j}(s,a))^{-1} \langle e_{S_{j+1}}, p_{j+1}(.|s,a)\rangle$$

is finite. Let M be the finite number given in (4.44). Let $\eta_n := 1$, and for $j = n-1, n-2, \ldots, 0$ define $\eta_j := (1 + \eta_{j+1} M_1)(1+M)$. Then for each $j \in \{0,1,\ldots,n-1\}$

$$e_{D_j} + \langle y_{j+1}^{\eta}, p_{j+1}(.|.)\rangle = e_{D_j} + \eta_{j+1}\langle e_{S_{j+1}}, p_{j+1}(.|.)\rangle$$
$$\leqq (1 + \eta_{j+1} M_1) e_{D_j}$$
$$\leqq (1 + \eta_{j+1} M_1)(1+M) e_{S_j} = \eta_j e_{S_j} = y_j^{\eta}.$$

Together with $y_n^{\eta} = \eta_n e_{S_n} = e_{S_n} = e_{D_n}$ this implies that $L_2(-y^{\eta}) \leqq -v^e$. Since $-y^{\eta} \in Y$ we proved $-y^{\eta} \in W(-v^e)$.

Suppose $y \in W(v^e)$. Then $y_n \leqq e_{D_n} = e_{S_n}$, and $y_j \leqq e_{D_j} + \langle y_{j+1}, p_{j+1}(.|.)\rangle$. Backward induction shows $y_j \leqq \eta_j e_{S_j} = y_j^{\eta}$, $j = n, n-1, \ldots, 0$, which had to be proved.

Finally, since $-y^{\eta} \in W(-v^e)$ we have for $v \neq 0$ that $-\|v\|.y^{\eta} \in W(-\|v\|.v^e)$, and because of $-\|v\|.v^e \leqq v$ we conclude $-\|v\|.y^{\eta} \in W(v)$; clearly, this result holds too if $v = 0$. Moreover, since $W(v^e) \subset y^{\eta} - \tilde{Y}^+$ and $v \leqq \|v\| v^e$ it follows that $W(v) \subset \|v\|.y^{\eta} - \tilde{Y}^+$ for $v \neq 0$; backward recursion shows that this inclusion is still true for $v = 0$. □

PROPOSITION 4.25. *Assume* (4.40), (4.41a,c), (4.44). *Then in* $SDP_2$ *the*

*supremum is attained. Moreover, the optimal value function* $\varphi_2$ *of* $SDP_2$ *is Lipschitz continuous:*

$$(4.50) \qquad |\varphi_2(v_1) - \varphi_2(v_2)| \leqq K.\|v_1 - v_2\| \quad \forall v_1,v_2 \in V.$$

PROOF. Recall that $\varphi_2(v) = \sup\{\langle y,\tilde{p}_0\rangle : Y \in W(c+v)\}$. Fix $v \in V$. Lemma 4.24 part b shows, that $W(c+v) \neq \emptyset$ so $\varphi_2(v) > -\infty$, and also, because of $\tilde{p}_0 \geqq 0$, that $\varphi_2(v) \leqq \|c+v\|.\langle y^\eta,\tilde{p}_0\rangle \leqq \|c+v\|.\eta_0.\langle e_{s_0},p_0\rangle < \infty$. Hence $\varphi_2(v)$ is finite. Because of that, there exists a sequence $(y^{(k)}, k = 1,2,\dots) \subset W(c+v)$ with $\varphi_2(v) - 1/k \leqq \langle y^{(k)},\tilde{p}_0\rangle \leqq \varphi_2(v)$. Because of Lemma 4.24 part a we may, without loss of generality, assume $y^{(1)} \leqq y^{(2)} \leqq y^{(3)} \leqq \dots$ (inequalities with respect to $\tilde{Y}^+$): if necessary, replace $y^{(k)}$ by $\sup_{1\leqq h\leqq k} y^{(h)}$; this does not disturb the assumption on $\langle y^{(k)},\tilde{p}_0\rangle$ since $\tilde{p}_0 \geqq 0$. On the other hand, we know from Lemma 4.24 part b that $y^{(k)} \leqq \|c+v\|y^\eta$. Hence, $\bar{y}_j(s_j) := \lim_{k\to\infty} y_j^{(k)}(s_j)$ is finite for all j and $s_j$. Each $\bar{y}_j$ is measurable, so that $\bar{y} := (\bar{y}_0,\bar{y}_1,\dots,\bar{y}_n) \in Y$ since $y^{(1)} \leqq \bar{y} \leqq \|c+v\|y^\eta$ with $y^{(1)},y^\eta \in Y$. The last condition enables us also to conclude by the dominated convergence theorem that $L\bar{y} \leqq c+v$ and that $\langle\bar{y},\tilde{p}_0\rangle = \lim_k \langle y^{(k)},\tilde{p}_0\rangle$. So we did show the existence of a $\bar{y} \in W(c+v)$ with $\langle\bar{y},\tilde{p}_0\rangle = \varphi_2(v)$.

Let us now prove the Lipschitz continuity of $\varphi_2$. Fix $v^1,v^2 \in V$ and set $\varepsilon := \|v^1-v^2\|$. Since the supremum in the definition of $\varphi_2(v^1)$ is attained, there exists a $y^1 \in Y$ with $\varphi_2(v^1) = \langle y^1,\tilde{p}_0\rangle$ and $L_2y^1 \leqq v^1$. Lemma 4.24 part b shows that $L_2(-y^\eta) \leqq -v^e$ so that $L_2(y^1 - \varepsilon y^\eta) \leqq v^1 - \varepsilon v^e \leqq v^2$; the last inequality follows from the definition of $\varepsilon$. Consequently $y^1 - \varepsilon y^\eta \in W(v^2)$ and therefore $\varphi_2(v^2) \geqq \langle y^1 - \varepsilon y^\eta,\tilde{p}_0\rangle = \varphi_2(v^1) - \varepsilon\langle y^\eta,\tilde{p}_0\rangle$. By exchanging $v^1$ and $v^2$ we find also $\varphi_2(v^1) \geqq \varphi_2(v^2) - \varepsilon\langle y^\eta,\tilde{p}_0\rangle$. Hence (4.50) is true with $K := \langle y^\eta,\tilde{p}_0\rangle = \eta_0.\langle e_{s_0},p_0\rangle < \infty$. □

Summarizing the results we have

THEOREM 4.26. *a. Assume* (4.40), (4.41a,c), (4.42) *and* (4.44). *Then*

$$(4.51) \qquad -\infty < \max SDP_2 \leqq \inf DPA \leqq \inf SDP = \inf SDP_1 < \infty.$$

*b. If a topology on* $V = \prod_{j=0}^n F(D_j)$ *can be defined satisfying*

(4.52) (i) $\|v\|$ *is continuous,*
(ii) *the topological dual space* $V^*$ *of* V *is* $X := \Pi_{j=0}^{n} M(D_j)$,

*then*

(4.53) $-\infty < \max \text{SDP}_2 = \inf \text{DPA} = \min \text{SDP} = \min \text{SDP}_1 < +\infty.$

*Moreover, the optimal solutions of* $\text{SDP}_1$ *and* $\text{SDP}_2$ *are completely characterized by feasibility and complementary slackness. As a consequence, the conditions* (4.46), *inspired by* DPA, *are necessary and sufficient for optimality in the* SDP *problem.*

PROOF. (a) (4.51) follows from the Propositions 4.20, 4.22, 4.23 and 4.25.
(b) Using (4.52) and Proposition 4.25 we derive from Theorem 4.5 part b that $\sup \text{SDP}_2 = \min \text{SDP}_1$. Proposition 4.18 part b shows that min SDP = $\min \text{SDP}_1$ if $\inf \text{SDP}_1 = \min \text{SDP}_1$. The remaining statements in (4.53) follow from (4.51). The last part of Theorem 4.26 is a consequence of Theorem 4.2, Proposition 4.22 and the discussion around (4.49). □

Theorem 4.26 looks nice but its weak point with regard to applications is, of course, that the existence of a topology on V which satisfies (4.52) has to be established. As in Section 4.5 we are urged to come up with the following compactness and continuity assumptions.

(4.54) $S_j$ and $A_j$ are compact Borel spaces, $j = 0,1,\ldots,n-1,(n,)$
$D_j$ is closed, $j = 0,1,\ldots,n-1$,
$c_j$ is continuous, $j = 0,1,\ldots,n$.

Since (4.54) implies that each $D_j$ is compact, each $c_j$ is bounded so that without loss we may replace each bounding function by 1.

THEOREM 4.27. *Assume that* (4.54) *holds, and replace in the definition of* $(\text{SDP}_1,\text{SDP}_2)$ $F(D_i)$ *by* $C(D_i)$, $i = 0,1,\ldots,n$, *and each bounding function by* 1. *Then* (4.40), (4.41a,c), (4.42), (4.44) *and* (4.52) *are true so that all results of Theorem 4.26 apply.*

PROOF. Direct. Use $C(D_i)^* = M(D_i)$. □

We derived Theorem 4.27 in the framework of dual linear programs; Theorem 4.5 part b is used to prove the stability of $SDP_2$. Since this stability implies that $SDP_1$ has optimal solutions, it is not surprising that we end up with compactness and continuity assumptions. Nevertheless, the assumptions (4.54) are not necessary for inf DPA = min SDP: in Section 6.4 we will discuss a rather general inventory control model with Euclidean state and action spaces, with $D_j$ closed but not compact and with $c_j$ lower semicontinuous and locally bounded below. Together with some assumptions which imply that only compact subsets of $D_j(s_j)$ have to be considered and which imply certain continuity properties of the transition probabilities $p_j$, these assumptions are sufficient to prove inf DPA = min SDP. The proof is based on a careful analysis of DPA.

In an attempt to get rid of the compactness assumption, let us analyze what can be proved if (4.54) is replaced by (4.40), (4.41a,c), (4.42), (4.44) and

$$e_{D_j} \in \mathcal{F}_c(D_j),\ c_j \in \mathcal{F}_c(D_j),\ j = 0,1,\ldots,n,$$

where $\mathcal{F}_c(D_j)$ is the subspace of $\mathcal{F}(D_j)$ consisting of functions which are continuous with respect to the topology in $D_j$ induced by the metric structure. If in $SDP_2$ the constraint space $V = \Pi_{j=0}^{n}\mathcal{F}(D_j)$ is replaced by $V_c := \Pi_{j=0}^{n}\mathcal{F}_c(D_j)$, then still its optimal value function is Lipschitz continuous with respect to the norm in $V_c$ (inherited from the norm in V). Hence, sup $SDP_2$ = min $\overline{SDP_1}$ where $\overline{SDP_1}$ is the conjugate dual problem of $SDP_2$. $\overline{SDP_1}$ is a linear program (see [23] for an explicit description). It resembles $SDP_1$. The basic difference is that its space of variables $\Pi_{j=0}^{n}\mathcal{F}_c(D_j)^*$ is larger than $X = \Pi_{j=0}^{n}\mathcal{M}(D_j)$, since $\mathcal{F}_c(D_j)^*$ contains all additive (rather than countably additive) set functions on $(D_j,\mathcal{D}_j)$ which satisfy certain boundedness conditions. Since integrals with respect to bounded additive set functions do not have all nice properties which are available in the countably additive case, the constraints in $\overline{SDP_1}$ are different from those in $SDP_1$, but nevertheless: on X there is a complete agreement between $\overline{SDP_1}$ and $SDP_1$. Therefore, a possibility to show sup $SDP_2$ = min $SDP_1$ would be to decompose the optimal solution of $\overline{SDP_1}$ into its countably additive part and its purely singular part, and to find conditions which guarantee that the countably additive part solves $SDP_1$. Ponstein [28] has studied such conditions.

However, his results refer to the decomposition of $L^*_\infty(S,\Sigma,\mu)$ rather than $L^*_\infty(S,\Sigma)$ which would be relevant for our problem. By the way, it is not to be expected that in general $\min \overline{SDP_1} = \min SDP_1$: Heilmann [12] provides an example (for an infinite horizon stationary dynamic programming problem) with $\min \overline{SDP_1} = \inf SDP_1$ where the infimum is not attained.

Up to now we have been preoccupied with proving stability: $\sup SDP_2 = \min SDP_1$. How about proving normality: $\sup SDP_2 = \inf SDP_1$? In the literature on duality theory we did not find sufficient conditions for normality which result into easily verifiable conditions on the Markovian decision process. On the other hand, it is known from dynamic programming theory, that "almost always" inf DPA = inf SDP holds. Typically, the proof has the following structure. In DPA, define for each $\varepsilon > 0$, each $j = n-1,\ldots,0$ and each $s \in S_j$ the set

$$D_j^\varepsilon(s) := \{a \in D_j(s)\colon f_j(s) + \varepsilon \geqq c_j(s,a) + \int_{S_{j+1}} f_{j+1}(t)p_{j+1}(dt|s,a)\}$$

where $f_j(s) + \varepsilon$ is replaced by $-\frac{1}{\varepsilon}$ if $f_j(s) = -\infty$. Then $D_j^\varepsilon(s_j) \neq \emptyset$. Use a measurable selection theorem in order to prove the existence of a deterministic policy $\delta$ with $\delta_j(s_j) \in D_j^\varepsilon(s_j)$ $\forall j$ $\forall s_j$. Then it follows by recursion that if each $f_j > -\infty$

$$\inf SDP \leqq \int v_0^\delta dp_0 \leqq \int f_0 dp_0 + n\varepsilon = \inf DPA + n\varepsilon$$

so that inf SDP $\leqq$ inf DPA. In this way, using the Jankov-von Neumann analytically measurable selection theorem [1 Propositions 7.49 and 7.50], Bertsekas and Shreve prove essentially the following result.

PROPOSITION 4.28. *Assume* (4.38), (4.42). *If the set of admissible policies is enlarged with all sequences of uniformly measurable transition probabilities, then*

$$\inf DPA = \inf SDP.$$

PROOF. Proposition 8.2 in [1]. In [1] the model is stationary, with terminal cost function $c_n = 0$. But by extending the number of stages with 1 one gets rid of the terminal cost function; and by state augmentation one reduces nonstationary models to stationary ones [1 Chapter 10]. The restriction to Markovian policies is justified in [1 Proposition 8.11]. □

Whereas we are not able to derive the dynamic programming result in Proposition 4.28 as a corollary of a duality result, it is interesting to notice that the reverse approach works.

PROPOSITION 4.29. *Assume* (4.40), (4.41a,c), (4.42), (4.44), *and replace in the definition of* SDP *each* $S_j$ *by the σ-algebra of universally measurable subsets of* $S_j$. *Then*

$$\max SDP_2 = \inf SDP_1.$$

PROOF. Combine (the discussion after) Proposition 4.23 with Theorem 4.27 and Proposition 4.28. □

In the remainder of this section we point out some relations to the existing literature which were not mentioned previously. For an extensive discussion on stochastic dynamic programming in general state and action spaces we refer to [14,34,35], the monograph [1] of Bertsekas and Shreve, and the references given there. The relevance of linear programming duality for dynamic programming has been recognized for a long time. We refer to [10] for an annotated bibliography. Mostly stationary models over an infinite horizon and under various objectives are considered. Complete results are known if state and action spaces are finite, e.g. [2,26]. Yamada [37] gives results for the average cost criterion using amongst others compact subsets of $\mathbb{R}^n$ as state and action spaces. For general state and action space in a stationary infinite horizon model Heilmann [9,11,12] formulates linear programming models and their duals, which are comparible with $SDP_2$ and $\overline{SDP}_1$. Hordijk [16] gives an exposition on nonstationary finite horizon models with finite action and state spaces. Not surprisingly, in this case every dynamic programming result can be derived easily from the finite-dimensional linear programming duality theory and the simplex method. Heilmann [13] considers the general case. Our model resembles that of Heilmann [13]. However, we consider a finite horizon. More importantly, we avoid to work with noncountably additive set functions, by choosing the dual pair a priori. Moreover, our framework allows for unbounded cost functions, due to the use of bounding functions. Whereas to our best knowledge bounding functions are not used in the literature on the linear programming approach to dynamic programming, they are not unknown to dynamic programmers. Just as in our approach, bounding functions have been introduced as a means to guarantee finiteness of the cost-to-go functions (and their convergence in the stationary infinite horizon case [36]). Moreover, extrapolation and approximation bounds might be improved by the use of bounding functions [15,17]. In all these papers the dynamic programming recursion is used as the basic tool, and therefore a bounding function on the *state space* is used to get a suitable "similarity transformation" of the state space. Typical assumptions are (in our notation)

(4.55) $\quad \sup_{(s,a)\in D_j} (e_{S_j}(s))^{-1} |c_j(s,a)| < \infty,$

(4.56) $\quad \sup_{(s,a)\in D_j} (e_{S_j}(s))^{-1} \int_{S_{j+1}} e_{S_{j+1}}(t) p_{j+1}(dt|s,a) < \infty.$

The linear programming approach led us to the introduction of bounding functions on the *action spaces*, too. It is interesting to notice, that (4.55) and (4.56) are equivalent to our assumptions (4.40), (4.41), (4.44) in the special case $e_{A_j} \equiv 1$ for all j. Therefore, our assumptions are more general, and it seems that they allow for the same results.

## 4.7 CONCLUSIONS

We analyzed several optimization problems in probabilities from a linear programming point of view. The similarity between the various problems is striking. Under weak finiteness conditions each of them corresponds to a linear program, and for this linear program a dual linear program can be formulated which is an approximation problem in a function space. Optimality of feasible solutions for both problems is achieved if complementary slackness holds. This application of the elementary duality theorem is so trivial, that some authors apply it without realizing that it has to do with linear programming duality; they do not formulate the dual problem and do not realize that they are looking for an optimal dual solution, too. In order to show that (asymptotical) complementarity slackness is necessary for optimality, one needs verifiable sufficient conditions for normality. For that reason we analyzed the possible application of a stability criterion provided by Rockafellar. In most cases it appeared to be profitable to consider the original problem in probabilities as the conjugate dual of the problem in function spaces. The optimal value function of the latter problem can be shown to be continuous in the norm topology, and this topology is compatible with the duality if the underlying spaces are compact metric, and the given functions are continuous. More general situations can also be covered if one is able to find a compatible weak topology, in which the optimal value function is bounded below. In a simple case we had success with this direct approach based on advanced duality theory. In general, however, it seems to be necessary to use the specific properties of the problem at hand in order to get the most general results. For example, one may use less strict finiteness conditions, and choose σ-algebra's which are adjusted to the given problem. Sometimes one needs advanced topological measure theory, and sometimes the topological considerations do not seem to be dominant in the derivation of the most general results. Anyhow, we feel that looking at optimization problems in probabilities as linear programming problems and studying their duals too, provides useful insight. This is not on strained terms with the natural fact, that using specific techniques for specific problems provides results which reach beyond the results of the abstract duality theory.

REFERENCES

1. D.P. BERTSEKAS, S.E. SHREVE (1978). *Stochastic Optimal Control: The Discrete-Time Case*, Academic Press, New York-San Francisco-London.
2. C. DERMAN (1970). *Finite State Markovian Decision Processes*, Academic Press, New York-San Francisco-London.
3. N. DUNFORD, J.T. SCHWARTZ (1957). *Linear Operators: Part I*, Interscience, New York.
4. Y. ERMOLIEV, A. GAIVORONSKI (1984). *Duality Relations and Numerical Methods for Optimization Problems on the space of Probability Measures with Constraints on Probability Measures*, IIASA Working paper WP-84-86, IIASA, Laxenburg.

5. A.V. FIACCO, K.O. KORTANEK (eds.) (1983). *Semi-Infinite Programming and Applications*. Lecture notes in economics and mathematical systems 215, Springer, Berlin-Heidelberg-New York-Tokyo.

6. N. GAFFKE, L. RÜSCHENDORF (1981). On a class of extremal problems in statistics. *Math. Operationsforsch. Statist. Ser. Optim. 12*, 123-135.

7. K. GLASHOFF, S.-Å. GUSTAFSON (1983). *Linear Optimization and Approximation*. Applied Mathematical Sciences 45, Springer, New York-Heidelberg-Berlin.

8. S.-Å. GUSTAFSON, K.O. KORTANEK (1983). Semi-infinite programming and applications. A.BACHEM, M. GRÖTSCHEL, B. KORTE (eds.). *Mathematical Programming: The State of the Art - Bonn 1982*, Springer, Berlin-Heidelberg-New York-Tokyo, 132-157.

9. W.-R. HEILMANN (1977). Stochastische dynamische Optimierung als Spezialfall linearer Optimierung in halbgeordneten Vectorräumen. *Manuscripta Math. 23*, 57-66.

10. W.-R. HEILMANN (1978). Solving stochastic dynamic programming problems by linear programming - an annotated bibliography. *Z. Oper. Res. Ser. A 22*, 43-53.

11. W.-R. HEILMANN (1978). Generalized linear programming in Markovian decision processes. *Bonner Mathematische Schriften 98*, 33-39.

12. W.-R. HEILMANN (1979). Solving a general discounted dynamic program by linear programming. *Z. Wahrsch. Verw. Gebiete 48*, 339-346.

13. W.-R. HEILMANN (1979). A linear programming approach to general nonstationary dynamic programming problems. *Math. Operationsforsch. Statist. Ser. Optim. 10*, 325-333.

14. K. HINDERER (1970). *Foundations of Nonstationary Dynamic Programming with Discrete Time Parameter*. Lecture notes in operations research and mathematical systems 33, Springer, Berlin-Heidelberg-New York.

15. K. HINDERER, G. HÜBNER (1977). On approximate and exact solutions for finite stage dynamic programs. H.C. TIJMS, J. WESSELS (eds.). *Proceedings of the Advanced Seminar on Markov Decision Theory*, Mathematical Centre Tracts 93, Mathematisch Centrum, Amsterdam.

16. A. HORDIJK (1979). From linear to dynamic programming via shortest paths. P.C. VAN BAAYEN, D. VAN DULST, J. OOSTERHOFF (eds.). *Proceedings of the Bicentennial Congress of the Wiskundig Genootschap, part II*, Mathematical Centre Tracts 101, Mathematisch Centrum, Amsterdam, 213-231.

17. G. HÜBNER (1980). Bounds and good policies in stationary finite stage Markovian decision problems. *Adv. in Appl. Probab. 12*, 154-173.

18. K. ISII (1960). The extrema of probability determined by generalized moments (I): Bounded random variables. *Ann. Inst. Statist. Math. 12*, 119-133.

19. S. KARLIN, W.J. STUDDEN (1966).*Tchebycheff Systems: with Applications in Analysis and Statistics*, Interscience, New York.

20. H.G. KELLERER (1984). Duality theorems and probability metrics. M. IOSIFESCU (ed.). *Proceedings of the Seventh Conference on Probability Theory, August 29-September 4, 1982, Braşov*, Romanian Academy of Sciences, Bucureşti, 211-220.

21. H.G. KELLERER (1984). Duality theorems for marginal problems. *Z. Wahrsch. Verw. Gebiete 67*, 399-432.

22. J.H.B. KEMPERMAN (1983). On the role of duality in the theory of moments. A.V. FIACCO, K.O. KORTANEK (eds.). *Semi-Infinite Programming and Applications*, Lecture notes in economics and mathematical systems 215, Springer, Berlin-Heidelberg-New York-Tokyo.

23. W.K. KLEIN HANEVELD (1979). *The Linear Programming Approach to Finite Horizon Stochastic Dynamic Programming*, Report 79-03-OR, Institute of Econometrics, University of Groningen, Groningen.

24. W.K.KLEIN HANEVELD (1984). *Abstract LP Duality and Bounds on Variables*, Report 84-13-OR, Institute of Econometrics, University of Groningen, Groningen.

25. V.L. LEVIN (1984). The problem of mass transfer in a topological space, and probability measures having given marginal measures on the product of two spaces. *Soviet Math. Dokl. 29 No 3*, 638-643.

26. H. MINE, S. OSAKI (1970). *Markovian Decision Processes*, American Elsevier, New York.

27. K.R. PARTHASARATHY (1967). *Probability Measures on Metric Spaces*, Academic Press, New York.

28. J. PONSTEIN (1981). On the use of purely finitely additive multipliers in mathematical programming. *J. Optim. Theory Appl. 33*, 37-55.

29. J. PONSTEIN (1980). *Approaches to the Theory of Optimization*, Cambridge University Press, Cambridge.

30. H. RICHTER (1957). Parameterfreie Abschätzung und Realisierung von Erwartungswerten. *Blätter Deutscher Gesellschaft Versicherungsmathematik 3*, 147-161.

31. R.T. ROCKAFELLAR (1970). *Convex Analysis*, Princeton University Press, Princeton NJ.

32. R.T. ROCKAFELLAR (1974). *Conjugate Duality and Optimization*, SIAM monograph, SIAM, Philadelphia Pa.

33. H.H. SCHAEFFER (1966). *Topological Vector Spaces*, Springer, Berlin-Heidelberg-New York.

34. S.E. SHREVE, D.P. BERTSEKAS (1978). Alternative theoretical frameworks for finite horizon discrete time stochastic optimal control. *SIAM J. Control Optim. 16*, 953-978.

35. S.E. SHREVE, D.P. BERTSEKAS (1979). Universally measurable policies in dynamic programming. *Math. Oper. Res. 4*, 15-30.

36. J. WESSELS (1977). Markov programming by successive approximations with respect to weighted supremum norm. *J. Math. Anal. Appl. 58*, 326-335.

37 K. YAMADA (1975). Duality theorem in Markovian decision problems. *J. Math. Anal. Appl. 50*, 579-595.

# CHAPTER 5

# ON INTEGRATED CHANCE CONSTRAINTS

## 5.1 INTRODUCTION

In Chapter 3 we introduced *integrated chance constraints* (ICCs) as a modeling tool for here-and-now stochastic programming problems; see (3.28). In the present chapter we analyze the mathematical properties of this new concept in some detail. Let us review its rationale. As indicated in Section 3.2, if in the constraints of a linear programming problem random coefficients occur with unknown realizations, then in order to have a unequivocal meaning of "feasibility" one has to make additional specifications. There are two well-known modeling techniques for this: in *chance-constrained programming* (CCP) the probability of infeasibility is restricted, and in *stochastic programming with recourse* (SPR) the effects of infeasibility are penalized. For convenience, we here consider "penalty cost" models as recourse models, see Remark 3.6. Several authors [12,10,11,3,2] established certain equivalences between CCP and SPR. Their results are not completely convincing, however; for example, CCP problems may be nonconvex whereas SPR problems are always convex [5 page 90]. Even if mathematical equivalence can be established there still are differences between CCP and SPR models which might be important for the model builder. Here we want to stress two major differences between CCP and SPR:

(i) In CCP "risk" is measured *qualitatively*, whereas in SPR it is done *quantitatively*, see Remark 3.9.

(ii) In CCP "risk aversion" is modeled by *constraints* whereas in SPR it is done by *penalties*.

These distinctions are useful to characterize integrated chance constraints: in ICC models "risk" is measured quantitatively, as in SPR, but "risk aversion" is modeled by constraints, as in CCP. ICCs can be very useful in practice. On the one hand it might be acceptable to have a constraint violated, even with high probability if necessary, if the amount by which it is violated is small enough. In particular this might be the case if the constraint represents a quantitative goal to be reached rather than a logical or technical necessity, as is often the case in practical linear programming models. In such cases the quantitative approach to infeasibility of SPR is appropriate. On the other hand, however, this does not mean that one has the obligation to introduce a penalty cost, or discrepancy cost [1], at the same time. It is quite conceivable that in certain circumstances it is more appropriate to fix an upper bound on the risk, as in CCP, rather than to penalize the risk, as in SPR. In particular this might be the case if it is difficult to specify numerical values for the costs of infeasibility, or if risk aversion is the main purpose of the model.

The contents of the present chapter can be summarized as follows. In Section 5.2 two different ICC formulations for single random constraints are studied. In the first type there is a fixed upper bound on the risk, and in the second one the upper bound depends on the functions and distributions involved. Both give rise to convex feasibility sets, which increase strictly and continuously with the risk aversion parameter. In these respects the ICCs behave better than the corresponding CCs. In Section 5.3 ICC formulations are given for the joint risk of infeasibility of a system of random inequalities. It appears that the analogue of the first type of single ICC has the same nice behaviour. In each of the three nice cases there is an intimate relation with SPR models, as shown in Section 5.4. This relationship is closely connected to duality theory, since it is based on the interpretation of penalties as Lagrange multipliers for ICCs. The conclusions are summarized in Section 5.5.

## 5.2 SINGLE INTEGRATED CHANCE CONSTRAINTS

We consider a linear programming model with random coefficients in the constraints, assuming that these coefficients are only known in distribution. In particular we are interested in the question under which condition a decision vector x is called feasible. In this section we consider one scalar constraint in $x \in \mathbb{R}^n$,

$$(5.1) \qquad \Sigma_{j=1}^{n} a_j(\omega)x_j \geqq b(\omega),$$

where $\omega := (a,b) = (a_1,\ldots,a_n,b)$ is a random vector with a known probability distribution. Defining $\eta(x) \equiv \eta(x,\omega) := \Sigma_{j=1}^{n} a_j x_j - b$, $(\eta(x))^- := \max(0,-\eta(x))$, we call the random variable $(\eta(x))^-$ the *shortage* determined by (5.1). The constraint (5.1) reflects the idea that one wants to avoid positive shortage, but because of the random character it is impossible in general or undesirable to exclude shortages completely. Depending on the *definition of risk* together with the specification of the maximal risk level (i.e. the *degree of risk aversion*) one gets different *risk constraints* as a translation of (5.1). In the most well-known case risk is understood as *probability of positive shortage*, resulting in a feasible set determined by a *chance constraint* (CC)

$$(5.2) \qquad X_0(\alpha) := \{x \in \mathbb{R}^n: E\ \mathrm{sgn}(\eta(x))^- \leqq \alpha\},\ \alpha \in [0,1].$$

Notice that $E\ \mathrm{sgn}(\eta(x))^- = P(\eta(x) < 0)$ so that

$$x \in X_0(\alpha) \iff P(\Sigma_{j=1}^{n} a_j x_j \geqq b) \geqq 1-\alpha.$$

The risk aversion parameter $\alpha$ denotes the maximum acceptable risk level; it has to be specified in advance. Notice that for this specification of risk the *amount of shortage* is not relevant; only its sign counts. In situations where this amount is important, it might be more appropriate to use the *mean shortage* $E(\eta(x))^-$ as a measure for risk. This leads to the following analogue of (5.2):

$$(5.3) \qquad X_1(\beta) := \{x \in \mathbb{R}^n\colon\ E(\eta(x))^- \leqq \beta\},\ \beta \in [0,\infty],$$

where the risk aversion parameter $\beta$ has to be chosen in advance. Because of (3.29) we call (5.3), and other constraints asking for upper bounds on $E(\eta(x))^-$, *integrated chance constraints* (ICCs). Whereas $\alpha$ in (5.2) is scale-free, $\beta$ in (5.3) depends on scale. If the "demand" b is fixed and positive, one may choose e.g. $\beta = \alpha b$ for a scale-free $\alpha \in [0,1]$.

It is also possible to specify the maximal accepted risk not as a fixed number $\beta$, but depending on the distribution of $\eta(x)$. For example, since a natural upper bound for $E(\eta(x))^-$ is given by $E|\eta(x)|$, we introduce, for $\alpha \in [0,1]$ to be specified in advance,

$$(5.4) \qquad X_2(\alpha) := \{x \in \mathbb{R}^n\colon\ E(\eta(x))^- \leqq \alpha . E|\eta(x)|\}.$$

In (5.4) also the mean surplus $E(\eta(x))^+ = E \max(0,\eta(x))$ is taken into account, since $E(\eta(x))^- + E(\eta(x))^+ = E|\eta(x)|$. Just like (5.3) also (5.4) is a direct analogue of (5.2), at least if $P(\eta(x) = 0) = 0$, as can be seen by rewriting the chance constraint as

$$E\ \mathrm{sgn}(\eta(x))^- \leqq \alpha . E\ \mathrm{sgn}|\eta(x)|.$$

In addition to (5.3) and (5.4) we call the attention to the integrated chance constraint of the type

$$(5.5) \qquad X_3(\gamma) := \{x \in \mathbb{R}^n\colon\ E(\eta(x))^- \leqq \gamma . P(\eta(x) < 0)\},\ \gamma \in [0,\infty],$$

introduced by Prékopa [7]. Here $\gamma$ is the maximal accepted value for the conditional mean shortage $E[(\eta(x))^- \mid \eta(x) < 0]$.

Obviously, each of the feasibility sets $X_0(\alpha)$, $X_1(\beta)$, $X_2(\alpha)$ and $X_3(\gamma)$ is nondecreasing in the risk aversion parameters $\alpha$, $\beta$, $\alpha$ and $\gamma$, respectively. They coincide for extreme specifications of these parameters:

$$X_0(0) = X_1(0) = X_2(0) = X_3(0) = \{x \in \mathbb{R}^n\colon\ P(\eta(x) < 0) = 0\},$$

(but, as a matter of fact, this *extremely safe set* is often empty, unfortunately), and, of course,

$$X_0(1) = X_1(\infty) = X_2(1) = X_3(\infty) = \mathbb{R}^n.$$

However, for intermediate values of the risk aversion parameters they may behave quite differently as the following examples show.

EXAMPLE 5.1. $n = 1$, $P((a,b) = (-1,1)) = P((a,b) = (1,1)) = \frac{1}{2}$.
The feasibility sets unequal to $\emptyset$ or $\mathbb{R}^n$ are:

$$\begin{aligned}
X_0(\alpha) &= \{|x_1| \geqq 1\} && \text{if } \alpha \in [\tfrac{1}{2},1),\\
X_1(\beta) &= \{|x_1| \leqq 2\beta - 1\} && \text{if } \beta \in [1,\infty),\\
X_2(\alpha) &= \{|x_1| \geqq (2\alpha - 1)^{-1}\} && \text{if } \alpha \in (\tfrac{1}{2},1),\\
X_3(\gamma) &= \{|x_1| < 1\} && \text{if } \gamma \in [1,2),\\
&= \{|x_1| \leqq \gamma-1\} && \text{if } \gamma \in [2,\infty).
\end{aligned}$$

EXAMPLE 5.2. $n = 1$, $P((a,b) = (-1,0)) = 2/3$, $P((a,b) = (1,0)) = 1/3$.

$$\begin{aligned}
X_0(\alpha) = X_2(\alpha) &= \{0\} && \text{if } 0 \leqq \alpha < 1/3,\\
&= (-\infty,0] && \text{if } 1/3 \leqq \alpha < 2/3,\\
&= \mathbb{R} && \text{if } 2/3 \leqq \alpha \leqq 1,\\
X_1(\beta) &= [-3\beta,3\beta/2] && \text{if } 0 \leqq \beta \leqq \infty,\\
X_3(\gamma) &= [-\gamma,\gamma] && \text{if } 0 \leqq \gamma \leqq \infty.
\end{aligned}$$

EXAMPLE 5.3. $n=2$, $P((a,b) = (-2,1,0)) = P((a,b) = (1,-2,0)) = \frac{1}{2}$.

$$\begin{aligned}
X_0(\alpha) &= \{x \in \mathbb{R}^2 : 2x_1 - x_2 \leqq 0 \textit{ and } 2x_2 - x_1 \leqq 0\}, \text{ if } 0 \leqq \alpha < \tfrac{1}{2},\\
&= \{x \in \mathbb{R}^2 : 2x_1 - x_2 \leqq 0 \textit{ or } 2x_2 - x_1 \leqq 0\}, \text{ if } \tfrac{1}{2} \leqq \alpha < 1,\\
X_1(\beta) &= \left\{x \in \mathbb{R}^2 : \begin{array}{l} 2x_1 - x_2 \leqq 2\beta,\ 2x_2 - x_1 \leqq 2\beta \\ \textit{and } x_1 + x_2 \leqq 2\beta \end{array}\right\}, \text{ if } 0 \leqq \beta \leqq \infty,\\
X_2(\alpha) &= \left\{x \in \mathbb{R}^2 : \begin{array}{l} (2-3\alpha)x_1 + (3\alpha-1)x_2 \leqq 0 \textit{ and} \\ (2-3\alpha)x_2 + (3\alpha-1)x_1 \leqq 0 \end{array}\right\}, \text{ if } 0 \leqq \alpha \leqq \tfrac{1}{2},\\
&= \left\{x \in \mathbb{R}^2 : \begin{array}{l} (2-3\alpha)x_1 + (3\alpha-1)x_2 \leqq 0 \textit{ or} \\ (2-3\alpha)x_2 + (3\alpha-1)x_1 \leqq 0 \end{array}\right\}, \text{ if } \tfrac{1}{2} \leqq \alpha \leqq 1,\\
X_3(\gamma) &= \{x \in \mathbb{R}^2 : 2x_1 - x_2 \leqq \gamma \textit{ and } 2x_2 - x_1 \leqq \gamma\} \cup\\
&\quad \{x \in \mathbb{R}^2 : 2x_1 - x_2 > 0,\ 2x_2 - x_1 > 0 \textit{ and } x_1 + x_2 \leqq 2\gamma\},\\
&\qquad \textit{if } 0 \leqq \gamma \leqq \infty.
\end{aligned}$$

See Figure 5.1 for an illustration.

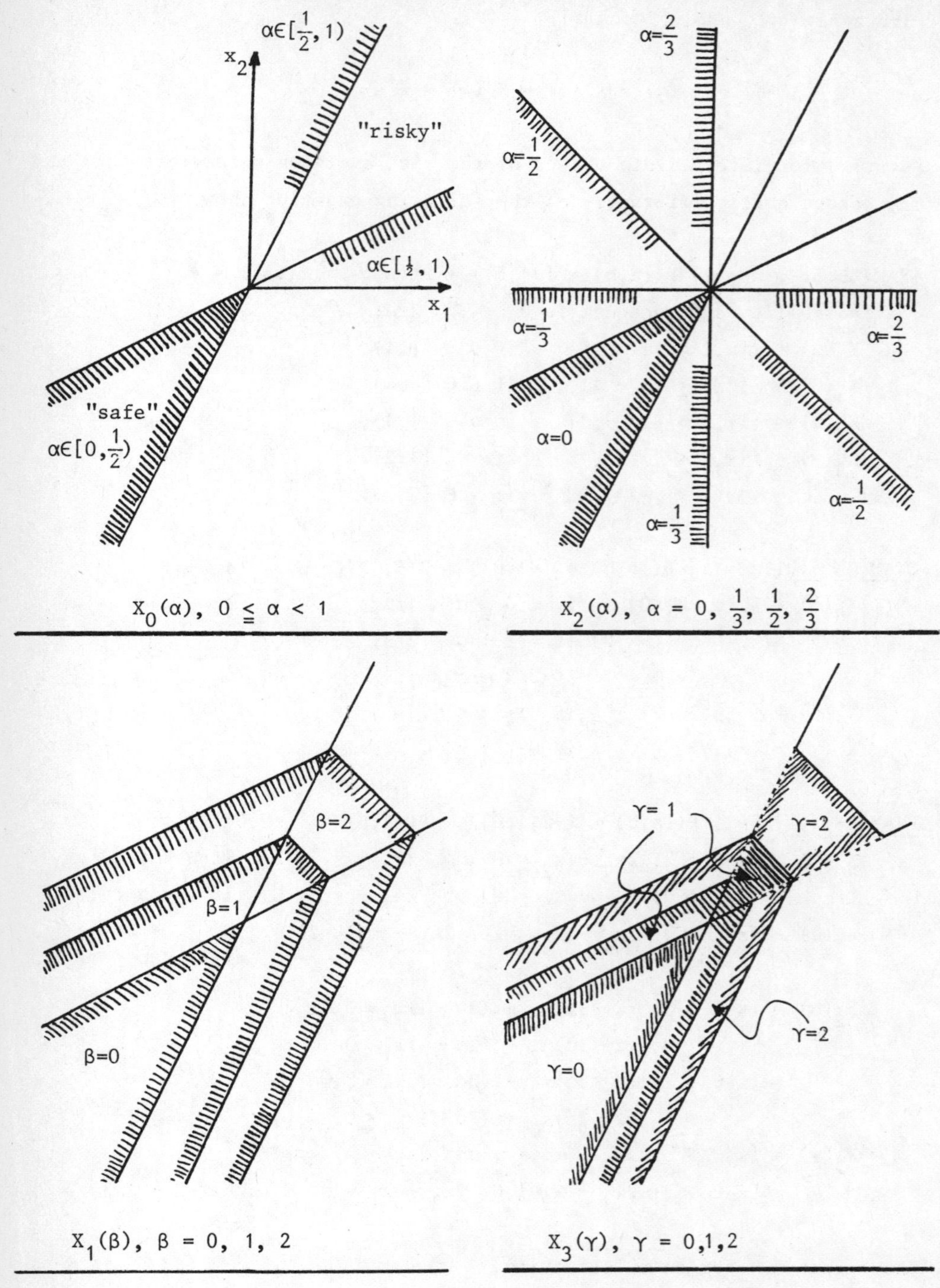


FIGURE 5.1. The feasibility sets of Example 5.3.

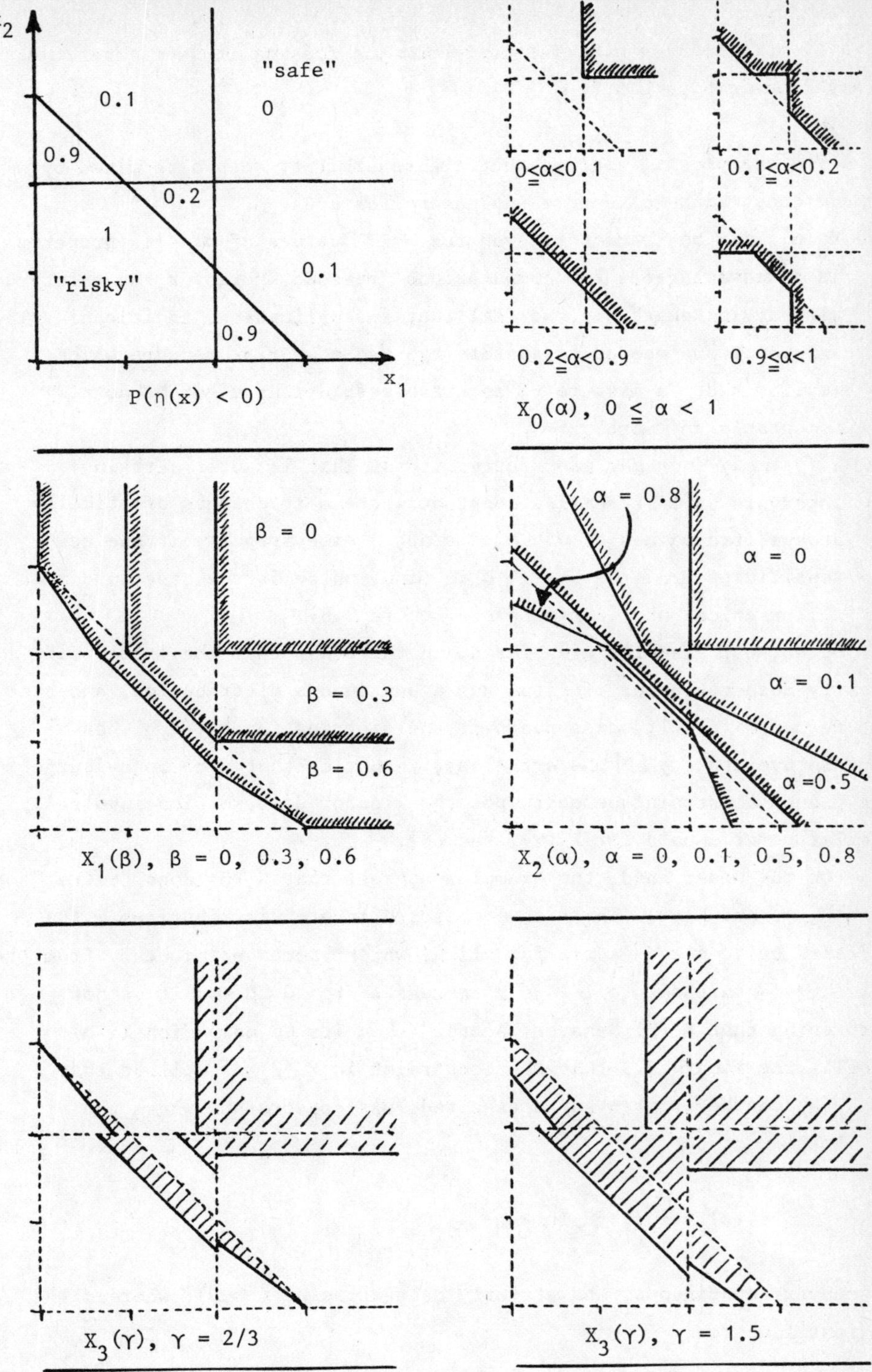


FIGURE 5.2. The feasibility sets of Example 5.4.

EXAMPLE 5.4. $n = 2$, $P((a,b) = (0,3,6)) = P((a,b) = (3,0,6)) = 0.1$, $P((a,b) = (2,2,6)) = 0.8$. The corresponding feasibility sets are visualized in Figure 5.2.

The examples illustrate that the feasibility sets determined by chance constraints have some unpleasant features.

(i) $X_0(\alpha)$ may be *nonconvex*, even for small values of $\alpha$. This property is disadvantageous for computational reasons. Moreover, as observed already in Remark 3.8, we feel that in applications additional arguments are needed to justify the use of a risk measure which implies that "a mixture of acceptable solutions might be not acceptable any more".

(ii) $X_0(\alpha)$ may *increase stepwisely* with $\alpha$. That is, over certain intervals of $\alpha$ it remains constant whereas it changes drastically at specific values of $\alpha$. This is not a nice property if one has sensitivity analysis w.r.t. $\alpha$ in mind, which is the case in almost all practical applications (see Remark 3.8).

Of course, our examples are biased, in the sense that the latter difficulty does not occur if (a,b) has a continuous distribution, and that the former difficulty disappears if the distribution of (a,b) has a log-concave density [7]. Nevertheless, we notice that nice behaviour of the chance constraint depends upon the kind of distribution involved. Similar remarks hold for $X_3(\gamma)$, see [8].

On the other hand, the examples suggest that $X_1(\beta)$ does "better" than $X_0(\alpha)$ and $X_3(\gamma)$. That is, even for discrete distributions $X_1(\beta)$ appeared to be a convex set for all $\beta$, which increases smoothly from the extremely safe set to $\mathbb{R}^n$ if $\beta$ increases from 0 to $+\infty$. It is not surprising that $X_1(\beta)$ behaves "better" than its CC companion $X_0(\alpha)$: recall from Remark 3.9 that the constraint in $X_1(\beta)$ as well as that in $X_0(\alpha)$ can be written as an expected loss constraint, where the ICC loss function

$$v_1(z) = z^-, \; z \in \mathbb{R},$$

is convex, continuous, and strictly decreasing on $(-\infty,0)$, whereas the CC loss function

$$v_0(z) = \text{sgn } z^-, \ z \in \mathbb{R},$$

does not possess these properties.

The examples suggest also that $X_2(\alpha)$ behaves "better" than $X_0(\alpha)$: it increases smoothly with $\alpha$ (except in Example 5.2) and it is convex for $\alpha \leqq \frac{1}{2}$, in spite of the discrete distributions. There is a basic difference between $X_2(\alpha)$ and $X_1(\beta)$. Let us call for a moment

$$R := \{x \in \mathbb{R}^n: \ P(\eta(x) < 0) = 1\}$$

the *extremely risky set*, since for each $x \in R$ the constraint (5.1) is violated with probability one. It is easily seen that such an extremely risky decision is not accepted as feasible by the chance constraint, as well as by the second integrated chance constraint:

$$X_0(\alpha) \cap R = X_2(\alpha) \cap R = \emptyset \text{ for all } \alpha < 1.$$

However, it is quite possible that $X_1(\beta) \cap R \neq \emptyset$, even for small values of $\beta$, see e.g. Figure 5.1. This illustrates that some values for x are accepted as feasible by the integrated chance constraint even if it is sure that shortage will occur; the reason is, of course, that the mean shortage is small then. From the same figure one sees that $X_3(\gamma)$ has the same property; this set, however, happens to be nonconvex for all $\gamma > 0$ in that case.

We now start the formal analysis of $X_1(\beta)$ and $X_2(\alpha)$. Throughout this section we assume

$$(5.6) \qquad E|a_j| < \infty, \ j = 1,\ldots,n, \ E|b| < \infty.$$

An uninteresting case is avoided by the assumption

$$(5.7) \qquad P(a=0) < 1.$$

If (5.7) does not hold then the support $S_{ab}$ of the random vector (a,b) is contained in the b-axis in $\mathbb{R}^{n+1}$. Sometimes it is appropriate to use the stronger condition

(5.8) $\quad S_{ab}$ is not a subset of *any* line in $\mathbb{R}^{n+1}$ through the origin.

This condition is not very strong; it excludes the possibility that the constraint (5.1) is homogeneous in one single random variable, i.e.

$$(5.9) \qquad \eta(x) = \omega_0 \cdot (\Sigma_{j=1}^n \hat{a}_j x_j - \hat{b})$$

for fixed $(\hat{a},\hat{b})$ and random $\omega_0$. Of course, if (5.9) holds then the analysis of the corresponding feasibility sets is straightforward, similar to that of Example 5.2.

The following sets will be used also

$$(5.10) \qquad \begin{aligned} M &:= \{x \in \mathbb{R}^n : P(\Sigma_{j=1}^n a_j x_j = b) < 1\}, \\ M^c &:= \{x \in \mathbb{R}^n : P(\Sigma_{j=1}^n a_j x_j = b) = 1\}. \end{aligned}$$

Obviously

$$(5.11) \qquad \begin{aligned} M = \emptyset &\Leftrightarrow M^c = \mathbb{R}^n \Leftrightarrow S_{ab} = \{0\}, \\ M = \mathbb{R}^n &\Leftrightarrow M^c = \emptyset \quad \Leftrightarrow b \text{ is not a linear function of } a. \end{aligned}$$

Also it is easily seen that

$$(5.12) \qquad x \in M^c \Leftrightarrow (x,-1) \perp S_{ab}$$

so that $M^c$ is a closed convex set.

LEMMA 5.5. *Assume* (5.6).

*a. The functions*

$$(5.13) \qquad f^-(x) := E(\eta(x))^-, \ f^+(x) := E(\eta(x))^+, \ f^o(x) := E|\eta(x)|, \ x \in \mathbb{R}^n,$$

*are finite, nonnegative, convex and Lipschitz continuous. For finite discrete distributions of* (a,b) *these functions are piecewisely linear, and for distributions of* (a,b) *with a density they are continuously differentiable, with*

(5.14) $\frac{\partial f^-}{\partial x_j}(x) = E[-a_j.\mathrm{sgn}(\eta(x))^-],\ x \in \mathbb{R}^n,\ j = 1,\ldots,n.$

*Moreover,*

(5.15) $\lim_{\lambda\to\infty} \frac{f^-(x + \lambda y) - f^-(x)}{\lambda} = E(\Sigma_{j=1}^n a_j y_j)^-,\ y \in \mathbb{R}^n.$

*b. Assume also* (5.8). *Then the function* $f^-(x)/f^o(x)$, *defined on* M, *attains all values in* $(\alpha_0,\alpha_1) \subset [0,1]$ *where*

(5.16a) $\alpha_0 := \inf_{x\in M}\{f^-(x)/f^o(x)\}$,

(5.16b) $\alpha_1 := \sup_{x\in M}\{f^-(x)/f^o(x)\}$.

PROOF. (a) Although a direct proof is not difficult, we may simply refer to more general results on SPR [5,6] since for i = −, + and o

$$f^i(x) = E[\min_{y\in\mathbb{R}^2}\{q_{i1}y_1 + q_{i2}y_2 : y_1 - y_2 = b - \Sigma_{j=1}^n a_j x_j,\ y_1, y_2 \geqq 0\}]$$

where

$$\begin{aligned}(q_{i1},q_{i2}) &:= (1,0) \text{ if } i = -,\\ &:= (0,1) \text{ if } i = +,\\ &:= (1,1) \text{ if } i = o,\end{aligned}$$

can be seen as second stages of simple recourse models. We will prove (5.15). For any pair (p,q) of real numbers, satisfying p = 0 or $|p| \geqq |q|$, the following equality holds:

(5.17) $(p+q)^- = p^- + q^- - \{q^-.\mathrm{sgn}\ p^+ + q^+.\mathrm{sgn}\ p^-\}$.

By the substitution $p := \lambda.\Sigma_{j=1}^n a_j y_j$ and $q := \eta(x)$ one derives from (5.17) that for any fixed $(a,b) \in \mathbb{R}^{n+1}$

$$\lim_{\lambda\to\infty}\lambda^{-1}\{(\eta(x + \lambda y))^- - (\eta(x))^-\} = (\Sigma_{j=1}^n a_j y_j)^-.$$

Moreover, $|\lambda^{-1}\{(\eta(x + \lambda y))^- - (\eta(x))^-\}| \leqq |\Sigma_{j=1}^n a_j y_j|$ for any $\lambda \neq 0$, and this majorant is integrable because of (5.6). Hence, the dominated convergence theorem of Lebesgue gives (5.15).
(b) Notice that $\{x \in \mathbb{R}^n: f^o(x) > 0\} = M$, and that $M \neq \emptyset$ because of (5.8). From $0 \leqq f^- \leqq f^o$ it follows that $0 \leqq \alpha_0 \leqq \alpha_1 \leqq 1$. Moreover, from (a) we know that $f^-/f^o$ is continuous on the open set M. Therefore, it will attain every value in $(\alpha_0,\alpha_1)$ if M is connected. We will show that (5.8) implies that M is connected. From (5.12) we know that $M^c$ is a convex set in $\mathbb{R}^n$. It separates M into disjoint parts if and only if it contains a hyperplane, and in that case the linear hull of $S_{ab}$ is at most one-dimensional. This is precisely the possibility which has been excluded by (5.8). □

Since the feasibility sets $X_1(\beta)$, $0 \leqq \beta \leqq \infty$, are level sets of $f^-(x) = E(\eta(x))^-$, their properties follow directly from Lemma 5.5 part a. For example, they are *closed* and *convex*. Obviously, $X_1(\beta)$ is nondecreasing with $\beta$. For small $\beta$ $X_1(\beta)$ may be empty: with

$$\beta_0 := \inf_{x\in\mathbb{R}^n} E(\eta(x))^-, \tag{5.18}$$

which is a finite number, we have $X_1(\beta) \neq \emptyset$ iff

(5.19)
either $\beta > \beta_0$
or $\beta = \beta_0$ and the infimum in (5.18) is attained.

On the other side of the range one has $X_1(\infty) = \mathbb{R}^n$. In the next theorem a general statement on the increase of $X_1(\beta)$ with $\beta$ will be shown. We use the following terminology. Any family $\{X(t),\ t \in T\}$ of closed subsets $X(t) \subset \mathbb{R}^n$ where the parameter t takes values in $T := [t_0,t_1] \subset [0,\infty]$ is called *nondecreasing* if

$$X(s) \subset X(t) \text{ if } s \in T,\ t \in T,\ s < t,$$

and it is called *strictly increasing* at $t \in T$ if it is nondecreasing and if

$$X(s) \neq X(t) \text{ for all } s \in T,\ s \neq t; \tag{5.20}$$

and it is called *continuously increasing* at $t \in T$ if it is non-decreasing and if

$$(5.21) \qquad \mathrm{cl}\ \cup_{s\in[t_0,t)} X(s) = X(t) = \cap_{s\in(t,t_1]} X(s).$$

Here it is understood that for $t = t_0$ only the second condition has to hold, and for $t = t_1$ only the first one.

THEOREM 5.6. Characterization of $X_1(\beta)$, $\beta \in [0,\infty]$. *Assume* (5.6).
*a. For any* $\beta \in [0,\infty]$ $X_1(\beta)$ *is a closed, convex subset of* $\mathbb{R}^n$. *It is nonempty iff* (5.19) *holds, and* $X_1(\beta) \neq \mathbb{R}^n$ *for all* $\beta < \infty$ *if* (5.7) *holds.*
*b.* $X_1(\beta)$ *is nondecreasing with* $\beta$. *The increase is continuous at all* $\beta \neq \beta_0$; *and it is strict at all* $\beta > \beta_0$ *if* (5.7) *holds.*
*c. Let* $\beta < \infty$ *with* $X_1(\beta) \neq \emptyset$. *Then*

$$(5.22) \qquad \left.\begin{array}{l} x + \lambda y \in X_1(\beta) \\ \forall x \in X_1(\beta),\ \forall\lambda \geqq 0 \end{array}\right\} \Leftrightarrow P(\Sigma_{j=1}^n a_j y_j \geqq 0) = 1.$$

*Therefore,* $X_1(\beta)$ *is bounded iff* $0 \in \mathrm{int\ conv}\ S_a$, *where* $S_a \subset \mathbb{R}^n$ *is the support of the distribution of* $a = (a_1,\ldots,a_n)$.

PROOF. (a) Follows from Lemma 5.5 part a. The last statement is a consequence of the fact that a bounded convex function is constant.
(b) (Continuous increase) Consider (5.21) with $X(t)$ replaced by $X_1(\beta)$. If $\beta < \beta_0$ (5.21) is obvious, since all three sets are empty. For $\beta > \beta_0$ (5.21) is a property of level sets, where for the first equality to be true it is used that $f^-$ is finite and convex [9 Corollary 7.6.1].
(Strict increase) Consider (5.20) with $X(t)$ replaced by $X_1(\beta)$. For $\beta = \infty$ (5.20) follows from (a). Fix $\beta \in (\beta_0,\infty)$. Because of (5.7) we know from (a) that $\emptyset \neq X_1(\beta) \neq \mathbb{R}^n$, so that $x_1,x_2 \in \mathbb{R}^n$ exist with $f^-(x_1) \leqq \beta$ and $f^-(x_2) > \beta$. Since $f^-$ is continuous on $\mathbb{R}^n$ each value in $[f^-(x_1),f^-(x_2)]$ is attained. Hence, for any $\beta' > \beta$ an $\hat{x} \in \mathbb{R}^n$ exists with $\beta < f^-(\hat{x}) \leqq \beta'$, so that $X_1(\beta') \neq X_1(\beta)$. For $\beta' < \beta$ there are two possibilities: either $X_1(\beta') = \emptyset$ and we have nothing to prove or $\exists x_3 \in \mathbb{R}^n$ with $f^-(x_3) \leqq \beta'$. In the latter case one proves $X_1(\beta') \neq X_1(\beta)$ similarly to the case $\beta' > \beta$.
(c) Since $f^-$ is finite and convex, all nonempty level sets $X_1(\beta)$ have

the same recession cone, namely $C = \{y \in \mathbb{R}^n: \lim_{\lambda\to\infty}\lambda^{-1}(f^-(x+\lambda y) - f^-(x)) \leqq 0\}$ where $x \in \mathbb{R}^n$ is arbitrary [9 Theorems 8.7 and 8.5]. Therefore, (5.22) follows from (5.15) since $E(\Sigma_{j=1}^n a_j y_j)^- \leqq 0$ iff $P(\Sigma_{j=1}^n a_j y_j < 0) = 0$. Also, $X_1(\beta)$ is bounded iff $C = \{0\}$ [9 Theorem 8.4]; equivalently iff $P(\Sigma_{j=1}^n a_j y_j \geqq 0) < 1$ for all $y \neq 0$. This is precisely the case if $\{0\}$ can not be separated from the support $S_a$ by a hyperplane in $\mathbb{R}^n$, and that is true iff 0 is an interior point of the convex hull of $S_a$. □

We will now analyze the second type of ICC, introduced in (5.4). Since $z = z^+ - z^-$, $|z| = z^+ + z^-$, we have

(5.23a) $X_2(\alpha) = \{x \in \mathbb{R}^n: E(\eta(x))^- \leqq \alpha.E|\eta(x)|\}$

(5.23b) $= \{x \in \mathbb{R}^n: (1-\alpha).E(\eta(x))^- \leqq \alpha.E(\eta(x))^+\}$

(5.23c) $= \{x \in \mathbb{R}^n: (1-2\alpha).E(\eta(x))^- \leqq \alpha.E\,\eta(x)\}$.

The representation (5.23b) shows that in (5.4) a positive mean shortage is only accepted if the corresponding mean surplus is large enough. Another representation uses M defined in (5.10):

(5.24) $X_2(\alpha) = M^c \cup \{x \in M: f^-(x)/f^o(x) \leqq \alpha\}$, $\alpha \in [0,1]$.

This shows for instance that, apart from the trivial case $M = \emptyset$, see (5.11), $X_2(\alpha) = \mathbb{R}^n$ iff $\alpha \geqq \alpha_1$ and that $X_2(\alpha) \neq \emptyset$ iff one of the following conditions holds:

(5.25)
(i) $M \neq \mathbb{R}^n$;
(ii) $M = \mathbb{R}^n$ and either $\alpha > \alpha_0$
or $\alpha = \alpha_0$ and the infimum in (5.16a) is attained.

Obviously, $X_2(\alpha)$ is nondecreasing with $\alpha$. Just as for $X_1(\beta)$, we will show that except for extreme specifications of the risk parameter the increase is strict and continuous. Moreover, $X_2(\alpha)$ is convex for all $\alpha \leqq \frac{1}{2}$:

THEOREM 5.7. Characterization of $X_2(\alpha)$, $\alpha \in [0,1]$. *Assume* (5.6) *and* (5.8).
*a. For any* $\alpha \in [0,1]$ $X_2(\alpha)$ *is a closed subset of* $\mathbb{R}^n$. *It is nonempty iff* (5.25) *holds, and* $X_2(\alpha) \neq \mathbb{R}^n$ *iff* $\alpha < \alpha_1$.
*b. If* $\alpha \leqq \frac{1}{2}$ *then* $X_2(\alpha)$ *is convex; it is even polyhedral if* (a,b) *has a finite discrete distribution. If* $\alpha > \frac{1}{2}$ *then* $X_2(\alpha)$ *is the complement in* $\mathbb{R}^n$ *of a convex set, hence nonconvex generally. For* $\alpha = \frac{1}{2}$ *the integrated chance constraint in* (5.4) *has a simple deterministic equivalent representation in terms of expected values of the coefficients:*

$$(5.26) \qquad X_2(\tfrac{1}{2}) = \{x \in \mathbb{R}^n:\ \Sigma_{j=1}^n \bar{a}_j x_j \geqq \bar{b}\}$$

*where* $(\bar{a}_1,\ldots,\bar{a}_n,\bar{b})$ *is the mean value of* $(a_1,\ldots,a_n,b)$.
*If* $P(b=0) = 1$ *then* $X_2(\alpha)$ *is a nonempty cone for all* $\alpha \in [0,1]$.
*c.* $X_2(\alpha)$ *is nondecreasing with* $\alpha$. *The increase is continuous at all* $\alpha \neq \alpha_0$; *it is strict at all* $\alpha \in (\alpha_0,\alpha_1)$.
*d. For* $\alpha \in (\frac{1}{2},1]$ $X_2(\alpha)$ *is nonempty and unbounded. For* $\alpha \in [0,\frac{1}{2}]$ *with* $X_2(\alpha) \neq \emptyset$

$$(5.27) \qquad \left.\begin{array}{l} x + \lambda y \in X_2(\alpha) \\ \forall x \in X_2(\alpha),\ \forall\lambda \geqq 0 \end{array}\right\} \Leftrightarrow E(\Sigma_{j=1}^n a_j y_j)^- \leqq \alpha . E|\Sigma_{j=1}^n a_j y_j|.$$

*Denote this condition on* $y \in \mathbb{R}^n$ *by* $y \in \hat{X}_2(\alpha)$. *Then any nonempty* $X_2(\alpha)$, $\alpha \in [0,\frac{1}{2}]$, *is bounded iff* $\hat{X}_2(\alpha) = \{0\}$. *In that case,* $0 \in \text{int conv } S_a$ *where* $S_a \subset \mathbb{R}^n$ *is the support of the distribution of* $a = (a_1,\ldots,a_n)$.

PROOF. (a) Because of (5.23c) we may write $X_2(\alpha) = \{x \in \mathbb{R}^n:\ h_\alpha(x) \leqq 0\}$ where

$$h_\alpha(x) := (1 - 2\alpha).f^-(x) - \alpha.E\,\eta(x) = f^-(x) - \alpha.f^o(x).$$

Since $|E\,\eta(x)| \leqq E|\eta(x)| < \infty$ it follows from Lemma 5.5 that $h_\alpha$ is finite and continuous, hence its level set $X_2(\alpha)$ is closed. The other statements have been discussed already; recall that assumption (5.8) implies that $M \neq \emptyset$, see (5.11).
(b) Since $E\,\eta(x) = \Sigma_{j=1}^n \bar{a}_j x_j - \bar{b}$ is linear in x and $f^-$ is convex we know that $h_\alpha$ is a convex function if $\alpha \leqq \frac{1}{2}$ and that it is a concave function if $\alpha \geqq \frac{1}{2}$. (5.26) follows from (5.23c). If $P(b=0) = 1$ then as well

$f^-$ as $f^o$ are positively homogeneous functions; hence $X_2(\alpha)$ is a cone in that case with $0 \in X_2(\alpha)$.

(c) (Continuous increase) Since for any fixed $x \in \mathbb{R}^n$ $h_\alpha(x)$ is a nonincreasing continuous function of $\alpha$, we have for any $\alpha \in [0,1)$

$$X_2(\alpha) = \cap_{\alpha' \in (\alpha,1]} X_2(\alpha').$$

Therefore, we only have to show

$$(5.28) \qquad X_2(\alpha) \subset \text{cl} \cup_{\alpha' \in [0,\alpha)} X_2(\alpha') \text{ if } \alpha > 0,\ \alpha \neq \alpha_0,$$

since the reverse inclusion follows from the monotonicity. If $0 < \alpha < \alpha_0$ then (5.28) is true since both sides are equal to $M^c$, and if $\alpha_1 < \alpha \leqq 1$ both are equal to $\mathbb{R}^n$. Hence, we only have to consider $\alpha_0 < \alpha \leqq \alpha_1$. Then both sides in (5.28) are nonempty. Suppose that (5.28) is not true. Then an $x_o \in X_2(\alpha)$ and a neighbourhood $N(x_o)$ exist such that $N(x_o) \cap X_2(\alpha') = \emptyset$ for all $\alpha' \in [0,\alpha)$. Since $h_\alpha(x)$ is continuous in $\alpha$ this implies that

$$(5.29) \qquad h_\alpha(x) \geqq 0 \ \forall x \in N(x_o), \text{ with equality for } x = x_o,$$

so that $x_o$ is a local minimum of $h_\alpha$. If $\alpha \leqq \frac{1}{2}$ then the convexity of $h_\alpha$ implies that the minimum is global, and therefore $h_\alpha(x) \geqq 0$ for all $x \in \mathbb{R}^n$ with equality for $x = x_o$. Consequently $\alpha = \alpha_0$. If $\alpha > \frac{1}{2}$ then $h_\alpha$ is a concave function with a local minimum in $x = x_o$. Hence, we must have then $h_\alpha(x) = 0 \ \forall x \in \mathbb{R}^n$. Consequently, $\alpha = \alpha_0$ $(=\alpha_1)$. In both cases we got a contradiction with the assumption $\alpha_0 < \alpha$. This completes the proof of (5.28).

(Strict increase) Because of (5.8) we know from Lemma 5.5 that $M \neq \emptyset$ and that $f^-(x)/f^o(x)$ attains all values in $(\alpha_0,\alpha_1)$. The proof that $X_2(\alpha)$ is strictly increasing at $\alpha \in (\alpha_0,\alpha_1)$ is now completely similar to the corresponding proof of Theorem 5.6 part b.

(d) Unless $\bar{a} = 0$ and $\bar{b} > 0$ we know from (5.26) that $X_2(\frac{1}{2})$ is nonempty and unbounded, so a fortiori this is true for $X_2(\alpha)$, $\alpha > \frac{1}{2}$. Suppose now $\bar{a} = 0$ and $\bar{b} > 0$. Then it follows from (5.23c) that

$$X_2(\alpha) = \{x \in \mathbb{R}^n:\ E(\eta(x))^- \geqq \frac{\alpha}{2\alpha - 1}\bar{b}\},\ \alpha > \tfrac{1}{2},$$

so that

$$X_2(\alpha) \supset \mathbb{R}^n \setminus X_1(\frac{\alpha}{2\alpha - 1}\bar{b}), \ \alpha > \tfrac{1}{2}.$$

Now (5.8) implies (5.7), so that because of Theorem 5.6 $\mathbb{R}^n \setminus X_1(\beta) \neq \emptyset$ for $\beta < \infty$. Hence, $X_2(\alpha) \neq \emptyset$, $\alpha > \frac{1}{2}$. Since a convex set in $\mathbb{R}^n$ can only have a bounded complement in $\mathbb{R}^n$ if it is equal to $\mathbb{R}^n$ itself, it follows also that $X_2(\alpha)$ is unbounded for $\alpha > \frac{1}{2}$.
Let us prove (5.27). Let $\alpha \in [0,\frac{1}{2}]$ and assume that $X_2(\alpha) \neq \emptyset$. Since $X_2(\alpha)$ is a level set of the convex function $h_\alpha$, a vector $y \in \mathbb{R}^n$, $y \neq 0$, is a direction of recession of $X_2(\alpha)$ iff

$$h'_\alpha(y) := \lim_{\lambda\to\infty} \frac{h_\alpha(x + \lambda y) - h_\alpha(x)}{\lambda} \leqq 0$$

for all $x \in X_2(\alpha)$ [9 Theorems 8.7 and 8.5]. From (5.15) it follows that

$$\begin{aligned} h'_\alpha(y) &= (1 - 2\alpha).E(\Sigma_{j=1}^n a_j y_j)^- - \alpha.E(\Sigma_{j=1}^n a_j y_j) \\ &= E(\Sigma_{j=1}^n a_j y_j)^- - \alpha.E|\Sigma_{j=1}^n a_j y_j|. \end{aligned}$$

This completes the proof of (5.27) and of the characterization of boundedness of $X_2(\alpha)$. Finally, the condition $E(\Sigma_{j=1}^n a_j y_j)^- > \alpha.E|\Sigma_{j=1}^n a_j y_j|$ for all $y \neq 0$ is stronger than $E(\Sigma_{j=1}^n a_j y_j)^- > 0$ for all $y \neq 0$, which is equivalent to $0 \in \text{int conv } S_a$ as shown in Theorem 5.6. □

REMARK 5.8. The choice $\alpha = \frac{1}{2}$ in $X_0(\alpha)$ reflects more or less a risk-neutral attitude: the median of $\eta(x)$ must be nonnegative. Similarly, $\alpha = \frac{1}{2}$ in $X_2(\alpha)$ corresponds to a risk-neutral attitude: the mean value of $\eta(x)$ must be nonnegative. In fact, the practice of neglecting randomness in the constraints may be interpreted as replacing (5.1) by $X_2(\frac{1}{2})$. Consequently, in $X_2(\alpha)$ one should specify $\alpha < \frac{1}{2}$ just as in usual CCs, if one is risk-averse. In unfortunate circumstances risk is unavoidable: $X_2(\alpha) = \emptyset$ for all $\alpha \in [0,\frac{1}{2}]$, see for instance Example 5.1.

REMARK 5.9. Many of the results in Theorems 5.6 and 5.7 remain true if the underlying constraint (5.1) is replaced by the *convex nonlinear* constraint in $x \in \mathbb{R}^n$

$$(5.1') \qquad g(x,\omega) \geqq 0$$

where $\omega$ is a random vector. If it is assumed that

$g(.,\omega)$ is concave in $x$ for almost all values of $\omega$,

(5.6') $\quad E_\omega|g(x,\omega)| < \infty \quad$ for all $x \in \mathbb{R}^n$,

(5.7') $\quad E_\omega|g(\hat{x},\omega)| > 0 \quad$ for at least one $\hat{x} \in \mathbb{R}^n$,

(5.8') $\quad \dfrac{E_\omega(g(x,\omega))^-}{E_\omega|g(x,\omega)|} \quad$ attains all values in the convex hull of its range,

then again $X_1(\beta)$, $\beta \geqq 0$, and $X_2(\alpha)$, $0 \leqq \alpha \leqq \frac{1}{2}$, are closed convex sets, strictly and continuously increasing with the risk aversion parameters $\beta$ and $\alpha$, respectively. Of course, here we identify $\eta(x)$ with $g(x,\omega)$. The proof is a repetition of the proof of the corresponding statements in Theorems 5.6 and 5.7, using the fact that also in the more general case the functions $E(\eta(x))^-$ and $-E\,\eta(x)$ are finite, convex and continuous.

Another generalization is possible. Recall that $X_1(\beta)$ can be written as

$$X_1(\beta) = \{x \in \mathbb{R}^n\colon\ E\, v_1(\eta(x)) \leqq \beta\}$$

where $v_1(z) = z^-$, $z \in \mathbb{R}$. The use of this particular loss function implies a uniform weighting of all "risks" $P(\eta(x) < t)$, $t \in (-\infty,0]$; see Remark 3.9. A more general weighting corresponds to the choice of an arbitrary loss function $v$, resulting in the integrated chance constraint

$$X_1'(\beta) := \{x \in \mathbb{R}^n\colon\ E\, v(\eta(x)) \leqq \beta\},\ \beta \in \mathbb{R}.$$

$X_1'(\beta)$ has similar properties as $X_1(\beta)$ if $v$ is convex.

It is well-known that in special cases the chance constraint in the definition of $X_0(\alpha)$ has a deterministic equivalent constraint which can be calculated easily. In particular this is true if only the right-hand side is random, or if the joint distribution is normal. Similar deterministic equivalents will be derived for the integrated chance constraints related to $X_1(\beta)$ and $X_2(\alpha)$.

Suppose first that $(a_1,\ldots,a_n)$ is deterministic. Then $E(\eta(x))^- = G(\Sigma_{j=1}^n a_j x_j)$ with $G(z) := E(z-b)^-$. By partial integration one derives, using the finiteness of $\bar{b}$,

$$(5.30) \qquad G(z) = \int_z^\infty (1-F(t))dt = \bar{b} - z + \int_{-\infty}^z F(t)dt, \; z \in \mathbb{R},$$

where $F(t) := P(b<t)$. The function G is finite, convex and continuous, with subgradients $\partial G(z) = [-1 + P(b<z), -1 + P(b\leqq z)]$. Its asymptotes for $|z| \to \infty$ are given by $G_0(z) = (z-\bar{b})^-$; in fact, $G(z) = G_0(z)$ iff $z \leqq b_0$ or $z \geqq b_1$, where $b_0 := \inf\{t: F(t) > 0\}$ and $b_1 := \sup\{t: F(t) < 1\}$. From this characterization of the function G one easily derives deterministic equivalents for $X_1(\beta)$ and $X_2(\alpha)$; they have the form $\{x: \Sigma_{j=1}^n a_j x_j \geqq k\}$ for a certain k, as is well-known for $X_0(\alpha)$.

<u>THEOREM 5.10</u>. <u>Only right-hand side random</u>. *Assume that* $(a_1,\ldots,a_n)$ *is deterministic and that* $E|b| < \infty$. *Then*

$$X_0(\alpha) = \{x \in \mathbb{R}^n: \Sigma_{j=1}^n a_j x_j \geqq k_\alpha^0\}, \; 0 < \alpha < 1,$$

*where* $k_\alpha^0 := \inf\{z: -\alpha \in \partial G(z)\}$ *i.e.* $k_\alpha^0 = \inf F^{-1}(1-\alpha)$.

$$X_1(\beta) = \{x \in \mathbb{R}^n: \Sigma_{j=1}^n a_j x_j \geqq k_\beta^1\}, \; 0 < \beta < \infty,$$

*where* $k_\beta^1$ *is the unique solution of* $G(k) = \beta$.

$$X_2(\alpha) = \{x \in \mathbb{R}^n: \Sigma_{j=1}^n a_j x_j \geqq k_\alpha^2\}, \; 0 < \alpha < 1,$$

*where* $k_\alpha^2$ *is the unique solution of* $(1 - 2\alpha)G(k) = \alpha(k-\bar{b})$.

*The functions* $\beta \to k_\beta^1$, $0 < \beta < \infty$, *and* $\alpha \to k_\alpha^2$, $0 < \alpha < 1$, *are continuous and strictly decreasing, and*

$$\lim_{\alpha\downarrow 0} k_\alpha^0 = \lim_{\beta\downarrow 0} k_\beta^1 = \lim_{\alpha\downarrow 0} k_\alpha^2 = b_1,$$

$$\lim_{\alpha\uparrow 1} k_\alpha^0 = \lim_{\alpha\uparrow 1} k_\alpha^2 = b_0,$$

$$\lim_{\beta\uparrow\infty} (k_\beta^1 - (\bar{b}-\beta)) = 0.$$

*Moreover*, $k_{\frac{1}{2}}^2 = \bar{b}$.

PROOF. The formula for $X_2(\alpha)$ is based on (5.23c). The statements about $k_\alpha^0$, $k_\beta^1$, $k_\alpha^2$ follow directly from the characterization of G, or from Theorems 5.6 and 5.7. □

The graph of the function G is sketched in Figure 5.3. Also the characteristic numbers $k_\alpha^0$, $k_\beta^1$ and $k_\alpha^2$ are indicated.

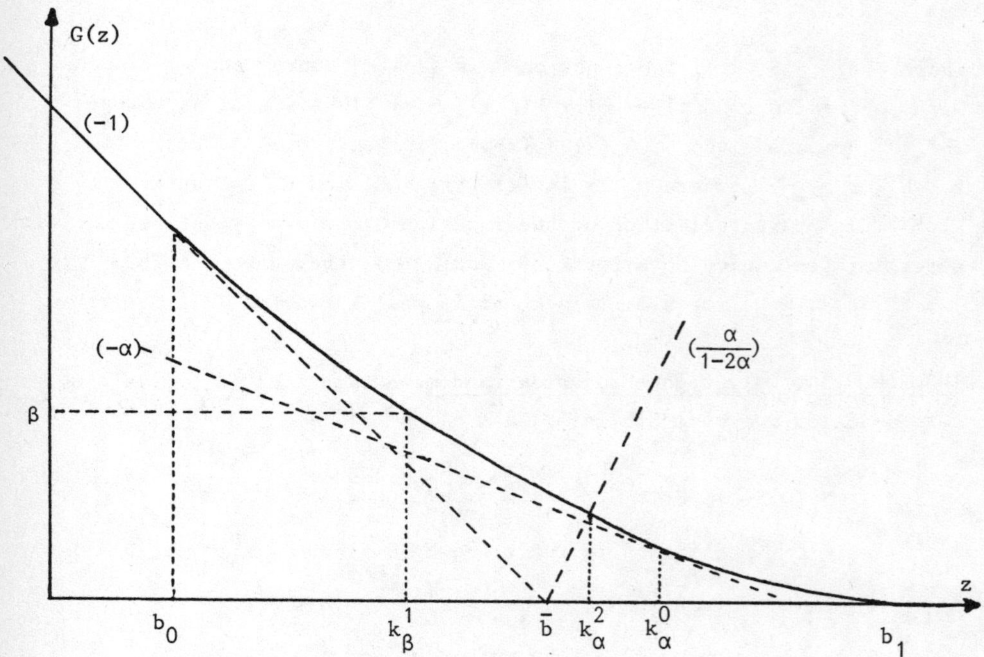


FIGURE 5.3. The function $G(z) = E(b-z)^-$. Slopes are indicated by (.).

Suppose now that $(a_1,\ldots,a_n,b)$ are normally distributed. In this case $\eta(x)$ has also a normal distribution, with mean value $\mu(x) = \Sigma_{j=1}^n \bar{a}_j x_j - \bar{b}$ and variance $\sigma^2(x) = \tilde{x}'V\tilde{x}$, with $\tilde{x} := (x,-1)$, and V the covariance matrix of (a,b). It is well-known [5] that $\sigma(x)$ is a convex function. If $\sigma(x) = 0$ then $E(\eta(x))^- = (\mu(x))^-$, and if $\sigma(x) > 0$

$$E(\eta(x))^- = \sigma(x)G_1(\frac{\mu(x)}{\sigma(x)}), \text{ where } G_1(z) := E(z-\xi)^-,$$

$\xi$ being a standard normal random variable. Obviously, the analysis of $G_1$ follows from that of G; from (5.30) we get

$$G_1(z) = \int_z^\infty (1 - \Phi_0(t))dt = - z + \int_{-\infty}^z \Phi_0(t)dt,$$

$\Phi_0$ being the distribution function of the standard normal distribution. Its density function $\varphi_0$ satisfies $\varphi_0'(z) = -z\varphi_0(z)$. Therefore $\frac{d}{dz}(z\Phi_0(z) + \varphi_0(z)) = \Phi_0(z)$, so that $G_1(z)$ can be expressed as

$$G_1(z) = - z + z\Phi_0(z) + \varphi_0(z), \; z \in \mathbb{R}.$$

THEOREM 5.11. Normal distribution for (a,b).

$$X_0(\alpha) = \{x \in \mathbb{R}^n: \; \mu(x) \geqq \kappa_\alpha^0 . \sigma(x)\}, \; 0 < \alpha < 1,$$

*where* $\kappa_\alpha^0 = \Phi_0^{-1}(1-\alpha)$, *i.e.* $G_1'(\kappa_\alpha^0) = -\alpha$.

$$X_1(\beta) = \{x \in \mathbb{R}^n: \; \mu(x) \geqq \kappa_{\beta/\sigma(x)}^1 . \sigma(x)\}, \; 0 < \beta < \infty,$$

*where* $\kappa_\beta^1$ *is the unique solution of* $G_1(k) = \beta$.

$$X_2(\alpha) = \{x \in \mathbb{R}^n: \; \mu(x) \geqq \kappa_\alpha^2 . \sigma(x)\}, \; 0 < \alpha < 1,$$

*where* $\kappa_\alpha^2$ *is the unique solution of* $(1 - 2\alpha)G_1(k) = \alpha k$.

$X_0(\alpha)$ *and* $X_2(\alpha)$ *are convex for* $\alpha \leqq \frac{1}{2}$; $X_1(\beta)$ *is convex for all* $\beta$. *The functions* $\alpha \to \kappa_\alpha^0$, $\beta \to \kappa_\beta^1$, $\alpha \to \kappa_\alpha^2$ *are continuous and strictly decreasing, with* $\kappa_{\frac{1}{2}}^0 = \kappa_{\frac{1}{2}}^2 = 0$, $\kappa_\beta^1 = 0$ *if* $\beta = (2\pi)^{-\frac{1}{2}}$ *and*

$$\lim_{\alpha\downarrow 0}\kappa_\alpha^0 = \lim_{\beta\downarrow 0}\kappa_\beta^1 = \lim_{\alpha\downarrow 0}\kappa_\alpha^2 = +\infty,$$

$$\lim_{\alpha\uparrow 1}\kappa_\alpha^0 = \lim_{\alpha\uparrow 1}\kappa_\alpha^2 = -\infty, \; \lim_{\beta\to\infty}(\kappa_\beta^1 + \beta) = 0.$$

PROOF. As the proof of Theorem 5.10. □

REMARK 5.12. Up to now we considered ICC formulations for an inequality constraint (5.1). Whereas the CC formulation of an *equality constraint* $\Sigma_{j=1}^n a_j x_j = b$ is useless, this is not true for ICCs. For example, one may define

$$(5.31) \qquad X_4(\beta_1,\beta_2) := \{x \in \mathbb{R}^n: \; E(\eta(x))^- \leqq \beta_1, \; E(\eta(x))^+ \leqq \beta_2\},$$

$$(5.32) \qquad X_5(\beta) := \{x \in \mathbb{R}^n: \; E|\eta(x)| \leqq \beta\}.$$

Both sets are convex, but they are empty if the risk aversion parameters are specified too small.

## 5.3 JOINT INTEGRATED CHANCE CONSTRAINTS

In this short section we consider m random constraints, $m \geqq 2$,

(5.33) $\quad A_i x := \Sigma_{j=1}^n a_{ij} x_j \geqq b_i, \; i = 1,\dots,m,$

where $(A,b) = (a_{ij}, b_i)$ has a known distribution with finite means. If for each constraint a separate degree of risk aversion is specified, (5.33) can be reformulated in terms of ICCs as

(5.34) $\quad x \in \tilde{X}_1(\beta_1,\dots,\beta_m) := \cap_{i=1}^m X_{1i}(\beta_i), \; 0 \leqq \beta_i \leqq \infty,$

(5.35) $\quad x \in \tilde{X}_2(\alpha_1,\dots,\alpha_m) := \cap_{i=1}^m X_{2i}(\alpha_i), \; 0 \leqq \alpha_i \leqq 1,$

where $X_{1i}(\beta_i)$ and $X_{2i}(\alpha_i)$ are defined as in (5.3) and (5.4). Of course, $\tilde{X}_1$ and $\tilde{X}_2$ have the same properties as $X_1$ and $X_2$.

A well-known alternative for separate CCs is the *joint* CC

(5.36) $\quad Y_0(\alpha) := \{x \in \mathbb{R}^n\colon \; P(Ax \geqq b) \geqq 1-\alpha\}, \; 0 \leqq \alpha \leqq 1.$

Since $Y_0(\alpha)$ can be formulated as

$$Y_0(\alpha) = \{x \in \mathbb{R}^n\colon \; E \operatorname{sgn} \max_i (\eta_i(x))^- \leqq \alpha\}$$

where $\eta_i(x) := \Sigma_{j=1}^n a_{ij} x_j - b_i$, the obvious generalization of $X_1(\beta)$ to joint ICC is

(5.37) $\quad Y_1(\beta) := \{x \in \mathbb{R}^n\colon \; E \max_i (\eta_i(x))^- \leqq \beta\}, \; 0 \leqq \beta \leqq \infty.$

Unlike $Y_0(\alpha)$, $Y_1(\beta)$ is *convex* for all distributions of (A,b) and for all values of the risk aversion parameter. In fact, the same statement is true for

$$Y_1'(\beta) := \{x \in \mathbb{R}^n\colon \; E\|(\eta(x))^-\| \leqq \beta\}, \; 0 \leqq \beta \leqq \infty,$$

$$Y_2''(\beta) := \{x \in \mathbb{R}^n: \ E\, v(\eta(x)) \leqq \beta\},\ \beta \in \mathbb{R},$$

where $\|.\|$ denotes *any* norm in $\mathbb{R}^n$, and v denotes *any* convex multiattribute loss function , respectively. Joint variants of $X_2(\alpha)$ are less obvious. One might define e.g. for $0 \leqq \alpha \leqq 1$,

$$Y_2(\alpha) := \{x \in \mathbb{R}^n: \ E \max_i (\eta_i(x))^- \leqq \alpha.E \max_i |\eta_i(x)|\},$$

$$Y_2'(\alpha) := \{x \in \mathbb{R}^n: \ (1-\alpha).E \max_i (\eta_i(x))^- \leqq \alpha.E \max_i (\eta_i(x))^+\},$$

$$Y_2''(\alpha) := \{x \in \mathbb{R}^n: \ (1-2\alpha).E \max_i (\eta_i(x))^- \leqq \alpha.E \min_i \eta_i(x)\},$$

each of which reduces to $X_2(\alpha)$ if m = 1. $Y_2(\alpha)$ and $Y_2'(\alpha)$ are nonconvex generally, but $Y_2''(\alpha)$ is convex for $\alpha \leqq \frac{1}{2}$.

It is also possible to define ICC feasibility sets for which no CC companion exists. For example, for weights $r_i > 0$ with $\Sigma_{i=1}^m r_i = 1$

$$(5.38) \qquad Z_1(\beta,r) := \{x \in \mathbb{R}^n: \ \Sigma_{i=1}^m r_i.E(\eta_i(x))^- \leqq \beta\},\ 0 \leqq \beta \leqq \infty,$$

convex for all $\beta$, and

$$(5.39) \qquad Z_2(\alpha,r) := \{x \in \mathbb{R}^n: \ \Sigma_{i=1}^m r_i.E(\eta_i(x))^- \leqq \alpha.\Sigma_{i=1}^m r_i.E|\eta_i(x)|\},\ 0 \leqq \alpha \leqq 1,$$

convex for all $\alpha \leqq \frac{1}{2}$.

## 5.4 INTEGRATED CHANCE CONSTRAINTS AND RECOURSE MODELS

Just as in ICCs, in simple recourse models "risk" is interpreted as "mean shortage". But risk aversion is specified in terms of penalty cost for shortages rather than by prescribing the maximal acceptable risk. It is clear, that there must be an intimate relation between both model types, expressed by Lagrange multipliers. It will be shown that for each convex ICC model an equivalent (often simple) recourse model exists. Hence, the criticism in [4], that CCP is a deficient modeling technique because of the poor correspondence between CCP and SPR, is not valid for convex ICCs.

Suppose that the finite convex objective function c(x) is to be minimized on the nonempty convex set X. Let (5.34) be part of the problem

$$\mathrm{ICC}_1(\beta) \qquad \text{minimize}_{x\in X}\{c(x): E(A_i x - b_i)^- \leqq \beta_i,\ i = 1,\dots,m\},\ \beta \geqq 0,$$

This convex program is strictly feasible if $\beta$ is not too small, that is: $\beta_i > \beta_{i0} := E(A_i\hat{x} - b_i)^-$, $\forall i$, for a suitable $\hat{x} \in X$. The Lagrangian problem

$$L_1(\lambda) \qquad \text{minimize}_{x\in X}\{c(x) + \Sigma_{i=1}^m \lambda_i . E(A_i x - b_i)^-\},\ \lambda \geqq 0,$$

is a simple recourse model, and we have

<u>THEOREM 5.12</u>. <u>Equivalence of $\mathrm{ICC}_1(\beta)$ and $L_1(\lambda)$</u>.
*a. If* $\beta_i > \beta_{i0} \forall_i$ *then there exist optimal Lagrange multipliers* $\lambda_i^0$ *for the constraints in* $\mathrm{ICC}_1(\beta)$, *and* $x^* \in X$ *solves* $\mathrm{ICC}_1(\beta)$ *iff it is feasible for it and if it is a solution of* $L_1(\lambda^0)$.
*b. If* $x^* \in X$ *solves* $L_1(\lambda)$ *for any* $\lambda \geqq 0$ *then it solves* $\mathrm{ICC}_1(\beta)$ *with* $\beta_i := E(A_i x^* - b_i)^-$.

<u>PROOF</u>. Follows directly from the Karush-Kuhn-Tucker Theorem; see e.g. Theorem 2.3 and the discussion there. The condition on $\beta_i$ implies that Slater's constraint qualification holds. □

Of course, the mathematical equivalence in Theorem 5.12 does not mean that both models are equivalent from a practical point of view, since the specification of penalty costs may be more difficult or less difficult than that of the risk aversion parameters $\beta$.

Consider now the separate ICCs (5.35). With obvious adjustments Theorem 5.12 part a remains true if $\mathrm{ICC}_1(\beta)$ and $L_1(\lambda)$ are replaced by $\mathrm{ICC}_2(\alpha)$ and $L_2(\lambda,\alpha)$, respectively:

$$\mathrm{ICC}_2(\alpha) \qquad \text{minimize}_{x\in X}\{c(x): E(A_i x - b_i)^- \leqq \alpha_i E|A_i x - b_i|,\ i = 1,\dots,m\},\quad 0 \leqq \alpha_i \leqq \tfrac{1}{2},$$

$$L_2(\lambda,\alpha) \qquad \text{minimize}_{x\in X}\{c(x) + \Sigma_{i=1}^m((1 - \alpha_i)\lambda_i E(A_i x - b_i)^- + (-\alpha_i)\lambda_i E(A_i x - b_i)^+)\},\ \lambda \geqq 0,\ 0 \leqq \alpha_i \leqq \tfrac{1}{2}.$$

Similarly for (5.39). For (5.31), (5.32) and (5.38) one easily formulates equivalent simple recourse models. That is, with obvious adjustments both parts of Theorem 5.12 hold.

Also for the joint integrated chance constraints (5.37) equivalence to a recourse model can be proved. In this case the recourse is fixed and complete, but not simple. Defining

$$\mathrm{ICC}_3(\beta) \qquad \text{minimize}_{x \in X}\{c(x) \colon E \max_i (A_i x - b_i)^- \leqq \beta\},\ \beta \geqq 0,$$

and

$$L_3(\lambda) \qquad \text{minimize}_{x \in X}\{c(x) + Q_\lambda(x)\},\ \lambda \geqq 0,$$

where $\quad Q_\lambda(x) := E \min_y\{qy \colon Wy = b - Ax,\ y \geqq 0\},$

$$\binom{q}{W} := \begin{pmatrix} \lambda & 0 \\ e & -I \end{pmatrix},\ e \in \mathbb{R}^m \text{ with } e_i = 1\ \forall i,$$

one easily verifies that $Q_\lambda(x) = \lambda.\ E \max_i (A_i x - b_i)^-$, so that Theorem 5.12 holds also for $\mathrm{ICC}_3(\beta)$ and $L_3(\lambda)$.

## 5.5 CONCLUSIONS

In many cases, ICCs may be an appropriate tool for modelling random linear constraints. ICCs are more appropriate than CCs if the underlying concept "risk := mean shortage" is more appropriate than "risk := probability of positive shortage". ICCs give rise to convex optimization problems, for all distributions of the random coefficients involved, matrix as well as righthand side. Moreover, the feasibility sets defined by ICCs change continuously and strictly with the risk aversion parameters, for all distributions. Models with ICCs are more appropriate than SPR models, if one is not able to specify the penalty costs for infeasibilities. Unlike CC models, ICC models exhibit a natural mathematical equivalence with SPR models, provided by Lagrange multipliers. Computation of mean shortages is not easy in general; however, in simple situations where deterministic equivalent formulations for CCs exist, the same is true for ICCs, and as far as Monte Carlo simulation is concerned, there does not seem to be much difference in difficulty either. On the other hand, one might expect that progress in computation in simple and fixed recourse models can be used for computation with ICCs.

REFERENCES

1. M.A.H. DEMPSTER (1968). On stochastic programming: I. Static linear programming under risk. *J. Math. Anal. Appl. 21*, 304-343.
2. S.J. GARSTKA (1980). The economic equivalence of several stochastic programming models. M.A.H. DEMPSTER (ed.). *Stochastic Programming*, Academic Press, New York-San Francisco-London, 83-91.
3. S.J. GARSTKA, R.J.-B. WETS (1974). On decision rules in stochastic programming. *Math. Programming 7*, 117-143.
4. A.J. HOGAN, J.G. MORRIS, H.E. THOMPSON (1981). Decision problems under risk and chance-constrained programming: Dilemmas in the transition. *Management Sci. 27*, 698-716.
5. P. KALL (1976). *Stochastic Linear Programming*, Springer, Berlin-Heidelberg-New York.
6. P. KALL, D.J. STOYAN (1982). Solving stochastic programming problems with recourse including error bounds. *Math. Operationsforsch. Statist. Ser. Optim. 13*, 431-447.
7. A. PREKOPA (1973). Contributions to the theory of stochastic programming. *Math. Programming 4*,202-221.
8. A. PREKOPA (1974). Programming under probabilistic constraints with a random technology matrix. *Math. Operationsforsch. Statist. Ser. Optim. 5*, 109-116.
9. R.T. ROCKAFELLAR (1970). *Convex Analysis*, Princeton University Press, Princeton NJ.
10. G.H. SYMONDS (1968). Chance-constrained equivalents of some stochastic programming problems. *Oper. Res. 16*, 1152-1159.
11. D.W. WALKUP, R.J.-B. WETS (1970). Stochastic programs with recourse: Special forms. H. KUHN (ed.). *Proceedings of the Princeton Symposium on Mathematical Programming*, Princeton University Press, Princeton NJ, 139-161.
12. A.C. WILLIAMS (1965). On stochastic linear programming. *SIAM J. Appl. Math. 13*, 927-940.

CHAPTER 6

# ON THE BEHAVIOUR OF THE OPTIMAL VALUE OPERATOR OF DYNAMIC PROGRAMMING

## 6.1 INTRODUCTION

In this chapter we deal with the dynamic programming algorithm DPA for the finite-horizon stochastic dynamic programming problem SDP introduced in Section 4.6. Roughly speaking, the following decision process is described by SDP. There is a discrete time stochastic system, and the state of the system evoluates in a Markovian way. At each stage of time the current state is observed, and using this information the decision maker has to select an action. Any action will influence the immediate cost of the corresponding stage as well as the probability distribution of the next state. The purpose is to minimize the expected costs,summed over all stages. For details on the SDP problem we refer the reader to Section 4.6. There we gave also a formal definition of a "policy"; a policy may be seen as a complete specification of all particular choices which possibly are made by the decision maker at any stage and at any state.

In Section 4.6 we also described the dynamic programming algorithm DPA for the SDP problem; it defines by backward recursion a sequence of so-called cost-to-go functions. A typical step in this recursion can be described by means of an operator T to be defined. T transforms a "cost" function $\bar{f}$ of the "next" state $\bar{s}$ to a "cost" function f of the "current" state s in the following way:

$$f(s) := (T\bar{f})(s) := \inf_{a\in D(s)}\{c(s,a) + \int_S f(\bar{s})p(d\bar{s}|s,a)\}.$$

Here D(s) is the set of feasible actions a, c is the "current" immediate cost function and $p(.|s,a)$ is the probability distribution of the "next" state. The operator T is called a (one-stage) *optimal value operator*: the value $T\bar{f}(s)$ represents the minimum expected cost one has to pay if one starts in state s, if the immediate cost is given by c and if $\bar{f}(\bar{s})$ represents the cost associated with ending up in state $\bar{s}$. Obviously, the optimal actions of the one-stage optimization problem described in the definition of $T\bar{f}(s)$ are given by the minimum set (which may be empty)

$$D^*(s) := \arg\min_{a\in D(s)}\{c(s,a) + \int_S \bar{f}(s)p(d\bar{s}|s,a)\}.$$

For any N-stage SDP problem let $T_n$ be the optimal value operator of stage n, n = 1,...,N. Then the dynamic programming algorithm for this problem can be written in a simple way as the backward recursion

$$\text{(DPA)} \qquad f_n := T_n f_{n+1}, \; n = N,\ldots,1,$$

which starts with the given terminal cost function $f_{N+1}$. As has been shown in Proposition 4.22 and the discussion after it, it is possible that DPA generates a (deterministic Markovian) policy; this is the case if all minimum sets $D_n^*(s)$ of DPA are nonempty, and if each $D_n^*(.)$ admits a measurable selection. Recall that any policy generated by DPA is optimal for SDP, uniformly for all distributions of the initial state.

Unfortunately, it is not true in general that DPA generates a policy at all. The main purpose of this chapter is to provide sufficient conditions on the data of a special but important class of SDP problems which guarantee that DPA generates a policy. That $D_n^*(s) \neq \emptyset$ for all s and all n is assured by lower semicontinuity and compactness conditions, basically, and the existence of measurable selections is assured with help of a measurable selection theorem for closed-valued Borel measurable multifunctions with a Euclidean range space. Moreover, it will appear that the same conditions imply that DPA behaves *properly*. Roughly speaking, we mean by this that each cost-to-go function $f_n$, defined in DPA, is a well-defined extended real Borel measurable function without additional provisions (Recall from Section 4.6 that in general one has to allow for uniformly measurable functions and to extend the definition of integrals using the provision $\infty - \infty := \infty$, in order to have DPA well-defined).

The contents of the present chapter are summarized as follows. In Section 6.2 we introduce our main result (Theorem 6.2). One of its implications for the optimal value operator T is given in Corollary 6.5. If all the optimal value operators $T_n$, $n = 1,\dots,N$, of an N-stage SDP problem satisfy the conditions of Corollary 6.5, then it is guaranteed that DPA generates an optimal (deterministic Markovian) policy for SDP. In Section 6.3 we give definitions and lemmas needed for the proof of the main theorem and its corollary in the end of this section. In Section 6.4 Theorem 6.2 is applied to finite-stage Markovian inventory control models. Due to the special properties of inventory control models the conditions for the existence of an optimal policy are weaker than those in Corollary 6.5, see Corollary 6.12 and Remark 6.15.

We conclude this introduction by pointing out the special class of SDP problems to which our results apply. Consider a general SDP problem as defined in Section 4.6. The following *assumptions* will be used throughout this chapter.

(i) Any transition probability $p_j(ds_{j+1}|s_j,a_j)$ which represents the distribution of the state $s_{j+1}$ at stage j+1 as a function of the state $s_j$ and the action $a_j$, both at stage j, can be described by a *system function* (state transformation function) $F_j$ and a vector $\xi_j$ of *disturbances* whose distribution $\Phi_j$ *does not depend on* $(s_j,a_j)$. That is

$$s_{j+1} = F_j(s_j,a_j,\xi_j)$$

and

$$p_j(B|s_j,a_j) := \Phi_j(\{\xi_j: F_j(s_j,a_j,\xi_j) \in B\}) \ \forall B.$$

Of course, because of the Markovian character of the transitions it is assumed that the disturbances of different stages are stochastically independent random vectors.

(ii) There are no state-dependent constraints on the actions, i.e. $D_j(s) = A_j \ \forall s \ \forall j$.

(iii) The cost function $c_j$ of any stage j may depend on the disturbances as well, i.e.

$$c_j = c_j(s_j,a_j,\xi_j) \ \forall j.$$

(iv) All state and vector action spaces are finite-dimensional Euclidean spaces.

Let us discuss these assumptions. Assumption (i) describes a well-known important special case of SDP. For example, it includes Markovian inventory control models as described in Section 6.4: then the "disturbances" represent the random "demands". It is, however, restrictive: in the general case of Section 4.6 one may be tempted to define $\xi_j := s_{j+1}$ with $F_j(s_j,a_j,\xi_j) := \xi_j$, but then the distribution of $\xi_j$ depends on $(s_j,a_j)$.

Assumption (ii) is not restrictive: without loss of generality one may include possible constraints $a_j \in D_j(s_j)$ in the cost function: just extend the domain of $c_j(s_j,a_j)$, which is

$$\{(s_j,a_j) \in S_j \times A_j: a_j \in D_j(s_j)\}$$

to the whole of $S_j \times A_j$, setting

$$c_j(s_j,a_j) := +\infty \text{ if } a_j \notin D_j(s_j).$$

Clearly, Assumption (iii) is not restrictive. On the other hand, cost functions of the type $c_j(s_j,a_j,\xi_j)$, are not really more general than functions of the type $c_j(s_j,a_j)$, since the former may be replaced by $\int c_j(s_j,a_j,z_j)\Phi_j(dz_j)$ without loss of generality.

Finally, Assumption (iv) is restrictive, of course. We need it because the measurable selection theorem which will be used is formulated in Euclidean spaces. We expect, however, that it can be generalized, but we did not work out this subject.

## 6.2 THE MAIN THEOREM

Suppose that $(Z,\mathcal{Z},\Phi)$ is a probability space, S a nonempty closed subset of a finite-dimensional Euclidean space, and A a nonempty closed subset of a finite-dimensional Euclidean space. S is called the state space and A is called the action space; both S and A are measure spaces with the Borel sets as measurable sets. Throughout this chapter with "measurable" in S and A is meant "Borel measurable". Suppose that $F: S\times A\times Z \to S$ (system function) and $c: S\times A\times Z \to \overline{\mathbb{R}}$ (one-stage cost function) are measurable with respect to the product $\sigma$-algebra. Here $\overline{\mathbb{R}}$ denotes the extended reals. For simplicity of notation we do not distinguish between the state spaces of different stages. However, every result to come remains true if the range space S of the system function F is replaced by another nonempty closed subset S' of a finite-dimensional Euclidean space. The *optimal value operator* T determined by S, A, $(Z,\mathcal{Z},\Phi)$, F and c is defined as follows. For "any" $\bar{f}: S \to \overline{\mathbb{R}}$ we set

$$f := T\bar{f}$$

if $f: S \to \overline{\mathbb{R}}$ is defined by

$$(6.1)\qquad f(s) := \inf_{a\in A}\int_Z g(s,a,z)\Phi(dz),\ s \in S,$$

where

$$(6.2)\qquad g(s,a,z) := c(s,a,z) + \bar{f}(F(s,a,z)),\ (s,a,z) \in S\times A\times Z.$$

Under appropriate conditions T transforms a large class of extended real functions on S to the extended real functions on S. Closely related to the operator T is the multifunction $D^*: S \prec A$ defined by

$$(6.3)\qquad D^*(s) := \arg\min_{a\in A}\int_Z g(s,a,z)\Phi(dz).$$

It is well-known that even under regularity conditions, the operator T does not preserve measurability [3]. In this chapter we give conditions in order that

(6.4)
a. $D^*$ is a nonempty-closed-valued measurable multifunction, and
b. f is lower semicontinuous.

The measurability of a closed-valued multifunction is defined in Section 6.3. If $D^*$ satisfies (6.4a) then there exists a measurable selection (i.e. a measurable function $a^*: S \to A$ such that $a^*(s) \in D^*(s)$ for all $s \in S$). Therefore, condition (6.4a) is important for the existence of optimal policies in SDP problems (See Sections 6.1 and 6.4). It is assured basically by lower semicontinuity and compactness conditions. Since we have iterative application of T in mind, property (6.4b) is important, together with the related question under which conditions the operator T preserves lower semicontinuity. The following result is well-known; e.g. Shreve [15 Theorem 3.4] and Hinderer [6 Theorem 17.12] give analogous results in a more general setting.

<u>THEOREM 6.1</u>. *Suppose that* $(s,a) \to g(s,a,z)$ *is lower semicontinuous,* $z \to g(s,a,z)$ *is measurable. If* A *is compact and* g *is nonnegative, then* f *is nonnegative and* (6.4) *holds.*

In this chapter we derive a more general theorem of this type. Of course, by replacing the nonnegativity of g by: $(s,a) \to \int g(s,a,z)\Phi(dz)$ is bounded from below, the conclusion of Theorem 6.1 remains valid, provided we read f bounded from below instead of f nonnegative. The same conclusion holds, if the compactness of A is replaced by the condition: $\{(s,a): \int g(s,a,z)\Phi(dz) \leq \gamma\}$ is compact for all $\gamma \in \mathbb{R}$. However, even under these conditions, Theorem 6.1 does not cover elementary cases like $S := Z := \mathbb{R}$, $A := \mathbb{R}^+$, $g(s,a,z) := s + a - z$ and $\Phi$ such that $\int z\Phi(dz) = \bar{\xi}$ is finite. In this case (6.4) holds trivially, but $f(s) = s - \bar{\xi}$ is not bounded from below on S. By adjusting the conditions on g in such a way that they only have to be true locally in a neighbourhood of any $s \in S$, we are able to derive a theorem which also covers the above elementary case. The nonnegativity of g in Theorem 6.1 is weakened to the "lower bound condition" LB, whereas the compactness condition is replaced by the "growth condition" G. These concepts will be defined with use of the following shorthand notation. For any

$h: S \times A \to \overline{\mathbb{R}}$ and any subset $B \subset S$ and any subset $C \subset A$ we define the functions $\pi_B h: A \to \overline{\mathbb{R}}$ and $\pi_C h: S \to \overline{\mathbb{R}}$ by

$$\pi_B h(a) := \inf_{s \in B} h(s,a), \ a \in A.$$

$$\pi_C h(s) := \inf_{a \in C} h(s,a), \ s \in S.$$

The same notation will be used for functions h with other domain spaces.

DEFINITION. *A function* $g: S \times A \times Z \to \overline{\mathbb{R}}$ *is said to satisfy the lower bound condition* LB *if for each bounded subset* $B \subset S$ *the function* $\pi_{B\times A} g$ *has an integrable minorant, i.e. a measurable real function on* Z *which does not exceed* $\pi_{B\times A} g$ *and has a finite integral with respect to* $\Phi$.

Of course, LB holds automatically if g is bounded below, or more generally, if g has an integrable minorant not depending on (s,a). However, LB admits also functions g which are not bounded from below as a function of s.

DEFINITION. *A function* $h: S \times A \to \overline{\mathbb{R}}$ *is said to satisfy the growth condition* G, *if for each bounded set* $B \subset S$ *the level sets of the function* $\pi_B h$, *defined by* $\{a \in A: \pi_B h(a) \leqq \alpha\}$, $\alpha \in \mathbb{R}$, *are bounded sets.*

Notice that the growth condition holds if for each $s_o \in S$ there exists a neighbourhood $B(s_o)$ such that $\pi_{B(s_o)} h$ has bounded level sets. Clearly, if $h(s,a) = \psi(s,a).\chi_C(a)$ for some function $\psi$ and $\chi_C$ the indicator function of a bounded set C or if h has bounded level sets, then the growth condition is satisfied automatically. However, G admits also functions h with level sets which are not bounded in s. In Lemma 6.8 we will prove that for lower semicontinuous functions h the growth condition is equivalent with: $\pi_B h$ is infcompact (defined in Section 6.3) for all compact subsets $B \subset S$.

In Section 6.3 we give the definition of a normal integrand. We can now state our main theorem.

THEOREM 6.2. *Suppose that* $g: S \times A \times Z \to \overline{\mathbb{R}}$ *is a normal integrand on* $(S \times A) \times Z$, *which satisfies the growth condition* G *a.e.* $\Phi$ *and the lower bound condition* LB. *Then the function* $f: S \to \overline{\mathbb{R}}$, *defined by*

$$f(s) := \inf_{a\in A} \int_Z g(s,a,z)\Phi(dz),$$

*is lower semicontinuous, and bounded from below on bounded subsets of* S. *Furthermore, for each* $s \in S$, *the set*

$$D^*(s) := \arg\min_{a\in A} \int_Z g(s,a,z)\Phi(dz)$$

*is a nonempty closed subset of* A, *compact if* $f(s) < \infty$. *Moreover, the multifunction* $D^*: S \multimap A$ *is measurable, so it admits a measurable selection.*

In comparison with Theorem 6.1 we replaced the condition "g is lower semicontinuous in (s,a) and measurable in z" by "g is a normal integrand on $(S \times A) \times Z$". The last condition is a little bit more restrictive (see Lemma 6.7a); we use it in order to circumvent measurability questions when working with the growth condition G, see the proof on Lemma 6.9b.

In the proof of Theorem 6.2 to be given in the next section we use elementary topological facts of $\mathbb{R}^n$ and well-known theorems of measure theory (especially, Fatou's lemma). Furthermore, properties of normal integrands as treated by Rockafellar [10] are used.

We explain with two counterexamples the "necessity" of the conditions G and LB.

<u>EXAMPLE 6.3</u>. $S := A := \mathbb{R}$, $(Z,\mathcal{Z},\Phi)$ arbitrary, and for all $z \in Z$: $g(s,a,z) := 0$ if $sa = 1$, $g(0,0,z) := 1$ and $g(s,a,z) := +\infty$ otherwise. Then every condition in Theorem 6.2 is satisfied, except G, but f is not lower semicontinuous at $s = 0$. Remark that g satisfies a "pointwise growth condition", i.e. for each fixed $s \in S$ (and $z \in Z$) the function $a \to g(s,a,z)$ has bounded level sets. So the weaker pointwise growth condition is not strong enough to assure the lower semicontinuity of f.

<u>EXAMPLE 6.4</u>. $(Z,\mathcal{Z},\Phi)$ is [0,1] endowed with its Borel sets and the Lebesgue measure, $S := [0,1]$, A is arbitrary, and $g(s,z) = g(s,a,z)$ is defined by: $g(0,0) := -\infty$, $g(s,z) := -s^{-1}$ if $s > 0$ and $0 \leq z \leq s$, whereas $g(s,z) := 0$ otherwise. Then every condition in Theorem 6.2 is satisfied

except LB. However, f is not lower semicontinuous at s = 0.

By verifying the conditions for g in Theorems 6.1 or 6.2, results on the operator T can be formulated. For example,

COROLLARY 6.5. *Let* T *be defined as in* (6.1) *and* (6.2). *In this definition assume that* $F: (S \times A) \times Z \to S$ *is a Caratheodory map (defined in Section* 6.3*), that* $c: (S \times A) \times Z \to \overline{\mathbb{R}}$ *is a normal integrand which satisfies the growth condition* G *a.e.*$\Phi$ *and that* $\pi_{S\times A}c$ *has an integrable minorant. Then* $f := T\bar{f}$ *is a lower semicontinuous function with a finite lower bound if* $\bar{f}$ *is so. Moreover, for* $D^*$ *defined in* (6.3) *all conclusions of Theorem* 6.2 *hold.*

The proof of Corollary 6.5 is postponed to the next section.

Notice that Corollary 6.5 can be applied iteratively in appropriate circumstances. Hence, if for an N-stage stochastic dynamic programming problem SDP the conditions of Corollary 6.5 are satisfied for the optimal value operator $T_n$ of *each* stage n = 1,...,N, and if the cost function related to the terminal state $s_{N+1}$ is lower semicontinuous with a finite lower bound, then *each* cost-to-go function defined by the DPA algorithm is lower semicontinuous, and this algorithm generates an optimal policy for SDP.

In Section 6.4 we will apply Theorem 6.2 in a similar way to finite-horizon Markovian inventory control models. Due to the special form of the system function in those models the conditions we end up with are weaker then those in Corollary 6.5, see Corollary 6.12. In Remark 6.17 we compare in some detail our results on Markovian inventory control models with those of Schäll [12,13,14]. Now we give a short exposition on the relation between our results and those in other related literature. Remark that the use of the LB condition brings us into the area of the so-called *negative dynamic programming* (see e.g. [16,6]). That is, for each initial state and each sequence of decisions the reward (= minus cost) has an integrable majorant. In the literature several sufficient conditions are given assuring lower semicontinuous cost-to-go functions and existence of optimal Borel policies, mostly in very general settings. Applied to our model in Section 6.4 they appear to be more restrictive. Hinderer [6 Theorem 17.12] assumes bounded cost functions and a compact action space. Shreve [15 Theorem 3.4] and Bertsekas and Shreve [2 Proposition 8.6, Corollary 8.6.1] have a compact action space (or infcompact cost functions). Freedman [5 Theorem 6] proves similar results if the costs are bounded above and the action space and the state space are compact. Rockafellar and Wets [11 Lemmas 2 and 3] prove similar results under convexity, infcompactness and summability conditions. For other related results we refer to Olsen [9] and Evstigneev [4]. Olsen gives conditions involving convexity guaranteeing that the dynamic programming approach solves the multistage stochastic program with recourse. Evstigneev proves analogous results by imposing rather simple conditions on the "static" formulation of the problem whereas our results impose conditions on the "dynamic" formulation. (For the meaning of "static" and "dynamic" formulations of multistage stochastic programming problems we refer to Section 3.3 or to [9]).

## 6.3 PROOF OF THE MAIN THEOREM

In this section we first recall the definitions and properties of lower semicontinuous and infcompact functions, normal integrands and Caratheodory maps for finite-dimensional Euclidean spaces (see e.g. [10,17]). Then additional lemmas are proved. These are used in the proofs of Theorems 6.1 and 6.2 and Corollary 6.5 which are given at the end of this section.

A function $f: A \to \overline{\mathbb{R}}$ is called *lower semicontinuous* (*lsc*) at $a \in A$ if $f(a) \leqq \liminf f(a_n)$ for all sequences $\{a_n\} \subset A$ converging to a. Equivalently, if the existence of a sequence $\{a_n\} \to a$ with $\{f(a_n)\} \to \mu < \infty$ implies that $f(a) \leqq \mu$. The function f is called lsc if it is lsc at each $a \in A$. Equivalently if the epigraph of f, $\{(a,\mu) \in A \times \mathbb{R}: f(a) \leqq \mu\}$ is closed, or, if for each $\mu \in \mathbb{R}$ the level set $\{a \in A: f(a) \leqq \mu\}$ is closed. The function f is called *infcompact* if these level sets are compact for all $\mu \in \mathbb{R}$. So an infcompact function is lsc. A function $f: A \times Z \to \overline{\mathbb{R}}$ is called a *normal integrand* on $A \times Z$ if its epigraph multifunction $E_f: Z \prec A \times \mathbb{R}$, determined by $E_f(z) := \{(a,\mu): f(a,z) \leqq \mu\}$, is closed-valued and measurable. That means that for each $z \in Z$ the function $a \to f(a,z)$ is lsc, and for each closed set $C \subset A \times \mathbb{R}$ it holds that $E_f^{-1}(C) := \{z \in Z: C \cap E_f(z) \neq \emptyset\}$ is measurable. A function $f: A \times Z \to \mathbb{R}^m$ is called a *Caratheodory map* if $a \to f(a,z)$ is continuous and $z \to f(a,z)$ is measurable.

<u>LEMMA 6.6.</u>

*a. Suppose* $f,g: A \to \overline{\mathbb{R}}$. *If* f *and* g *are lsc, so is* f+g. *If* f *is lsc and bounded from below, and* g *is infcompact, then* f+g *is infcompact. If* f *is lsc and* g *is infcompact, then* $f \geqq g$ *implies that* f *is infcompact.*
*b. Suppose* $f: A \to \overline{\mathbb{R}}$ *is lsc and* C *is a nonempty compact subset of* A, *then* f *attains its minimum on* C *in a nonempty compact subset of* C. *In particular, if* $f \not\equiv +\infty$ *is infcompact,* $\min_C f$ *is attained in a nonempty compact set. If furthermore* $f > -\infty$ *then* $\min f > -\infty$.
*c. Suppose* $f: S \times A \to \overline{\mathbb{R}}$ *is lsc and* $C \subset A$ *is compact. Then* $\pi_C f$ *is lsc.*

Recall the following definitions. Let $(Z,\mathcal{Z})$ be a measurable space, and $\Lambda$ the family of all nonnegative $\sigma$-finite measures $\lambda$ on $\mathcal{Z}$. For any $\lambda \in \Lambda$ the completion of $\mathcal{Z}$ with respect to $\lambda$ is defined by $\mathcal{Z}_\lambda :=$ $\{B \subset Z: \exists B_1, B_2 \in \mathcal{Z}$ such that $(B \setminus B_1) \cup (B_1 \setminus B) \subset B_2$ and $\lambda(B_2) = 0\}$. The measurable space $(Z,\mathcal{Z})$ is called *complete* if $\mathcal{Z} = \bigcap_{\lambda \in \Lambda} \mathcal{Z}_\lambda$.

<u>LEMMA 6.7.</u>

*a. Suppose that* $f: A \times Z \to \overline{\mathbb{R}}$ *is such that* $a \to f(a,z)$ *is lsc, and let* $\mathcal{B}$ *be the σ-algebra of Borel sets of* A. *If* f *is a normal integrand, then* f *is* $\mathcal{B} \otimes \mathcal{Z}$*-measurable (in particular:* $z \to f(a,z)$ *is* $\mathcal{Z}$*-measurable). The converse is true if the measurable space* $(Z,\mathcal{Z})$ *is complete.*
*b. If* $f,g: A \times Z \to (-\infty,\infty]$ *are normal integrands, then so is* f+g.
*c. Suppose that* $f: A \times Z \to \overline{\mathbb{R}}$ *is a normal integrand. Then* $\pi_A f$ *is a measurable extended real function on* $(Z,\mathcal{Z})$, *and the multifunction* $D: Z \rightrightarrows A$ *given by* $D(z) := \{a \in A: f(a,z) = \pi_A f(z)\}$ *has closed values, and is measurable. In particular,* $\operatorname{dom} D := \{z \in Z: D(z) \neq \emptyset\} \in \mathcal{Z}$ *and* D *admits a measurable selection.*
*d. Suppose that* $f: A \times Z \to \overline{\mathbb{R}}$ *is a normal integrand, and* $\varphi: S \times Z \to A$ *is a Caratheodory map. Then* $h: S \times Z \to \overline{\mathbb{R}}$ *given by* $h(s,z) := f(\varphi(s,z),z)$ *is a normal integrand, too.*
*e. Suppose that* $f: S \times A \to \overline{\mathbb{R}}$ *is lsc. Then it is a normal integrand, i.e.* $E_f(s) = \{(a,\mu) \in A \times \mathbb{R}: f(s,a) \leqq \mu\}$ *defines a closed-valued (Borel) measurable multifunction. In particular, each lsc function* $f: A \to \overline{\mathbb{R}}$ *is measurable.*

*f. Suppose that* $g\colon (S\times A)\times Z\to\overline{\mathbb{R}}$ *is a normal integrand and that* B *is a compact subset of* S. *Then* $\pi_B g$ *is a normal integrand on* $A\times Z$.

PROOF. For the statements (a), (b), (c), (d) we refer to [10: 2A,2M, 2K,1C and 2N]. Claim (e) holds since $A\times\mathbb{R}$ is $\sigma$-compact and for each compact set $C\subset A\times\mathbb{R}$ the set $E_f^{-1}(C)$ is closed, so measurable. To prove (f), remark first of all that Lemma 6.6c implies that $a\to\pi_B g(a,z)$ is lsc. In order to prove measurability, consider $E_g(z) = \{(s,a,\mu)\in S\times A\times\mathbb{R}\colon g(s,a,z)\leqq\mu\}$. Since g is a normal integrand $E_g^{-1}(H)\in Z$ for all closed $H\subset S\times A\times\mathbb{R}$. Now $E_{\pi_B g}(z) = \{(a,\mu)\in A\times\mathbb{R}\colon \pi_B g(a)\leqq\mu\} = \{(a,\mu)\colon \exists s\in B,\ g(s,a,z)\leqq\mu\}$, the latter inequality being true since the infimum is attained (Lemma 6.6b). For arbitrary $K\subset A\times\mathbb{R}$ we have $E_{\pi_B g}^{-1}(K) = \{z\in Z\colon \exists(a,\mu)\in K\cap E_{\pi_B g}(z)\} = \{z\in Z\colon \exists(s,a,\mu)\in B\times K\cap E_g(z)\} = E_g^{-1}(B\times K)$, and this set is measurable for all closed K. □

Lemma 6.8. *Let* $h\colon S\times A\to\overline{\mathbb{R}}$ *be lsc.*
*a.* h *satisfies the growth condition* G *if and only if* $\pi_B h\colon A\to\overline{\mathbb{R}}$ *is infcompact for all compact* $B\subset S$.
*b. If* h *satisfies* G *then* $\pi_A h\colon S\to\overline{\mathbb{R}}$ *is lsc.*

PROOF. (a) Remark first that in G "each bounded set $B\subset S$" may be replaced by "each compact set B", since each compact set is bounded, and each bounded set is contained in a compact set, and $\{a\in A\colon \pi_B h(a)\leqq\mu\}$ is increasing with B. Furthermore, B compact and h lsc imply $\pi_B h$ lsc (Lemma 6.6.c).
(b) Consider $\{(s_n,a_n)\}\subset S\times A$ with $\lim s_n=\bar{s}$ and $\lim h(s_n,a_n)=\beta<\infty$. Then with $B=\{s_n\colon n=1,2,\dots\}$ and $\alpha=\beta+1$ (if $\beta>-\infty$) or $\alpha=0$ (if $\beta=-\infty$) we have for sufficiently large n that $a_n\in\{a\in A\colon \exists s\in B,\ h(s,a)\leqq\alpha\}$. Since B is bounded and h satisfies G the sequence $\{a_n\}$ must be bounded. By passing to a subsequence we may assume that $\{a_n\}$ converges to $\bar{a}\in A$. Since h is lsc, $h(\bar{s},\bar{a})\leqq\lim h(s_n,a_n)=\beta$. We conclude that for each $s\in S$ and each sequence $\{(s_n,a_n)\}\subset S\times A$ such that $\{s_n\}$ converges to s it holds that $\pi_A h(s)\leqq\liminf h(s_n,a_n)$. We now prove that this condition is sufficient in order that $\pi_A h$ is lsc (actually, it is necessary too). Consider any sequence $\{s_n\}\to s$. Suppose first that a subsequence $\{s_{n_k}\}$ exists with $\pi_A h(s_{n_k})=-\infty$ for all k. Then $\{a_{n_k}\}\subset A$ exist

with $h(s_{n_k},a_{n_k}) \leqq -k$, which proves with the previous conclusion that $\pi_A h(s) = -\infty$ actually; at such an $s \in S$, $\pi_A h$ is lsc trivially. Suppose therefore without loss of generality that $\pi_A h(s_n) > -\infty$ for all n. Then $\{a_n\} \subset A$ exist with $\pi_A h(s_n) \leqq h(s_n,a_n) \leqq \pi_A h(s_n) + n^{-1}$, and the conclusion implies $\pi_A h(s) \leqq \liminf h(s_n,a_n) = \liminf \pi_A h(s_n)$. So $\pi_A h$ is lsc at s.□

LEMMA 6.9. *Suppose that* $\Phi$ *is a probability on* $(Z,\mathcal{Z})$ *and that* $g\colon (S\times A)\times Z \to \overline{\mathbb{R}}$ *satisfies the lower bound condition* LB *whereas* $(s,a) \to g(s,a,z)$ *is lower semicontinuous and* $z \to g(s,a,z)$ *is measurable.*
*a. Then the function* $h\colon S\times A \to \overline{\mathbb{R}}$ *given by*

$$h(s,a) := \int_Z g(s,a,z)\Phi(dz)$$

*is lower semicontinuous, and bounded from below on* $B\times A$ *for each bounded set* $B \subset S$.
*b. Suppose that* g *is a normal integrand on* $(S\times A)\times Z$*, actually. If* g *satisfies the growth condition* G *a.e.* $\Phi$*, then h satisfies* G.

PROOF. (a) Suppose $\{(s_n,a_n)\} \to (s,a) \in S\times A$, then

$$\begin{aligned}\liminf h(s_n,a_n) &= \liminf \int g(s_n,a_n,z)\Phi(dz)\\ &\geqq \int \liminf g(s_n,a_n,z)\Phi(dz)\\ &\geqq \int g(s,a,z)\Phi(dz) = h(s,a),\end{aligned}$$

the inequalities being true because of Fatou's lemma (applicable since $C := \{s_n : n=1,2,\dots\}$ is bounded and $\pi_{C\times A}g$ has an integrable minorant) and the lower semicontinuity of $(s,a) \to g(s,a,z)$. Therefore, h is lsc. Since g satisfies LB we have that $\pi_{B\times A}h > -\infty$ for all bounded $B \subset S$.
(b) Because of Lemma 6.8a we have to prove that for any compact $B \subset S$ the infcompactness of $a \to \pi_B g(a,z)$ a.e. $\Phi$ implies that $\pi_B h$ is infcompact. Fix a compact $B \subset S$. Remark that $g(s,a,z) \geqq \pi_B g(a,z)$ for all $(s,a,z) \in B\times A\times Z$. Both sides of the inequality sign are measurable for fixed (s,a) (Lemmas 6.7f and 6.7a). After integration, $h(s,a) \geqq \int \pi_B g(a,z)\Phi(dz)$ (both integrals exist $> -\infty$ since g satisfies LB), so for all $a \in A$

$$\pi_B h(a) \geqq \int \pi_B g(a,z)\Phi(dz).$$

We already proved that h is lsc, so with Lemma 6.6c it follows that $\pi_B h$ is lsc. With reference to Lemma 6.6a we conclude that $\pi_B h$ is infcompact if we show that $\int \pi_B g(a,z)\Phi(dz)$ is infcompact. First of all, since g is a normal integrand and B compact, $\pi_B g$ is a normal integrand, so $\int \pi_B g(a,z)\Phi(dz)$ is lsc (Lemma 6.9a with $\pi_B g$ instead of g). It is even infcompact: suppose on the contrary the existence of $\alpha \in \mathbb{R}$, $\{a_n\} \subset A$ such that $\lim_n \|a_n\| = \infty$ and $\int \pi_B g(a_n,z)\Phi(dz) \leqq \alpha$. Applying Fatou's lemma (possible since $\pi_{B\times A} g$ has an integrable minorant because g satisfies LB) we get the contradiction

$$\infty > \alpha \geqq \lim\inf \int \pi_B g(a_n,z)\Phi(dz) \geqq \int \lim\inf \pi_B g(a_n,z)\Phi(dz) = \int \infty d\Phi,$$

the equality being true since $a \to \pi_B g(.,z)$ has been given to be infcompact a.e. $\Phi$. □

PROOF OF THEOREM 6.2. Since g is a normal integrand satisfying LB we conclude from Lemma 6.9a that the function $h: S \times A \to \overline{\mathbb{R}}$ defined by $h(s,a) := \int_Z g(s,a,z)\Phi(dz)$ is lsc. Moreover $\pi_{B\times A} h > -\infty$ for all bounded sets $B \subset S$, so $\pi_B f > -\infty$. Furthermore, since $(s,a) \to g(s,a,z)$ is lsc and satisfies G a.e. $\Phi$ it follows from Lemma 6.8a that $a \to \pi_B g(a,z)$ is infcompact for all compact $B \subset S$ a.e. $\Phi$. Applying Lemma 6.9b we conclude that h satisfies G. With Lemma 6.8b it follows that $f = \pi_A h$ is lsc. Consider now the multifunction $D^*$. Since $B = \{s\}$ is compact, the function $a \to \pi_B h(a) = h(s,a)$ is infcompact for each s, so from Lemma 6.6b it follows that $D^*(s) = \arg\min_{a\in A} h(s,a)$ is a nonempty closed subset of A, even compact unless $h(s,a) = +\infty$ for all $a \in A$. The function h is lsc, so a fortiori a normal integrand (Lemma 6.7e). Therefore, $D^*$ is a measurable multifunction and admits a measurable selection (Lemma 6.7c). □

PROOF OF THEOREM 6.1. This is a simplified duplicate of the proof of Theorem 6.2 and will therefore not be given. Note that h satisfies G since A is compact, so the use of Lemma 6.9b can be circumvented. □

PROOF OF COROLLARY 6.5.

Define $g: S \times A \times Z \to \overline{\mathbb{R}}$ by $g(s,a,z) := c(s,a,z) + \bar{f}(F(s,a,z))$. The second term defines a normal integrand since F is a Caratheodory map and $\bar{f}$ is lsc (Lemma 6.7e and d), and therefore g is a normal integrand (Lemma 6.7b). Let $\bar{\gamma} := \pi_S \bar{f}$ then $\bar{\gamma} > -\infty$. Since for any subset $B \subset S$

$$\{a \in A \mid \pi_B g(a,z) \leqq \alpha\} \subset \{a \in A \mid \pi_B c(a,z) \leqq \alpha - \bar{\gamma}\},$$

g satisfies G a.e. $\Phi$. Let $\varphi$ be an integrable minorant for $\pi_{S\times A}c$, then so is $\varphi + \bar{\gamma}$ for $\pi_{S\times A}g$. So g satisfies LB, and Theorem 6.2 applies. Moreover, $\pi_S f \geqq \int \varphi d\Phi + \bar{\gamma} > -\infty$. □

## 6.4 MARKOVIAN INVENTORY CONTROL MODELS

Consider the following inventory control model. A shop deals with M items. During N successive time periods customers demand stochastic amounts $\xi_n \in \mathbb{R}^M$, $n = 1,2,\ldots,N$. At stage n (i.e. the beginning of period n) the stock at hand (state) $s_n \in \mathbb{R}^M$ is observed, and additional amounts are ordered by the management by choosing the production variables $a_n \in \mathbb{R}^P$. The costs involved are $c_n(s_n,a_n,\xi_n)$. At the next stage the resulting stocks are $s_{n+1} = F_n(s_n,a_n,\xi_n)$. After the N-th period there are costs $c_{N+1}(s_{N+1})$ caused by the final stock $s_{N+1}$. The optimization problem is: how to control the production variables at each stage in order to minimize the expected value of the total costs. Whereas at stage n the past demands $\xi_1,\ldots,\xi_{n-1}$ are known to the decision maker, the only available information on future demands is their probability distribution. We call this inventory control model Markovian, since the cost functions $c_n$ and the system functions $F_n$ depend on the past only via the current state $s_n$, and last but not least, since the demands in different periods are supposed to be stochastically independent. Remark that possible production constraints can be implemented by redefining $c_n(s_n,a_n,\xi_n) := +\infty$ for not allowed values of $(s_n,a_n)$.

The above model is a general formulation of most of the current inventory control models. For example, various variants of the classical Arrow, Karlin and Scarf-model [1] have $M = P = 1$, $F_n(s,a,\xi) = s+a-\xi$ in the case of backlogging of unsatisfied demand, or $F_n(s,a,\xi) = (s+a-\xi)^+$ in the case of lost sales, and the cost function is specified by $c_n(s,a,z) = \alpha^{n-1}.\{K.\psi(a) + c(a) + h((s+a-\xi)^+) + q((s+a-\xi)^-)\}$. Here $z^+ := \max(0,z)$, $z^- := \max(0,-z)$ for any $z \in \mathbb{R}$, $\alpha \in [0,1]$ is the discount factor, K is the set-up cost, $\psi(a) := \infty$ if $a < 0$, $\psi(0) := 0$,

$\psi(a) := 1$ if $a > 0$, c represents the variable production cost and h and q are the not necessarily linear holding and shortage costs.

We make the following assumptions. S is a nonempty closed Borel set of $\mathbb{R}^M$; A is a nonempty closed Borel set of $\mathbb{R}^P$, $\Phi_n$ is the probability distribution of $\xi_n$, with support $Z_n \subset \mathbb{R}^M$, the functions $F_n: S \times A \times Z_n \to S$, $c_n: S \times A \times Z_n \to (-\infty,\infty]$, $c_{N+1}: S \to (-\infty,\infty]$ are Borel measurable.

A sequence of functions $\delta := (\delta_n,\ n = 1,\dots,N)$ is called a *policy* if $\delta_n: S \times Z_1 \times \dots \times Z_{n-1} \to A$ is Borel measurable. So under policy $\delta$ the action $\delta_n(s,\xi_1,\dots,\xi_{n-1})$ is chosen at stage n if the initial state is s and the past demands appeared to be $\xi_1,\dots,\xi_{n-1}$. This definition reflects the condition that the controls must be nonanticipative. We write for short $\xi^{n-1} := (\xi_1,\dots,\xi_{n-1})$, so that $\delta_n = \delta_n(s,\xi^{n-1})$. The set of all policies is denoted by $\Delta$. The actual total costs resulting from initial state s and demands $\xi := (\xi_1,\dots,\xi_N)$ under $\delta$ are then given by

$$V^\delta(s,\xi) = \sum_{n=1}^{N} c_n(s_n,a_n,\xi_n) + c_{N+1}(s_{N+1})$$

with $s_1 = s$, and for $n = 1,\dots,N$: $a_n = \delta_n(s,\xi^{n-1})$ and $s_{n+1} = F_n(s_n,a_n,\xi_n)$. Clearly, $\xi \to V^\delta(s,\xi)$ is Borel measurable. The expected loss of $\delta \in \Delta$ and initial state $s \in S$ is

$$v^\delta(s) = \int V^\delta(s,\xi)\Phi(d\xi)$$

where $\Phi = \Phi_1 \otimes \Phi_2 \otimes \dots \otimes \Phi_N$ denotes the probability distribution of $\xi$. The optimization problem related with the Markovian inventory control model is then:

(MIC) find $\delta^* \in \Delta$ such that $v^{\delta^*}(s) = \min_{\delta\in\Delta} v^\delta(s)$ for all $s \in S$.

A policy $\delta^*$ which satisfies this condition is called *optimal*. The function $s \to v^*(s) := \inf_{\delta\in\Delta} v^\delta(s)$ is called the *value function*. Remark that MIC in fact determines a class of control problems parametrized by the initial state $s \in S$. Problem MIC is meaningless unless all integrals exist. Therefore, we assume that MIC is *well-defined*, i.e. for each $s \in S$ there exists a function $\xi \to p(s,\xi)$ which is Borel measurable and integrable with respect to $\Phi$, such that for all $\delta \in \Delta$, $V^\delta(s,\xi) \geqq p(s,\xi)$ a.e. $\Phi$. If MIC is well-defined, then the integral defining $V^{\delta}(s)$ can be calculated as iterative integral (Fubini).

In this section we give sufficient conditions in order that problem MIC is well-defined and can be solved by dynamic programming. The dynamic programming solution method calculates the cost-to-go functions $f_n: S \to \overline{\mathbb{R}}$ by means of the backward recursion

$$\text{(DPA)}\quad \begin{aligned} f_{N+1}(s) &:= c_{N+1}(s), && s \in S,\\ &\text{and for } n = N, N-1, \dots, 1: && \\ g_n(s,a,z) &:= c_n(s,a,z) + f_{n+1}(F_n(s,a,z)), && (s,a,z) \in S \times A \times Z_n,\\ f_n(s) &:= \inf_{a \in A} \int_{Z_n} g_n(s,a,z)\Phi_n(dz), && s \in S,\\ D_n^*(s) &:= \arg\min_{a \in A} \int_{Z_n} g_n(s,a,z)\Phi_n(dz), && s \in S. \end{aligned}$$

In general it is not sure that this method works. For example: if $f_{n+1} \not> -\infty$ then $g_n(s,a,z) = +\infty - \infty$ may occur; if $f_{n+1}$ is not Borel measurable then it is not sure that $g_n$ can be integrated. We say that the algorithm DPA *behaves properly* if for each $n = N, \dots, 1$ the functions $f_n$ are defined, $> -\infty$, and measurable, whereas measurable functions $a_n^*: S \to A$ exist such that $a_n^*(s) \in D_n^*(s)$, $\forall s \in S$. It is clear that the results in Section 6.2 are useful to guarantee proper behaviour.

We use the following notation. Let $\delta \in \Delta$, $s \in S$, then $s_n^{\delta}(s,\xi)$ denotes the state at stage n, incurred by initial state s, policy $\delta$ and demand realization $\xi$. So if $\delta = (\delta_n, n = 1, \dots, N)$ then $s_1^{\delta}(s,\xi) = s$, $s_{n+1}^{\delta}(s,\xi) = F_n(s_n^{\delta}(s,\xi), \delta_n(s,\xi^{n-1}), \xi_n)$, $n = 1, \dots, N$. Since by definition $\delta_n(s,\xi^{n-1})$ does not depend on $(\xi_n, \dots, \xi_N)$, also $s_n^{\delta}(s,\xi)$ does not depend on $(\xi_n, \dots, \xi_N)$. Therefore we may write $s_n^{\delta}(s,\xi^{n-1})$ for $s_n^{\delta}(s,\xi)$. Furthermore,

$$V^{\delta}(s,\xi) = \sum_{n=1}^{N} c_n(s_n^{\delta}(s,\xi^{n-1}), \delta_n(s,\xi^{n-1}), \xi_n) + c_{N+1}(s_{N+1}^{\delta}(s,\xi)).$$

A policy $\delta = (\delta_1, \dots, \delta_N)$ is called *Markovian* if for each n $\delta_n$ only depends on $s_n^{\delta}(s,\xi^{n-1})$, i.e. if $(s,\xi^{n-1})$ and $(\bar{s},\bar{\xi}^{n-1})$ are such that $s_n^{\delta}(s,\xi^{n-1}) = s_n^{\delta}(\bar{s},\bar{\xi}^{n-1})$ then $\delta_n(s,\xi^{n-1}) = \delta_n(\bar{s},\bar{\xi}^{n-1})$. The following result indicates the importance of proper behaviour of DPA for the MIC. It is of the same nature as Proposition 4.22.

THEOREM 6.10. *Suppose that the problem* MIC *is well-defined and that its backward recursion* DPA *behaves properly. Then the following statements hold.*

*a. There exists an optimal policy* $\delta^*$ *for* MIC *which is Markovian.*

*b. The value function* $V^*$ *of* MIC *equals* $f_1$.

*c. A policy* $\delta \in \Delta$ *is optimal for* MIC *if and only if for all* $s \in S$ *and all* $n = 1,2,\ldots,N$ *it holds that* $\delta_n(s,\xi^{n-1}) \in D_n^*(s_n^\delta(s,\xi^{n-1}))$, *a.s.*.

PROOF. Fix $s \in S$, $\delta \in \Delta$. Since MIC is well-defined we may calculate $V^\delta(s)$ as iterative integral. Since $c_n^\delta(s_n^\delta(s,\xi^{n-1}),\delta_n(s,\xi^{n-1}),\xi_n)$ does not depend on $(\xi_{n+1},\ldots,\xi_N)$ we may write

$$V^\delta(s) = \int_{Z_1}\{c_1 + \int_{Z_2}\{c_2 + \ldots \int_{Z_N}\{c_N + c_{N+1}\}d\Phi_N\} \ldots d\Phi_2\, d\Phi_1.$$

In the integral

$$\int_{Z_N}\{c_N(s_N^\delta(s,\xi^{N-1}),\delta_N(s,\xi^{N-1}),\xi_N) + c_{N+1}(F_N(s_N^\delta(s,\xi^{N-1}), \delta_N(s,\xi^{N-1}),\xi_N))\}\Phi_N(d\xi_N)$$

the vector $(s_N^\delta(s,\xi^{N-1}),\delta_N(s,\xi^{N-1}))$ does not depend on $\xi_N$, so this integral is at least $f_N(s_N^\delta(s,\xi^{N-1})) = f_N(F_{N-1}(s_{N-1}^\delta(s,\xi^{N-2}),\delta_{N-1}(s,\xi^{N-2}), \xi_{N-1})$ by definition of ($f_{N+1}$ and) $f_N$. By induction, $V^\delta(s) \geqq f_1(s)$ for all $s \in S$, all $\delta \in \Delta$. We will prove the claims of the theorem from this induction.

(a) and (b) We shall construct a Markovian policy $\delta^* \in \Delta$ for which holds: $V^{\delta^*}(s) = f_1(s)$ for all $s \in S$. Then clearly $\delta^*$ is optimal, and the value function of MIC is given by $f_1$. Fix for each n and each $s \in S$ an $a_n^*(s) \in D_n^*(s)$ such that $s \to a_n^*(s)$ is measurable. This is possible since DPA behaves properly. Then $\delta^* = (\delta_1^*,\ldots,\delta_N^*)$ is defined iteratively: $\delta_1^*(s) := a_1^*(s)$, and $\delta_n^*(s,\xi^{n-1}) := a_n^*(s_n^{\delta^*}(s,\xi^{n-1}))$ for $n = 1,\ldots,N$. Note that this is a correct definition since $s_n^{\delta^*}(s,\xi^{n-1})$ is determined by $(s,\xi^{n-1}$ and) $\delta_1^*,\ldots,\delta_{n-1}^*$. $\delta^*$ is a Markovian policy. Furthermore, $V^{\delta^*}(s) = f_1(s)$ for all $s \in S$. This can be easily verified in the above induction proof since for $\delta = \delta^*$ and each n actually the equality holds in

$$\int_{Z_n}\{c_n(s_n^\delta(s,\xi^{n-1}),\delta_n(s,\xi^{n-1}),\xi_n) + f_{n+1}(F_n(s_n^\delta(s,\xi^{n-1}), \delta_n(s,\xi^{n-1}),\xi_n))\}\Phi_n(d\xi_n) \geqq f_n(s_n^\delta(s,\xi^{n-1})).$$

(c) The "if" part follows as in part (a). The "only if" part follows by contradiction. If for any $\delta \in \Delta$, $\exists s \in S$, $\exists n \in \{2,\ldots,N\}$ such that the set $\{\xi\colon \delta_n(s,\xi^{n-1}) \notin D_n^*(s_n^\delta(s,\xi^{n-1}))\}$ has a strictly positive probability, then following the above induction proof one sees easily that $V^\delta(s) > f_1(s)$ so $\delta$ is not optimal. If $\delta_1(s) \notin D_1^*(s)$ this condition follows immediately. □

The interpretation of the functions $f_n$ is clear from the proof of Theorem 6.10. Define the optimization problem from stage n onward as follows

$$(\mathrm{MIC}_n) \qquad V_n^*(s) := \inf_{\delta\in\Delta_n} V_n^\delta(s) := \inf_{\delta\in\Delta_n} \int V_n^\delta(s,\xi)\ \Phi(d\xi)$$

where $\delta \in \Delta_n$ iff $\delta = (\delta_n,\delta_{n+1},\ldots,\delta_N)$ with: $\delta_i\colon S \times Z_n \times \ldots \times Z_{i-1} \to A$ measurable, and where $V_n^\delta(s,\xi)$ is the total cost in stages $n, n+1,\ldots,N$ incurred by having $s_n = s$ and using policy $\delta \in \Delta_n$. Actually, the proof of Theorem 6.10 does not hold only for MIC = $\mathrm{MIC}_1$, but for all $\mathrm{MIC}_n$. In particular, $f_n = V_n^*$, the value function of $\mathrm{MIC}_n$.

We now apply Theorem 6.2 to Markovian inventory control models.

<u>COROLLARY 6.11</u>. *Suppose that the problem* MIC *is well-defined, and suppose that* $c_{N+1}\colon S \to \overline{\mathbb{R}}$ *is lower semicontinuous,* $c_n\colon (S \times A) \times Z_n \to \overline{\mathbb{R}}$ *is a normal integrand for* $n = 1,\ldots,N$ *and* $F_n\colon (S \times A) \times Z_n \to S$ *is a Caratheodory map. Consider the backward recursion* DPA *of* MIC. *If for each* $n = N,N-1,\ldots,1$ *the function* $g_n$ *satisfies the growth condition* G *a.e.* $\Phi_n$ *and the lower bound condition* LB, *then* DPA *behaves properly and Theorem 6.10 applies.*

<u>PROOF</u>. Consider DPA. Since $c_{N+1}$ is lsc, $F_N$ is Caratheodory and $c_N$ is a normal integrand, $g_N$ is a normal integrand (Lemma 6.7). Since $g_N$ satisfies G a.e. $\Phi_N$ and LB Theorem 6.1 implies that $D_N^*$ is a nonempty-closed-valued measurable multifunction (having a measurable selection) and that $f_N$ is lsc (in particular: measurable) and $> -\infty$. Using the same argument inductively for $n = N-1,\ldots,1$ we conclude that DPA behaves properly. □

The only condition in Corollary 6.11 which does not fit in the dynamic programming scheme, is that MIC must be well-defined. In fact, this condition is closely related with the condition LB for all $g_n$. We shall tie these conditions together (and with the growth condition G) to get a result which seems to be a practical set of sufficient conditions to guarantee that an optimal (Markovian) policy exists that can be found by dynamic programming. Suppose that functions $h_n: S \times Z_n \times Z_{n+1} \times \ldots \times Z_N \to \overline{\mathbb{R}}$ exist, $n = 1,2,\ldots,N$, and $h_{N+1}: S \to \overline{\mathbb{R}}$ satisfying

$$\text{(DPA')} \qquad \begin{aligned} h_{N+1}(s) &\leqq c_{N+1}(s), \text{ and} \\ h_n(s,\xi_n,\ldots,\xi_N) &\leqq \inf_{a \in A}\{c_n(s,a,\xi_n) + h_{n+1}(F_n(s,a,\xi_n), \\ &\qquad \xi_{n+1},\ldots,\xi_N)\} \end{aligned}$$

for all $s \in S$, $\xi_n \in Z_n$ and $n = 1,2,\ldots,N$. If $\xi \to h_1(s,\xi)$ is integrable for each s, then clearly MIC is well-defined. On the other hand, if $\xi \to h_n(s,\xi_n,\ldots,\xi_N)$ is integrable for each n and s, it follows by induction that

$$\begin{aligned} g_n(s,a,z) &\geqq c_n(s,a,z) + \int h_{n+1}(F_n(s,a,z),\xi_{n+1},\ldots,\xi_N)\Phi_{n+1}(d\xi_{n+1}) \\ &\qquad \ldots \Phi_N(d\xi_N) \\ &\geqq \int h_n(s,z,\xi_{n+1},\ldots,\xi_N)\Phi_{n+1}(d\xi_{n+1}) \ldots \Phi_N(d\xi_N). \end{aligned}$$

From the second inequality we conclude that if for each bounded subset $B \subset S$ the function $\pi_B h_n: Z_n \times \ldots \times Z_N \to \overline{\mathbb{R}}$ has an integrable minorant (or itself is integrable) then $g_n$ satisfies LB. Furthermore, if for each $n = 1,\ldots,N$ and for each bounded subset $B \subset S$ the function $a \to \bar{h}_n^B(a,z)$, defined by

$$\begin{aligned} \bar{h}_n^B(a,z) := \pi_B c_n(a,z) + \inf_{s \in B} \int h_{n+1}(F_{n+1}(s,a,z),\xi_{n+1},\ldots,\xi_N) \\ \Phi_{n+1}(d\xi_{n+1})\ldots\Phi_N(d\xi_N) \end{aligned}$$

has bounded level sets for almost every z (with respect to $\Phi_n$), then it follows from the first inequality that $g_n$ satisfies G a.e. $\Phi_n$. So we proved our main result on MIC problems:

COROLLARY 6.12. *Suppose that a Markovian inventory problem* MIC *has been given, with* $c_{N+1}: S \to \overline{\mathbb{R}}$ *lower semicontinuous,* $c_n: (S \times A) \times Z_n \to \overline{\mathbb{R}}$ *a normal integrand and* $F_n: (S \times A) \times Z_n \to S$ *a Caratheodory map,* $n = 1,\ldots,N$. *Suppose that for each* $s \in S$ *measurable functions* $(\xi_n,\ldots,\xi_N) \to h_n(s,\xi_n,\ldots,\xi_N)$, $n = 1,\ldots,N$, *exist satisfying* DPA' *such that, for each bounded* $B \subset S$, $\pi_B h_n$ *has an integrable minorant and* $\bar{h}_n^B(.,z)$ *has bounded level sets, a.e.* $\Phi_n$. *Then* MIC *is well-defined and its backward recursion* DPA *behaves properly, so Theorem 6.10 applies.*

If $c_n \geqq 0$ all n and A compact then we may take $h_n = 0$ in Corollary 6.12. However, Corollary 6.12 covers also problems with $c_n$ not bounded from below nor with bounded level sets.

EXAMPLE 6.13. $S = \mathbb{R}$, $A = \mathbb{R}^+$, $Z_n = \mathbb{R}$, $\Phi_n$ an arbitrary distribution on $Z_n$ with finite first absolute moment $\xi_n$, $F_n(s,a,z) := s + a - z$, $c_n(s,a,z) := k_n a + p_n.(s+a-z)^+ + q_n.(s+a-z)^-$ and $c_{N+1}(s) := -k_{N+1}s$. So unsatisfied demands are *backlogged*, production costs, holding costs, shortage costs and salvage values are linear. We claim that this model satisfies the conditions of Corollary 6.12 if for each n: $k_n + p_n + \ldots + p_N > k_{N+1}$. Intuitively, it is clear that this is a reasonable condition: if $k_n + p_n + \ldots + p_N < k_{N+1}$ then producing unlimited amounts at stage n would be optimal (costs = $-\infty$). Note that the values of $q_n$ are irrelevant for the condition, caused by the fact that each production must be nonnegative.

To prove the claim, remark that $c_n$, $F_n$ are continuous, so only suitable functions $h_n$ have to be constructed. By replacing $p_N$ by $p_N - k_{N+1}$, $q_N$ by $q_N + k_{N+1}$, we may assume that $k_{N+1} = 0$. We prove that

$$h_{N+1} := 0$$

and

$$h_n(s,\xi_n,\ldots,\xi_N) := \sum_{j=n}^{N} p_j.(s - \xi_n - \xi_{n+1} \ldots - \xi_j) + - \sum_{j=n}^{N} (p_j + q_j)^-.(s - \xi_n - \xi_{n+1} \ldots - \xi_j)^-$$

satisfy all conditions of Corollary 6.12. DPA' is satisfied since for $a \geqq 0$

$$c_n(s,a,z) = k_n.a + p_n.(s+a-z) + (p_n + q_n).(s+a+z)^-$$
$$\geq (k_n + p_n).a + p_n.(s-z) - (p_n + q_n)^-.(s-z)^-$$

so

$$c_n(s,a,\xi_n) + h_{n+1}(F_n(s,a,\xi_n),\xi_{n+1},\ldots,\xi_N) \geq (k_n + \sum_{i=n}^{N} p_i).a +$$
$$+ \sum_{j=n}^{N} p_j.(s - \xi_n - \xi_{n+1} \ldots - \xi_j) - (p_n + q_n)^-.(s - \xi_n)^- +$$
$$- \sum_{j=n+1}^{N} (p_j + q_j)^-.(s + a - \xi_n - \xi_{n+1} \ldots - \xi_N)^-$$
$$\geq h_n(s,\xi_n,\ldots,\xi_N),$$

the last inequality being true since $k_n + p_n + \ldots + p_N > 0$, $a \geq 0$ and $\min_{a \geq 0} -(x+a)^- = -x^-$. It is easy to verify that $h_n$ satisfies the other conditions of Corollary 6.12: for compact $B \subset S$ the function $\pi_B h_n$ is integrable since $\int |\xi_n| \Phi_n(d\xi_n) < \infty$ for all n, and $\bar{h}_n^B(a_n,z) \geq (k_n + \sum_{i=n}^{N} p_i).a_n + \varphi_n^B(z)$ for some real function $\varphi_n^B$ not depending on $a_n$, so that $\{a_n \geq 0 : \bar{h}_n^B(a_n,z) \leq \alpha\}$ is bounded for all $z \in Z$, $\alpha \in \mathbb{R}$.

<u>EXAMPLE 6.14</u>. Suppose that the backlogging assumption in the previous example is omitted; that means that unsatisfied demand is lost and only nonnegative inventory has to be considered. The only changes are

$$S := \mathbb{R}^+, \quad F_n(s,a,z) := (s+a-z)^+.$$

Without loss of generality we may assume that $k_{N+1} = 0$; replace $p_{N+1}$ by $p_{N+1} - k_{N+1}$, if necessary. Just like in the backlogging case, it can be proved that the functions $h'_{N+1} := 0$ and

$$h'_n(s,\xi_n,\ldots,\xi_N) := \sum_{j=n}^{N} p_j.(s - \xi_n - \xi_{n+1} \ldots - \xi_j) +$$
$$- \sum_{j=n}^{N} (r_j)^-.((\ldots((s - \xi_n)^+ - \xi_{n+1})^+ \ldots -\xi_{j-1})^+ - \xi_j)^-$$

satisfy all conditions of Corollary 6.12, where $r_j := q_j + \sum_{i=j}^{N} p_i$.

REMARK 6.15. The Examples 6.13 and 6.14 are more general than they seem to be. For the assumptions of Corollary 6.12 are still satisfied if $c_n$ is a normal integrand with $c_n(s,a,z) \geqq k_n a + p_n.(s+a-z)^+ + q_n.(s+a-z)^- + \gamma_n$, where the only condition on the constants $k_n$, $p_n$, $q_n$ and $\gamma_n$ is that $k_n + p_n + \ldots + p_N > k_{N+1}$, $n = 1,\ldots,N$. These lower bound conditions on the functions $c_n$ are rather weak. For example, since A is not assumed to be compact it is allowed that $a \to \bar{c}_n(s,a) = \int c_n(s,a,z)\Phi_n(dz)$ is not bounded from below on A. However, possible unbounded low immediate costs in period n are more than nullified by unbounded high costs in future periods. So an important feature of our assumptions is that the lower bound conditions for the cost functions of different periods are tied together. Of course this type of conditions works because of the special form of the state transformation function F: $F(s,a,z) = s+a-z$ or $(s+a-z)^+$. Since these specifications of F and those of the lower bound of $c_n$ are typical for inventory control models, we formulated Corollary 6.12 specially for this type of problems.

Notice that the above conditions on $c_n$ imply that $(s,a) \to \bar{c}_n(s,a)$ is lsc (same proof as Lemma 6.9a). Possible set-up costs can be included since for $K > 0$ the function $a \to K.\psi(a)$, with $\psi(a) := \infty$ if $a < 0$, $\psi(0) := 0$, $\psi(a) := 1$ if $a > 0$, is lsc. It is clear that also other types of discontinuity can be handled within the framework of Corollary 6.12.

REMARK 6.16. In this section we gave conditions which guarantee that the backward recursion DPA generates an optimal (Markovian) policy for the MIC problem. Such policies are characterized by means of measurable selections $a_n^*(s) \in D_n^*(s)$. As is well-known, the DPA approach to the MIC problem provides more structural information on optimal policies if additional convexity and linearity assumptions are made. In particular one may have *level policies*, where

$$a_n^*(s) = (y_n^* - s)^+$$

for certain optimal target inventory levels, $y_n^*$, $n = 1,\ldots,N$, or, if set-up costs are included, *(s,S)-policies*, where

$$\begin{aligned} a_n^*(s) &= (S_n^* - s)^+ \text{ if } s < s_n^* \\ &= 0 \qquad\qquad \text{otherwise} \end{aligned}$$

for certain optimal target inventory levels $S_n^*$ and critical levels $s_n^*$, n = 1,...,N. For a discussion on the existence of optimal (s,S)-policies we refer to [14]. In [7,8] we give sufficient conditions for the existence of optimal level policies in MIC problems, not only for the backlogging case but also for the non-backlogging case. Similar results are derived in Chapter 8 for the *non-Markovian* inventory control model with backlogging and with linear cost functions.

REMARK 6.17. We now compare our results with recent developments in stochastic dynamic programming, as exemplified by the work of Schäl [12,13,14]. In [14] a finite horizon Markovian inventory control model is studied. The subject is not the existence of optimal Markovian policies, as it is in Corollary 6.12, but it is the existence of optimal Markovian policies of a certain type, namely *(s,S)-policies*. As might be expected, the conditions on $\bar{c}_n(s,a) := \int c_n(s,a,z)\Phi_n(dz)$ are more restrictive than ours (Remark 6.15): apart from the set-up cost only continuous costs are considered, and the production costs are linear; moreover, there is a set of generalized convexity conditions which do the job. In [12] a decision model over an infinite horizon is considered, which is Markovian but also stationary (except for the discount factor) and conditions are given in order that optimal or ε-optimal *stationary* plans exist. For that reason it is not of much use for our finite horizon MIC model. In [13] Schäl gives two sets of conditions, W and S, which guarantee the existence of an optimal policy in a general nonstationary dynamic programming problem. Corollary 6.12 is only formulated for a less general problem: finite instead of infinite horizon, finite-dimensional Euclidean spaces as state and action spaces instead of general standard Borel spaces. Furthermore, the transition probability measure $Q_n$ from $S \times A$ to S has the special form: $Q_n(s,a,B) = \Phi_n(\{z \in Z_n : F_n(s,a,z) \in B\})$. It is not surprising that the conditions of Corollary 6.12 cannot be derived from the general conditions W: A compact, Q w-continuous and $r_n \in \hat{C}(S \times A)$, or S: A compact, Q(s,.) s-continuous and $r_n \in \hat{Q}(S \times A)$, where $r_n(s,a) = -\int c_n(s,a,z)\Phi_n(dz)$. For under the assumptions of Corollary 6.12 A is not necessarily compact, and although under the conditions of Corollary 6.12 Q is w-continuous, in general it is not s-continuous. Furthermore, $r_n$ does not need to be an element of $\hat{C}(S \times A)$: it is upper semicontinuous, but generally not bounded from above. For the same fact it is not necessarily true that

$r_n \in \hat{Q}(S \times A)$. Remark 6.15 shows the existence of models to which Corollary 6.12 can be applied but not the results of [13].

REMARK 6.18. If S is countable the theorems reduce a lot. Then without any further constraints $f_n: S \to \overline{\mathbb{R}}$ is measurable and (if S has no cluster points, what we assume) even lsc. Likewise, each selection of a multi-function $D_n^*: S \twoheadrightarrow A$ is measurable. So DPA behaves properly iff all $f_n > -\infty$ and $D_n^*(s) \neq \emptyset$ for all n and s. Furthermore, G and LB can be formulated pointwisely in s. For example:

THEOREM 6.2'. *Suppose* $S = \{0,1,2,\ldots\}$. *Suppose that for each* $s \in S$, $(a,z) \to g(s,a,z)$ *is a normal integrand. If furthermore* $a \to g(s,a,z)$ *has bounded level sets a.e.* $\Phi$ *and* $z \to g(s,a,z)$ *has an integrable minorant uniformly in* $a \in A$, *both for each* $s \in S$, *then* (6.4) *holds.*

COROLLARY 6.12'. *Suppose that a Markovian inventory control model* MIC *has been given, with* $S = A = \{0,1,2,\ldots\}$, *and let for each* $(s,a) \in S \times A$ *and each* $n = 1,2,\ldots,N$ *the functions* $z \to c_n(s,a,z)$ *and* $z \to F_n(s,a,z)$ *be measurable.*
*Suppose that for each* $s \in S$ *measurable functions* $(\xi_n,\ldots,\xi_N) \to h_n(s,\xi_n,\ldots,\xi_N)$ *exist,* $n = 1,\ldots,N$ *satisfying* DPA', *with finite integrals* $\bar{h}_n(s)$, *such that* $a \to c(s,a,z) + \bar{h}_n(F_n(s,a,z))$ *has bounded level sets a.e.* $\Phi_n$. *Then* MIC *is well-defined and its backward recursion* DPA *behaves properly, so Theorem 6.10 applies.*

REFERENCES

1. K.J. ARROW, S. KARLIN, H. SCARF (1958). *Studies in the Mathematical Theory of Inventory and Production*, Stanford University Press, Stanford Ca.
2. D.P. BERTSEKAS, S.E. SHREVE (1978). *Stochastic Optimal Control; the Discrete Time Case*, Academic Press, New York.
3. D. BLACKWELL, D. FREEDMAN, M. ORKIN (1974). The optimal reward operator in dynamic programming. *Ann. Probab.* 2, 926-941.
4. I.V. EVSTIGNEEV (1976). Measurable selection and dynamic programming. *Math. Oper. Res.* 1, 267-272.
5. D.A. FREEDMAN (1974). The optimal reward operator in special cases of dynamic programming problems. *Ann. Prob.* 2, 942-949.

6. K. HINDERER (1970). *Foundations of Non-stationary Dynamic Programming with Discrete Time Parameter*. Lecture notes in operations research and mathematical systems 33, Springer, Berlin-Heidelberg-New York.

7. W.K. KLEIN HANEVELD (1977). *Markovian Inventory Control Models*, Report Institute of Econometrics OR 7702, University of Groningen.

8. W.K. KLEIN HANEVELD (1978). *Markovian Inventory Control Models; Improvements and Extensions*. Report Department of Mathematics, University of Kentucky, Lexington Ky.

9. P. OLSEN (1976). When is a multistage stochastic programming problem well-defined? *SIAM J. Control Optim. 14*, 518-527.

10. R.T. ROCKAFELLAR (1976). Integral functionals, normal integrands and measurable selection. J.P. GOSSEZ, E.J. LAMI DOZO, J. MAWHIN, L. WAELBROECK (eds.). *Nonlinear Operators and the Calculus of Variation*, Lecture notes in mathematics 543, Springer, Berlin-Heidelberg-New York, 157-207.

11. R.T. ROCKAFELLAR, R.J.-B. WETS (1976). Nonanticipativity and $L_1$-martingales in stochastic optimization problems. *Math. Programming Stud. 6*, 170-187.

12. M. SCHÄL (1972). On continuous dynamic programming with discrete time parameter. *Z. Wahrsch. Verw. Gebiete 21*, 279-288.

13. M. SCHÄL (1975). On dynamic programming: compactness of the space of policies. *Stochastic Process. Appl. 3*, 345-364.

14. M. SCHÄL (1976). On the optimality of (s,S)-policies in dynamic inventory models with finite horizon. *SIAM J. Appl. Math. 30*, 528-537.

15. S. SHREVE (1977). *Dynamic Programming in Complete Separable Spaces*. Ph.D. thesis, University of Illinois at Urbana-Champaign, Urbana Ill.

16. R.E. STRAUCH (1966). Negative dynamic programming. *Ann. Statist. 37*, 871-890.

17. R.J.-B. WETS (1976). *Grundlagen Konvexer Optimierung*. Lecture notes in economics and mathematical systems 137, Springer, Berlin-Heidelberg-New York.

CHAPTER 7

# ROBUSTNESS AGAINST DEPENDENCE IN PERT

## 7.1 INTRODUCTION

A typical *minimax* model of stochastic programming can be described as

$$\text{minimize}_{T \in \mathcal{T}} \sup_{H \in \mathcal{H}} E_H f(T,\xi).$$

Here T is a vector of decision variables, restricted to a fixed feasible set $\mathcal{T} \subset \mathbb{R}^N$. The function f to be minimized depends not only on T but also on the unknown realization of an n-dimensional random vector $\xi$. For that reason the expected value of f is chosen as the objective function. However, the distribution function H of $\xi$ is not completely known to the decision maker: it may be any element of a family $\mathcal{H}$ of distribution functions. The supremization in the "inner problem" indicates that a *worst-case* approach is adopted: the decision on T in the "outer problem" is based on the *most unfavorable distributions* in $H$.

In this chapter we deal with such a minimax model under the additional specification that $H$ describes the situation that the decision maker is uncertain with respect to the *dependence* between the components $\xi_i$, $i = 1,\ldots,n$, of $\xi$. That is, it is assumed that each $\xi_i$ has a known distribution function $F_i$ and that

$$H := H(F_1,\ldots,F_n)$$

where $H(F_1,\ldots,F_n)$ is the family of all joint distributions on $\mathbb{R}^n$ which have the prescribed marginals $F_1,\ldots,F_n$. Under this specification of $H$ we call any optimal solution of (the outer part of) the minimax problem *robust against dependence*; this term is borrowed from Rüschendorff [36] who uses it in a more traditional way.

With $H := H(F_1,\ldots,F_n)$ the inner problem of the minimax problem is precisely a *marginal problem* as described in Section 4.5, except that infimization is replaced by supremization. There we noticed already, that a marginal problem is a generalization of the well-known transportation problem. For that reason, one might call it a *distribution problem*, since this term covers the case of probability distributions as well as that of physical distributions. From Section 4.5 we know, that the marginal problem can be dualized. Its dual problem involves a one-sided approximation of the "criterion function" $f(T,.)$ by a separable function. By means of duality theory it is possible to characterize optimal (i.e. worst-case) joint distributions by means of a condition on their supports: the complementary slackness conditions indicate, that these distributions must be restricted to the subset of the range space of the random vector $\xi$ on which the approximation in the dual problem is exact.

The contents of the present chapter can now be summarized as follows. In Section 7.2 we deal again with the marginal problem and its dual. Special attention is paid to conditions under which the so-called NW Corner Rule Solution is optimal. The name refers to the well-known device to get a starting basic feasible solution of the transportation problem. In the remaining sections an extensive exposition on the *minimax approach to project planning* is given. A project is described as a set of activities together with a set of precedences between them which is represented by a network. Each activity needs time, and together with the precedence relations these times determine the minimum completion time of the project. In Section 7.3 we characterize critical activities; by definition, for such activities no delay can be tolerated without delaying the project completion. The results of this section are used in the final section, where it is assumed that the durations of the activities are random variables. At stake is the decision of making a promise on the project completion time. In practical situations only very partial information on the joint distribution is available. PERT is a heuristic for this problem that is often used, but that has a poor probabilistic basis. We show in Section 7.4 that the minimax model of stochastic programming defined above can be applied successfully. In this case $T \in \mathbb{R}$ is the promise on the project completion time, $\xi$ is the vector of random durations of the activities, and $f(T,\xi)$ is the delay of the project beyond T. First the inner problem is studied under the assumption, that $H = H(F_1,\ldots,F_n)$, that is, that it is a marginal problem. Due to the special form of its criterion function it is possible to reduce the dual of the marginal problem to a finite-dimensional convex minimization problem; its variables can be interpreted as promises with respect to the completion time of the activities separately. By means of the optimal dual solution the most unfavorable distributions are characterized completely. This is done in an appealing way by dualizing the reduced again, with a *flow problem* defined on the PERT network as a result. Optimal flows appear to be so-called criticality numbers of the most unfavorable distributions. Next the results on the inner problem are applied to the outer problem. After that, the assumption of complete knowledge of the marginal distributions is dropped. It is shown, that even if only partial information on the marginals is available, as is not unusual in project planning, the minimax approach can be used. Finally some numerical results are given, showing that the minimax approach to project planning is computationally feasible.

## 7.2 THE MARGINAL PROBLEM AND ITS DUAL

Let n be a natural number, $n \geqq 2$, let, for $i = 1,\ldots,n$, $X_i$ be a Borel subset of $\mathbb{R}$ and $\mu_i$ a probability measure on the Borel sets of $\mathbb{R}$ such that $\mu_i(X_i) = 1$. One might restrict $X_i$ to be the support of $\mu_i$, or its convex hull, but this is not necessary. Defining $X := \Pi_{i=1}^n X_i$, let

$M(\mu_1,\mu_2,\dots,\mu_n)$ be the family of all probability measures $\mu$ on the Borel sets of X which have the marginals $\mu_1,\dots,\mu_n$. Finally, let $f\colon X \to \mathbb{R}$ be any Borel measurable function. Then the *marginal problem* (or *distribution problem*) on X with criterion function f and prescribed marginals $\mu_i$, $i = 1,\dots,n$, is by definition

$$(P) \qquad \text{maximize}_{\mu\in M(\mu_1,\dots,\mu_n)} \int_X f\,d\mu.$$

Formulated in terms of n random variables $\xi_i$ and their distribution functions $F_i$, $F_i(x_i) := P(\xi_i < x_i) = \mu_i((-\infty,x_i))$, $x_i \in \mathbb{R}$, P can be rewritten as

$$(P') \qquad \text{maximize}_{H\in H(F_1,\dots,F_n)}\ E_H f(\xi_1,\dots,\xi_n)$$

where, as in Section 7.1, $H(F_1,\dots,F_n)$ denotes the family of all n-dimensional distribution functions H with marginals $F_1,\dots,F_n$. Moreover, by transforming the criterion function, it is possible to reformulate P' in such a way that the marginals are uniform distributions. Indeed, let $F_i^{-1}$ be an arbitrary version of the inverse function of $F_i$. That is, $F_i^{-1}$ is any real function on (0,1) satisfying

$$F_i^{-1}(t) = x_i \ \Rightarrow\ F_i(x_i) \leqq t \leqq F_i(x_i + 0).$$

Then it is easy to show, that the distribution function of $F_i^{-1}(\omega_i)$ is $F_i$ if $\omega_i$ is a random variable with uniform distribution $U = U(0,1)$ on (0,1). Consequently, the distribution of $(F_1^{-1}(\omega_1),\dots,F_n^{-1}(\omega_n))$ is feasible for P'. Moreover, *every* $H \in H(F_1,\dots,F_n)$ can be represented as the distribution of $(F_1^{-1}(\omega_1),\dots,F_n^{-1}(\omega_n))$ for a random vector $(\omega_1,\dots,\omega_n)$ with uniform marginals on (0,1) [42,35], so that P' is equivalent to

$$(P'') \qquad \text{maximize}_{G\in H(U,\dots,U)}\ E_G f(F_1^{-1}(\omega_1),\dots,F_n^{-1}(\omega_n)).$$

In fact, it is even possible to reformulate P" in terms of only one single U(0,1) random variable $\omega_0$:

$$(P''') \qquad \text{maximize}_{\varphi}\ E_{\omega_0} f(\varphi_1(\omega_0),\varphi_2(\omega_0),\dots,\varphi_n(\omega_0)),$$

where $\varphi_i$ is an arbitrary so-called *rearrangement* of $F_i^{-1}$ [35]. For our purposes the formulations P and P' are the most appropriate; we will switch the notations freely.

It is clear that the marginal problem P is trivial if the criterion function f is *separable* on X, that is, if $f(x) = \Sigma_{i=1}^n f_i(x_i)$ $\forall x = (x_1,\ldots,x_n) \in X$. Then $\int f d\mu$ depends only on $\mu$ via the marginals $\mu_i$ which are fixed. Therefore a separable function f does not discriminate between the feasible solutions and does not deserve the name criterion function. If f is not separable on X, then still separable functions are important in the analysis of problem P. For instance, if f' is another criterion function which satisfies

$$f' - f \text{ is separable and integrable}$$

then f' and f are equivalent criterion functions, in the sense that replacing f by f' in P has only the effect of adding a finite constant to the objective function. A simple condition on f which guarantees that P is well-defined and that its optimal value sup P is finite employs integrable separable functions:

$$|f(x)| \leqq \Sigma_{i=1}^n e_i(x_i), \ \forall x = (x_1,\ldots,x_n) \in X, \tag{7.1}$$

where $e_i\colon X_i \to [0,\infty)$ is a Borel measurable function with $\int_{X_i} e_i d\mu_i < \infty$ $\forall i$.

Notice that the previous remark shows, that under (7.1) one may assume without loss of generality that $f \geqq 0$, or that $f \leqq 0$: just replace f by $f + \Sigma e_i$ or by $f - \Sigma e_i$, respectively. Since each $\mu_i$ is a finite measure, one may also without loss of generality assume that the "bounding functions" $e_i$ in (7.1) satisfy $e_i(x_i) \geqq 1$ $\forall x_i \in X_i$; as a consequence, (7.1) is equivalent to the boundedness conditions (4.34) and (4.35) that have been employed in Section 4.5 on the linear programming approach to the marginal problem.

As has been shown in Section 4.5 the natural dual problem of the marginal problem P involves an approximation of the criterion function f by integrable separable functions. The dual problem can be described as follows. Let $Y_i$ denote the linear space of real Borel functions on $X_i$ which are integrable with respect to $\mu_i$, and let Y be the linear space of all *separable* real functions on X determined by $Y_i$, $i = 1, \ldots, n$:

$$Y := \{y \in \mathbb{R}^X\colon\ y(x_1,\ldots,x_n) = \Sigma_{i=1}^n y_i(x_i)\ \forall x \in X,\ y_i \in Y_i\ \forall i\}.$$

Then the dual problem is

$$\text{(D)} \qquad \text{minimize}_{y\in Y}\left\{\Sigma_{i=1}^n \int_{X_i} y_i\, d\mu_i : \begin{array}{l}\Sigma_{i=1}^n y_i(x_i) \geq f(x_1,\ldots,x_n) \\ \text{for all } (x_1,\ldots,x_n) \in X\end{array}\right\}.$$

The following strong duality theorem holds.

THEOREM 7.1. Duality for P and D.
*Suppose that the criterion function* f *is upper semicontinuous, and that* $|f|$ *is bounded by an integrable separable function. Then the following statements are true.*

*a. Both* P *and* D *have optimal solutions, and they have the same finite optimal value:*

$$\text{(7.2)} \qquad -\infty < \max P = \min D < +\infty.$$

*b. The probability measure* $\mu^*$ *on* X *is optimal for* P *and the separable function* $y^* \in Y$ *is optimal for* D *if and only if the following conditions hold.*

*(i) Primal feasibility* : $\mu^*$ *has the required marginals* $\mu_i$, $i = 1, \ldots, n$.

*(ii) Dual feasibility* : $y^*$ *is an upper bound for* f *on* X.

*(iii) Complementary slackness:* $\mu^*(X^*) = 1$, *where*

$$\text{(7.3)} \qquad X^* := \{x \in X\colon\ \Sigma_{i=1}^n y_i^*(x_i) = f(x_1,\ldots,x_n)\}.$$

PROOF. (a) As is seen from (7.1), $|\sup P| < \infty$. The other statements follow from Kellerer's results [19,20] summarized at the end of Section 4.5.

(b) This is a direct consequence of the fact that sup P = inf D, see Theorem 4.2. □

For the sake of completeness we call the attention to the slight differences between the dual pair (P,D) of the present section and the corresponding pair $(MA_1, MA_2)$ discussed in Section 4.5. First of all, $MA_1$ is a minimization problem whereas its counterpart P is a maximization problem. As a consequence, $MA_2$ involves a lower approximation of the criterion function whereas in D this function is approximated from above. There is a second, less trivial, distinction. Following Kellerer, in D we defined $Y_i$ to be the space of $\mu_i$-integrable functions, whereas in $MA_2$ only the subspace of $Y_i$ is considered consisting of all functions $y_i$ that satisfy $\sup_{x_i \in X_i} |y_i(x_i)|/e_i(x_i) < \infty$ for $e_i$ the bounding function used in (7.1).

For $n := 2$, $X_1 := \{s_1,\dots,s_a\}$, $X_2 := \{t_1,\dots,t_b\}$, $p_i := \mu_1(\{s_i\})$, $q_j := \mu_2(\{t_j\})$, $x_{ij} := \mu(\{s_i,t_j\})$, $c_{ij} := f(s_i,t_j)$, the marginal problem P reduces to the well-known *transportation problem*

$$(P_o) \qquad \text{maximize}_{x \in \mathbb{R}^{ab}} \left\{ \Sigma_{i=1}^{a} \Sigma_{j=1}^{b} c_{ij} x_{ij} : \begin{array}{l} \Sigma_{j=1}^{b} x_{ij} = p_i,\ i = 1,\dots,a, \\ \qquad\qquad x_{ij} \geqq 0\ \forall i,j \\ \Sigma_{i=1}^{a} x_{ij} = q_j,\ j = 1,\dots,b, \end{array} \right\}$$

and the dual D becomes then

$$(D_o) \qquad \text{minimize}_{y_1 \in \mathbb{R}^a, y_2 \in \mathbb{R}^b} \{\Sigma_{i=1}^{a} p_i y_{1i} + \Sigma_{j=1}^{n} q_j y_{2j} : y_{1i} + y_{2j} \geqq c_{ij}\ \forall i,j\}.$$

In different degrees of generality, the dual pair (P,D) has been studied extensively in probability theory. For example, if f is the indicator function of $\{x \in \mathbb{R}^2 : |x_1 - x_2| \leqq \varepsilon\}$ problem P is a major tool in proving a famous theorem of Strassen: the minimum Ky Fan distance between two random variables with given distributions equals the Prokhorov distance of their distributions [39,37,12,15]; related results are contained in [19,20,40,41]. Closely related is the problem of the existence of probability measures under conditions including prescribed one-dimensional marginals [16,17,18,39,11]. For the theory of variance reduction P gives an appropriate framework with e.g. $f(\xi_1,\dots,\xi_n) := -(\Sigma_{i=1}^{n}(\xi_i - E(\xi_i)))^2$. Several inequalities can be derived using P and D [10,33,34,32,1,42,23]. In the literature mentioned often a statement related to P is derived by solving D explicitly. In Section 7.4 we will follow the same line in deriving results for the minimax approach applied to a stochastic programming model for a project planning problem.

We conclude this section with some remarks on two special feasible solutions for P. The first one is the joint distribution specified by *independence*: $\mu^{IND}(\Pi_{i=1}^{n} B_i) := \Pi_{i=1}^{n} \mu_i(B_i)$. It is interesting to realize, that the independent solution can only be optimal for P in trivial cases. Indeed, Theorem 7.1 shows, that $\mu^{IND}$ is optimal for P iff Borel sets $S_i \subset X_i$, $i = 1,\ldots,n$, exist such that $\mu_i(S_i) = 1$ $\forall i$ and f is separable on $\Pi_{i=1}^{n} S_i$. In such a case every feasible solution for P is in fact optimal. If the criterion function f is not separable one may expect that the optimal solutions of P show strong dependence between the marginals.

One other special feasible solution for P deserves attention. Let $\mu^o$ be the probability measure on X defined by the distribution function

$$(7.4) \qquad H^o(z_1,\ldots,z_n) := \min_{i=1,\ldots,n} F_i(z_i), \; z = (z_1,\ldots,z_n) \in \mathbb{R}^n.$$

It is easy to verify that $H^o$ indeed is a distribution function with marginals $F_i$. We call $\mu^o$ the *NW (Corner Rule) Solution* or the *NW (Corner Rule) Distribution* of P, since in the case of the transportation problem $P_o$ it reduces to the well-known basic feasible solution determined by the North West Corner Rule. In terms of P" the NW Distribution corresponds to the uniform distribution on the main diagonal in the unit cube, that is $\omega_1 = \omega_2 = \ldots = \omega_n$. In other words, $H^o$ is the distribution function of the random vector

$$(7.5) \qquad (F_1^{-1}(\omega), F_2^{-1}(\omega), \ldots, F_n^{-1}(\omega)), \; \omega \text{ is } U(0,1).$$

In P''' the NW Distribution is described by $\varphi_i = F_i^{-1}$ $\forall i$: no real rearrangement is made.

Since $\mu^o$ depends only on the marginals but not on the criterion function f, it is not to be expected to be optimal for P in general. However, it is possible to give conditions on f which guarantee that $\mu^o$ is optimal for all specifications of $\mu_i$, $i = 1,\ldots,n$. The function f is called *L-super-additive* [25] if for all $x \in \mathbb{R}^n$, all unit vectors $e_i$ and $e_j$, $i \neq j$, and all nonnegative scalars $\alpha$ and $\beta$

$$f(x + \alpha e_i + \beta e_j) + f(x) \geq f(x + \alpha e_i) + f(x + \beta e_j).$$

<u>THEOREM 7.2</u>. *If* $\mu_1,\ldots,\mu_n$ *have first moments and* f *is L-super-additive then the NW Distribution* $\mu^o$ *is optimal for* P.

<u>PROOF</u>. See [35]. □

The well-known result that the NW Distribution maximizes the correlation coefficient of two random variables can be formulated as a consequence of this theorem: take $f(x_1,x_2) = x_1.x_2$. For the transportation problem $P_o$ the NW Solution is, as can easily verified,

$$(7.6) \qquad x^o_{kl} = (\min(\Sigma^k_{i=1}p_i,\Sigma^l_{j=1}q_j) - \max(\Sigma^{k-1}_{i=1}p_i,\Sigma^{l-1}_{j=1}q_j))^+,$$

where $z^+ := \max(0,z)$.

<u>COROLLARY 7.3</u>. *The NW Solution (7.6) is optimal for* $P_o$ *for all values of* $p_i \geq 0$, $q_j \geq 0$ *with* $\Sigma^a_{i=1}p_i = \Sigma^b_{j=1}q_j$ (=1) *if and only if for each* $k = 1,\ldots,a-1$ *and for each* $l = 1,\ldots,b-1$

$$(7.7) \qquad c_{kl} + c_{k+1,l+1} \geq c_{k+1,l} + c_{k,l+1}.$$

<u>PROOF</u>. Sufficiency follows from Theorem 7.2 (a direct proof based on Theorem 7.1 is also not difficult). Necessity can be proved by contradiction: if (7.7) does not hold for $(k,l) = (s,t)$ then (7.6) is not optimal if $p_s = p_{s+1} = \frac{1}{2}$ and $q_t = q_{t+1} = \frac{1}{2}$. □

As a typical application of Corollary 7.3 let us interpret $P_o$ as a *dynamic production* model. Then $p_i$ is the production capacity in period i, $q_j$ is the demand in period j, and $x_{ij}$ is the production in period i destinated for demand in period j, with corresponding unit cost $c_{ij}$. Both indices i and j are ordered in a natural way, namely increasing with time. The NW Solution can then be interpreted as a FIFO type of strategy: the oldest demands have highest priority. According to Corollary 7.3 it gives the minimum cost solution for all values of the capacity parameters $p_i$ and demand parameters $q_j$ (provided the problem is balanced: $\Sigma_i p_i = \Sigma_j q_j$) if for all (k,l) the cost vector c satisfies

$$c_{kl} + c_{k+1,l+1} \leqq c_{k+1,l} + c_{k,l+1}.$$

This is true for example if $c_{ij} = d + h(i-j)$, h convex. On the other hand, under the same conditions for c the total cost is maximized under the LIFO strategy. The corresponding solution can be called the NE Solution since high values of j are coupled with low values of i and vice versa. By the way, generalizations of the NE Solution to general n are not as obvious as in the NW case [33].

From (7.5) useful formulas for $\mu^o$ and $H^o$ can be derived. For example, let $-\infty \leqq a_i < b_i \leqq \infty$, $-\infty \leqq z_i \leqq \infty$, $I, I_j \subset \{1,\ldots,n\}$ then the probability of each rectangle under $H^o$ can be expressed in terms of the marginal distribution functions:

$$P_{H^o}(a_i \leqq \xi_i < b_i,\ i \in I) = (\min_I F_i(b_i) - \max_I F_i(a_i))^+,$$

$$P_{H^o}(a_i < \xi_i \leqq b_i,\ i \in I) = (\min_I F_i(b_i + 0) - \max_I F_i(a_i + 0))^+,$$

$$(7.8) \qquad P_{H^o}(\xi_i < z_i, i \in I_1;\ \xi_i \leqq z_i, i \in I_2;\ \xi_i \geqq z_i, i \in I_3;\ \xi_i > z_i, i \in I_4) =$$

$$= (\min\{\min_{I_1} F_i(z_i), \min_{I_2} F_i(z_i + 0)\} +$$

$$- \max\{\max_{I_3} F_i(z_i), \max_{I_4} F_i(z_i + 0)\})^+.$$

Hence, for any $z \in \mathbb{R}^n$ and any $h,k \in \{1,\ldots,n\}$ we have

$$P_{H^o}(\xi_h < z_h, \xi_k \geqq z_k) \cdot P_{H^o}(\xi_h \geqq z_h, \xi_k < z_k) = 0,$$

$$(7.9) \qquad P_{H^o}(\xi_h < z_h, \xi_k > z_k) \cdot P_{H^o}(\xi_h \geqq z_h, \xi_k \leqq z_k) = 0,$$

$$P_{H^o}(\xi_h \leqq z_h, \xi_k > z_k) \cdot P_{H^o}(\xi_h > z_h, \xi_k \leqq z_k) = 0.$$

This is a characteristic property of the NW Distribution. It holds also for arbitrary disjoint index sets $I_1$ and $I_2$ instead of $\{h\}$ and $\{k\}$. Hence, $H^o$ has positive probability mass in at most one of each pair of opposite translated orthants in $\mathbb{R}^n$, except if we are concerned with the positive and negative orthants. For that case we have a characteristic maximality property:

THEOREM 7.4. *For arbitrary* $z \in \mathbb{R}^n$ *and* $I_1, I_2 \subset \{1,\ldots,n\}$, *define* $A_1$ *and* $A_2$ *by*

$$A_1 := \{x \in \mathbb{R}^n\colon\ x_i < z_i \text{ for } i \in I_1 \text{ and } x_i \leq z_i \text{ for } i \in I_2\},$$

$$A_2 := \{x \in \mathbb{R}^n\colon\ x_i > z_i \text{ for } i \in I_1 \text{ and } x_i \geq z_i \text{ for } i \in I_2\}.$$

*Then, for arbitrary marginal distributions* $\mu_i$,

$$\max_{\mu \in M(\mu_1,\ldots,\mu_n)} \mu(A_k) = \mu^o(A_k),\ k = 1,2.$$

*Consequently, the NW Distribution maximizes the probability of all translated positive or negative orthants.*

PROOF. By directly using (7.8), or from Theorem 7.2 by showing that the indicator function of $A_k$ is L-super-additive. □

If we call a distribution H with marginals $F_1,\ldots,F_n$ *positively quadrant-dependent* if $H(z_1,\ldots,z_n) \geq \Pi_{i=1}^n F_i(z_i)$ for all $z \in \mathbb{R}^n$ [24], then Theorem 7.4 characterizes the NW Distribution as the *maximally* positively quadrant-dependent distribution compatible with the fixed marginals.

Finally, we remark that from Theorem 7.2 it follows that the NW Distribution is optimal for P if f(x) of f(-x) is the distribution function of a finite measure on the Borel sets of $\mathbb{R}^n$. The same result is contained in [32]. In [22] we gave a proof based on Theorem 7.4.

## 7.3 CRITICAL ACTIVITIES IN A PROJECT NETWORK

In this section we deal with some aspects of deterministic project planning. The definitions to be given and the properties to be discussed serve primarily as a preparation for the next section in which we consider stochastic versions of the project planning model.

As usual, a project is represented by an acyclic connected directed graph with one source and one sink. The arcs of the graph correspond to the activities which have to be completed in order that the project is finished, and the graph describes the precedences between the activities. Let n be the number of activities, and p the number of paths from source to sink; the activities will be indicated by i, $i = 1,\dots,n$, and the paths by j, $j = 1,\dots,p$. Let $A = (a_{ji})$ be the $p \times n$ incidence matrix of the graph, i.e.

$$a_{ji} = 1 \text{ if activity } i \text{ is contained in path } j,$$
$$a_{ji} = 0 \text{ if not.}$$

The index set of all activities of path j is denoted by $B_j$,

$$B_j := \{i \in \{1,\dots,n\}: a_{ji} = 1\},\ j = 1,\dots,p.$$

Obviously, $B_1,\dots,B_p$ are different nonempty sets adding up to $\{1,\dots,n\}$. The index set of all paths through activity i is denoted by

$$D_i := \{j \in \{1,\dots,p\} : a_{ji} = 1\},\ i = 1,\dots,n.$$

Each $D_i$ is nonempty and $\cup_i D_i = \{1,\dots,p\}$. For $x = (x_1,\dots,x_n) \in \mathbb{R}^n$ the (minimum) *project completion time* determined by x is given by

$$\text{(7.10)} \qquad R(x) := \max_{j=1,\dots,p} (Ax)_j = \max_{j=1,\dots,p} \Sigma_{i \in B_j} x_i,\ x \in \mathbb{R}^n.$$

Here $x_i$ is to be interpreted as the duration of activity i, $i = 1,\dots,n$. If the maximum is attained by j, path j is called a *critical path* with respect to x. We denote the set of all critical paths with respect to x by

$$\text{(7.11)} \qquad CP(x) := \arg\max_{j=1,\dots,p} (Ax)_j,\ x \in \mathbb{R}^n.$$

Then the set of all *critical activities* with respect to x is

$$\text{(7.12)} \qquad CA(x) := \cup\{B_j : j \in CP(x)\},\ x \in \mathbb{R}^n.$$

By definition, each activity on a critical path is critical. The converse is also true, in the sense that

$$\text{(7.13)} \qquad j \in CP(x) \Leftrightarrow i \in CA(x) \text{ for all } i \in B_j.$$

This is due to the fact, that if two critical paths j and k contain the same activity i one gets another critical path by replacing in j all the activities before (or after) i by those of k. The term "critical" is explained by the following characterization, which is a direct consequence of the definitions. Let for $x \in \mathbb{R}^n$ and $i = 1,\dots,n$ the set $Z_i(x)$ be defined by

$$Z_i(x) := \{z \in \mathbb{R}^n:\ z_i > x_i,\ z_k = x_k \text{ for each } k \neq i\}$$

then we have

$$\text{(7.14)}\qquad \begin{aligned} i \in CA(x) &\Leftrightarrow R(z) > R(x)\ \forall z \in Z_i(x), \\ j \in CP(x) &\Leftrightarrow R(z) > R(x)\ \forall z \in \cup_{i \in B_j} Z_i(x). \end{aligned}$$

That is, a delay of any critical activity implies a delay of the project completion.

An alternative characterization of criticality can be given in terms of *flows* in the project network. By definition, a flow in the network is any vector $\pi = (\pi_1,\dots,\pi_n) \in \mathbb{R}^n$ (whose i-th element $\pi_i$ is interpreted as the flow in arc i) which satisfies the following conditions:

(i) $\pi_i \geqq 0\ \forall i$, i.e. the flow in any arc is in the direction of that arc;

(ii) the flow is conserved at each node except for the source and the sink.

Of course, for any flow in the network the total output of the source is equal to the total input of the sink; this number is called the *total flow*. Any flow in the network can be decomposed into flows in the paths from source to sink. Therefore, a representation of all flows in the network with a total flow $\alpha \geqq 0$ is given by

$$\text{(7.15)}\qquad \Pi(\alpha) := \{\pi \in \mathbb{R}^n:\ \exists \lambda \in \Lambda(\alpha),\ \pi = \lambda A\}$$

where

$$\text{(7.16)}\qquad \Lambda(\alpha) := \{\lambda \in \mathbb{R}^p: \lambda \geqq 0,\ \Sigma_{j=1}^{p} \lambda_j = \alpha\}.$$

Here $\pi$ and $\lambda$ are row vectors, and $\lambda_j$ is interpreted as the flow in path

j. Obviously,

(7.17) $\max_{\pi\in\Pi(1)} \pi x = \max_{\lambda\in\Lambda(1)} \lambda Ax = R(x),\ x \in \mathbb{R}^n.$

Hence, if $x_i$ is interpreted as the "unit reward" of the flow in arc i, then R(x) is the maximum reward to be gained when selecting any flow in the network with total flow equal to 1. We call $\pi^o$ and $\lambda^o$ optimal flows if in (7.17) the maxima are attained at $\pi = \pi^o$ and $\lambda = \lambda^o$, respectively; and we conclude from (7.17)

$$(7.18)\quad \begin{aligned} j \in CP(x) &\Leftrightarrow \exists \text{ optimal flow } \lambda^o \in \Lambda(1) \text{ with } \lambda^o_j > 0, \\ i \in CA(x) &\Leftrightarrow \exists \text{ optimal flow } \pi^o \in \Pi(1) \text{ with } \pi^o_i > 0. \end{aligned}$$

So we have two characterizations of criticality: the first one (7.14) can be interpreted in terms of consequential delays of the activities, whereas the second one (7.18) is expressed in terms of optimal flows in the network. It is interesting to notice, that (7.14) and (7.18) are related to each other by *duality*. In order to see this, notice that R(x) is the common optimal value of the following dual pair of linear programs

$$(LP_1)\quad \text{minimize}_{\substack{y\in\mathbb{R} \\ z\in\mathbb{R}^n}} \left\{(1\ 0)\binom{y}{z} : \begin{pmatrix} e & -A \\ 0 & I \end{pmatrix}\binom{y}{z} \geqq \binom{0}{x}\right\},$$

$$(LP_2)\quad \text{maximize}_{\substack{\lambda\in\mathbb{R}^p \\ \pi\in\mathbb{R}^n}} \left\{(\lambda\ \pi)\binom{0}{x} : \begin{array}{c} (\lambda\ \pi)\begin{pmatrix} e & -A \\ 0 & I \end{pmatrix} = (1\ 0) \\ \lambda \geqq 0,\ \pi \geqq 0 \end{array}\right\},$$

where $e := (1,1,\ldots,1)^T \in \mathbb{R}^p$ and I is the $n \times n$ unity matrix. Indeed, $LP_2$ is precisely the flow problem (7.17). By eliminating y, one easily derives that its dual $LP_1$ is equivalent to

$$(7.19)\quad \text{minimize}_{z\in\mathbb{R}^n} \{R(z) : z \geqq x\}.$$

Since R is a nondecreasing function, it follows that $(y^o, z^o)$ is optimal for $LP_1$ iff

$$y^o = R(x),\ z^o \text{ is optimal for } (7.19).$$

One optimal solution is just $z^o = x$. Other optimal solutions are more interesting, however, since they give information with respect to the possible slacks for the various activities. For example, the first statement of (7.14) can be rewritten as

$$i \in CA(x) \Leftrightarrow z_i^o = x_i \text{ for } all \text{ optimal solutions } z^o.$$

Comparison of this result with (7.18) reveals that (7.14) and (7.18) are dual characterizations of criticality: just recall from linear programming duality (see e.g. Chapter 2) that $z^o$ is optimal for $LP_1$ and for (7.19) and that $(\lambda^o,\pi^o)$ are optimal for $LP_2$ and (7.17) if, together with $y^o = R(x)$, $\lambda^o = \pi^o A$, the following conditions hold: for $j = 1,\ldots,p$ and $i = 1,\ldots,n$, respectively

$$\begin{aligned} &\lambda_j^o \geqq 0,\ R(x) \geqq (Az^o)_j,\ \lambda_j^o(R(x) - (Az^o)_j) = 0,\\ &\pi_i^o \geqq 0,\quad z_i^o \geqq x_i,\quad \pi_i^o(z_i^o - x_i) = 0. \end{aligned}$$

We conclude this section with the description of some properties of the project completion time function $R(x)$. These properties will appear to be useful in the next section. Corresponding to each path in the network we introduce an orthant in $\mathbb{R}^n$:

$$(7.20)\qquad K_j := \left\{x \in \mathbb{R}^n:\ \begin{matrix} x_i \geqq 0 \text{ if } i \in B_j \\ x_i \leqq 0 \text{ if } i \notin B_j \end{matrix}\right\},\ j = 1,\ldots,p.$$

That is, $x-z \in K_j$ iff $x_i \geqq z_i$ for all $i \in B_j$ and $x_i \leqq z_i$ for all $i \notin B_j$.

<u>LEMMA 7.5</u>. *For all* $x,z \in \mathbb{R}^n$

$$(7.21)\qquad R(x) \leqq R(z) + \Sigma_{i=1}^n (x_i - z_i)^+.$$

*Equality holds iff* $x-z \in \cup_{j\in CP(z)} K_j$. *If* $x-z \in K_j$ *and* $j \in CP(z)$ *then* $j \in CP(x)$ *too.*

<u>PROOF</u>. The inequality (7.21) follows from

$$(7.22)\qquad \Sigma_{i\in B_j} x_i \leqq \Sigma_{i\in B_j} z_i + \Sigma_{i=1}^n (x_i - z_i)^+ \quad \forall x\ \forall z\ \forall j.$$

In (7.22) equality holds iff $x-z \in K_j$. Therefore, $j \in CP(z)$ and $x-z \in K_j$ imply equality in (7.21), together with $j \in CP(x)$. On the other hand, if equality holds in (7.21) then for any $j \in CP(x)$

$$(7.23)\qquad \Sigma_{i\in B_j} x_i = R(x) = R(z) + \Sigma_{i=1}^n (x_i - z_i)^+ \geqq \Sigma_{i\in B_j} z_i + \Sigma_{i=1}^n (x_i - z_i)^+$$

so that equality must be true in (7.22) and (7.23). Hence $x-z \in K_j$ and $j \in CP(z)$. □

If it is important that the project is finished within time T, the function $(R(x) - T)^+$ is relevant; it represents the possible *delay* of the project beyond time T, as a function of the durations x of the activities. For the delay function we have a characterization which is similar to that of R in Lemma 7.5. It uses, in addition to (7.20), the following definitions

$$(7.24)\qquad K_0 := \{x \in \mathbb{R}^n:\ x \leqq 0\}$$

and

$$(7.25)\qquad \begin{array}{lll} CP_T(z) := \{0\} & \text{if } R(z) < T, \\ CP_T(z) := \{0\} \cup CP(z) & \text{if } R(z) = T, \\ CP_T(z) := CP(z) & \text{if } R(z) > T. \end{array}$$

<u>LEMMA 7.6</u>. *For all* $x,z \in \mathbb{R}^n$ *and each* $T \in \mathbb{R}$

$$(7.26)\qquad (R(x) - T)^+ \leqq (R(z) - T)^+ + \Sigma_{i=1}^n (x_i - z_i)^+.$$

*Equality holds if and only if*

$$(7.27)\qquad x-z \in \cup_{j\in CP_T(z)} K_j.$$

<u>PROOF</u>. (7.26) is a direct consequence of (7.21). It remains to show, that for any $z \in \mathbb{R}^n$ $X_T(z) = W_T(z)$, where

(7.28) $\quad X_T(z) := \{x \in \mathbb{R}^n\colon\ x-z \in K_j \text{ for a } j \in CP_T(z)\},$

(7.29) $\quad W_T(z) := \{x \in \mathbb{R}^n\colon\ (R(x) - T)^+ = (R(z) - T)^+ + \Sigma_{i=1}^n (x_i - z_i)^+\}.$

First we show that $X_T(z) \subset W_T(z)$. Let $x \in X_T(z)$. If $R(z) \leqq T$ and $x \in z + K_0$ this inclusion is trivial since $x \leqq z$ implies $R(x) \leqq R(z) \leqq T$ so that each term in (7.29) vanishes. If $R(z) \geqq T$ and a $j \geqq 1$ exists with $j \in CP(z)$ and $x \in z + K_j$ then $R(x) = R(z) + \Sigma_{i=1}^n (x_i - z_i)^+$ (Lemma 7.5), so that $R(x) \geqq T$ and $(R(x) - T)^+ = (R(z) - T)^+ + \Sigma_{i=1}^n (x_i - z_i)^+$. In all cases we conclude that $x \in W_T(z)$.

In order to prove the reverse inclusion, let $x \in W_T(z)$. Suppose first that $R(z) < T$. Then $(R(x) - T)^+ = \Sigma_{i=1}^n (x_i - z_i)^+$. From (7.21) we know, however, that $R(x) \leqq R(z) + \Sigma_{i=1}^n (x_i - z_i)^+ < T + \Sigma_{i=1}^n (x_i - z_i)^+$. Consequently, $R(x) - T < 0$ and $\Sigma_{i=1}^n (x_i - z_i)^+ = 0$ so that $x \in z + K_0$. Suppose now that $R(z) = T$. If $x-z \in K_0$ we have nothing to prove, so assume $\Sigma_{i=1}^n (x_i - z_i)^+ > 0$. Since $x \in W_T(z)$ it follows then that $R(x) > T$, and that equality must hold in (7.21), so that $x \in z + \cup_j \{K_j\colon j \in CP(z)\}$. Finally suppose $R(z) > T$. Then $x \in W_T(z)$ implies also equality in (7.21), since now $(R(x) - T)^+ = R(x) - T$ because of $(R(x) - T)^+ \geqq (R(z) - T)^+ > 0$. Hence, in all cases $x \in X_T(z)$. □

## 7.4 A MINIMAX APPROACH TO PROJECT PLANNING

7.4.1. Introduction. From now on we will study *stochastic* project planning problems. That is, it is assumed that a project has been specified completely, except for the fact that the durations of the activities are unknown. The duration of activity i is supposed to be a random variable $\xi_i$, $i = 1,\ldots,n$. Then the project completion time

$$\tau := R(\xi),$$

with $\xi := (\xi_1,\ldots,\xi_n)$ is a random variable too. Its distribution is determined by the distribution of $\xi$, since the function R, defined in (7.10), is completely specified because the graph of the project is assumed to be fixed. One would like to be able to calculate the probability distribution of $R(\xi)$; in particular, $E\,R(\xi)$, $P(R(\xi) > T)$

and $E(R(\xi) - T)^+$ for certain values of T might be important quantities for a decision maker who has to make a "promise" T on the project completion time. Also *criticality indices* as $P(j \in CP(\xi))$, $P(j \in CP(\xi), R(\xi) > T)$ and $P(i \in CA(\xi))$, $P(i \in CA(\xi), R(\xi) > T)$ are important in the analysis of the random case. However, there are two practical difficulties. First of all, the joint distribution of $(\xi_1,\ldots,\xi_n)$ has to be specified completely. Moreover, even if this is possible, the amount of calculations is prohibitive for networks of a reasonable size, see e.g. [26,7]. Under the assumption of independence of $\xi_1,\ldots,\xi_n$ it is possible to derive bounds for the expectation of $R(\xi)$ [9,6,1] and for its distribution function [21,38,30,14,4] from the marginal distributions. We will, however, *not* assume independence. Instead we apply the *minimax* approach that has been introduced in the beginning of this chapter. It is assumed that for each $\xi_i$ separately the complete distribution function $F_i$, $F_i(t) := P(\xi_i < t)$, $t \in \mathbb{R}$, is known, that has a finite mean value $\bar{\xi}_i$ and that is restricted to $[a_i,b_i]$ where

$$a_i := \lim_{\alpha \downarrow 0} F_i^{-1}(\alpha) \geqq 0,$$

$$b_i := \lim_{\alpha \uparrow 1} F_i^{-1}(\alpha) \leqq +\infty.$$

(The notation $[u,v]$ has to be interpreted as $[u,\infty)$ if $v = \infty$). But there is *no* knowledge with respect to the dependence between the random durations of the activities. This situation has also been studied by Meilijson and Nádas [27,29]. Hoewever, unlike their approach ours is characterized by the explicit use of duality theory; this makes our results more transparant and more general. Moreover, we explicitly consider a *decision* problem: we suppose that one has to make a "promise" T on the unknown completion time $\tau$ of the project. Our starting point is the simple recourse model

$$\text{(7.30)} \qquad \text{minimize}_{T \in \mathbb{R}} \{\alpha T + E(\tau - T)^+\}, \ \alpha \in [0,1].$$

That is, it is assumed that reduction of T is profitable at a rate $c \geqq 0$, but that, if the promise is not kept, the delay $(\tau-T)^+$ is penalized with $q \geqq c$ per time unit; then in (7.30) $\alpha$ is interpreted as $c/q$. Of course, if the distribution function $\Phi$ of $\tau$ would be known explicitly, the optimal solution $T^*$ of (7.30), which can be interpreted as a "production model" ($T$ := production, $\tau$ := demand, $c$ := unit production cost, $q$ := unit shortage cost), could be found simply by $T^* \in \Phi^{-1}(1-\alpha)$, that is

$$(7.31) \qquad P(\tau > T^*) \leqq \alpha \leqq P(\tau \geqq T^*).$$

The main problem of this section can now be described as the minimax formulation of (7.30):

$$(OP(\alpha)) \qquad \text{minimize}_{T\in\mathbb{R}} \{\alpha T + \sup_{H\in\mathcal{H}(F_1,\ldots,F_n)} E_H f(\xi)\}, \ \alpha \in [0,1],$$

where

$$(7.32) \qquad f(x) = f_T(x) := (R(x) - T)^+, \ x \in \mathbb{R}^n,$$

is the *project delay function*. As already has been indicated before, the "inner problem" of OP($\alpha$),

$$(IP(T)) \qquad \text{maximize}_{H\in\mathcal{H}(F_1,\ldots,F_n)} E_H f_T(\xi), \ T \in \mathbb{R},$$

is a marginal problem of the type P' (or P) of Section 7.2. This inner problem has independent value, and it will be studied first. In fact, the analysis is aimed at IP(T) for the most part, since the outer problem is simple if the inner problem is mastered.

The contents of the present section can now be summarized as follows. First, in Subsection 7.4.2, we apply Duality Theorem 7.1 to the inner problem to get its dual ID(T). The dual problem is reduced to a finite-dimensional convex programming problem, to be denoted by $\widetilde{ID}(T)$. It appears to be a simple recourse formulation of a problem related to (7.19); its variables can be interpreted as "promises" with respect to the activity durations $\xi$. Just as in the deterministic case $\widetilde{ID}(T)$ can be dualized. This dualization will be executed in Subsection 7.4.3. As a result of the second dualization we get a flow problem denoted by $\widetilde{IP}(T)$. Not surprisingly, $\widetilde{IP}(T)$ appears to be a reduction of the original marginal problem IP(T): the optimal flows in $\widetilde{IP}(T)$ are precisely the criticality numbers of the most unfavorable distributions in IP(T), as is indicated in Subsection 7.4.4. There we also give a complete characterization of all most unfavorable distributions. In Subsection 7.4.5 the successive reformulations of the inner problem are applied to the "outer" problem OP($\alpha$), with the sequence of equivalent problems OP($\alpha$), OD($\alpha$), $\widetilde{OD}(\alpha)$ and $\widetilde{OP}(\alpha)$ as a result. In Subsection 7.4.6 the assumption of complete knowledge of the marginal distributions, which in practical situations often is too demanding, will be relaxed. Also in that case the most unfavorable joint distributions are characterized completely, and the corresponding criticality numbers are calculated explicitly. In Subsection 7.4.7 it is shown that the minimax approach is tractable from a computational point of view.

Notations as OP($\alpha$) and IP(T) suggest that we interpret the outer problem as well as the inner problem as a parametrized family of optimization problems. This is the case, indeed. We will be interested in their optimal values as a function of the parameter:

$$\hat{h}(\alpha) := \inf OP(\alpha),\ \alpha \in [0,1],$$
$$h(T) := \sup IP(T),\ T \in \mathbb{R}.$$

By extending the definition of $\hat{h}(\alpha)$ we write

$$(7.33)\qquad \hat{h}(\alpha) := \inf_{T\in\mathbb{R}} \{\alpha T + h(T)\},\ \alpha \in \mathbb{R},$$

so that $\hat{h}$ is the *concave conjugate* function of h (see Chapter 2 for the definition). Since it will appear that h is a finite convex function, we have also the dual representation

$$(7.34)\qquad h(T) = \sup_{\alpha\in\mathbb{R}} \{-\alpha T + \hat{h}(\alpha)\},\ T \in \mathbb{R}.$$

For the proof of (7.34) and for more details we refer the reader to Proposition 7.14. The conjugacy relation between h and $\hat{h}$ is very useful for translating results on the inner problem to results on the outer problem, as will be seen.

Before starting the formal analysis we want to make a remark on the minimax model OP($\alpha$). Whereas the simple recourse model (7.30) is practically equivalent to the chance-constrained (CC) model

$$\text{minimize}_T\{T: P(\tau > T) \leqq \alpha\},$$

see (7.31), it is not true that OP(α) is equivalent to the minimax CC model

(7.35) $$\text{minimize}_T\{T: \sup_{H\in\mathcal{H}(F_1,\dots,F_n)} P_H(\tau > T) \leqq \alpha\}.$$

If $T^*$ and $H^*$ solve OP(α) then

$$E_H(\tau - T^*)^+ \leqq E_{H^*}(\tau - T^*)^+ \quad \forall H \in \mathcal{H}(F_1,\dots,F_n);$$

in this sense the distribution one is looking for in the inner problem of OP(α) is the most unfavorable one in $\mathcal{H}(F_1,\dots,F_n)$. However, $H^*$ is not necessarily most unfavorable in the sense of (7.35). Of course, $P_{H^*}(\tau > T^*) \leqq \alpha$, see (7.31), but the existence of a $H \in \mathcal{H}(F_1,\dots,F_n)$ with $P_H(\tau > T^*) > \alpha$ is not excluded.

7.4.2. <u>The inner problem, its dual and its reduced dual</u>. We will now study the inner problem IP(T). It is a marginal problem of the type P' (or P) discussed in Section 7.2, with $X_i := \mathbb{R}$. Its criterion function is the project delay function f, defined in (7.32). Since f satisfies the conditions of Duality Theorem 7.1 (just take $e_i(x_i) := 1 + |x_i|$, $x_i \in \mathbb{R}$, $i = 1,\dots,n$), we conclude from this theorem that most unfavorable distributions $H^*$ exist, and that they can be analyzed by means of the dual problem of IP(T)

(ID(T)) $$\text{minimize}_{y\in Y}\{E\,y: y \geqq f\},$$

where Y is the space of all separable real Borel functions on X, of which the components $y_i$ are integrable with respect to $F_i$. Notice that Theorem 7.1 implies that ID(T) has optimal solutions, and that

$$h(T) = \inf \text{ID}(T),\ T \in \mathbb{R}.$$

We will show that ID(T) can be reduced to a convex program in $\mathbb{R}^n$.

To that purpose we define for $i = 1,\dots,n$

$$\tilde{Y}_i := \{y_i \in \mathbb{R}^{X_i} : \exists w_i \in \mathbb{R}, \exists z_i \in \mathbb{R} \text{ such that } \forall x_i \in X_i \; y_i(x_i) = w_i + (x_i - z_i)^+\}$$

and

$$\tilde{Y} := \{y \in \mathbb{R}^X : y = \Sigma_{i=1}^n y_i,\ y_i \in \tilde{Y}_i,\ i = 1,\dots,n\}.$$

Since $\int |y_i(x_i)| dF_i(x_i) < \infty$ for all $y_i \in \tilde{Y}_i$, $\tilde{Y}$ is a subset of Y. Since for arbitrary but fixed $(x_1,x_2,\dots,x_{i-1},x_{i+1},\dots,x_n)$ the criterion function f, considered as a function of $x_i$ only, belongs to $\tilde{Y}_i$, it is not surprising that in ID(T) only upper bounds $y \in \tilde{Y}$ for f are relevant.

PROPOSITION 7.7. *Let* f *be defined by* (7.32) *and* (7.10). *Then for any* $y \in Y$, $y \geqq f$ *there exists a* $\tilde{y} \in \tilde{Y}$ *such that* $y \geqq \tilde{y} \geqq f$.

PROOF. Fix $y \in Y$ with $y \geqq f$. Then

$$(7.36) \qquad \infty > y_1(x_1) \geqq f(x) - \Sigma_{i=2}^n y_i(x_i) > -\infty, \ \forall x \in \mathbb{R}^n.$$

Denote the vector $(x_2,\dots,x_n) \in \mathbb{R}^{n-1}$ by $v_1$. Then, because of the special form of f, $f(x) - \Sigma_{i=2}^n y_i(x_i)$ can be written as $w_1(v_1) + (x_1 - z_1(v_1))^+$ for certain $w_1(v_1) \in \mathbb{R}$, $z_1(v_1) \in \mathbb{R}$ not depending on $x_1$. With $\tilde{y}_1$ defined by

$$(7.37) \qquad \tilde{y}_1(x_1) := \sup_{v_1 \in \mathbb{R}^{n-1}} \{w_1(v_1) + (x_1 - z_1(v_1))^+\},\ x_1 \in \mathbb{R},$$

it follows from (7.36) that

$$(7.38) \qquad \infty > y_1(x_1) \geqq \tilde{y}_1(x_1) \geqq f(x) - \Sigma_{i=2}^n y_i(x_i) > -\infty, \ \forall x \in \mathbb{R}^n.$$

We will show that $\tilde{y}_1 \in \tilde{Y}_1$. Define

$$\tilde{w}_1 := \sup_{v_1 \in \mathbb{R}^{n-1}} w_1(v_1),$$

$$\tilde{z}_1 := -\sup_{v_1 \in \mathbb{R}^{n-1}} (w_1(v_1) - z_1(v_1)) + \tilde{w}_1.$$

From (7.36) it follows that both suprema are finite. Moreover,

$$\begin{aligned}\tilde{y}_1(x_1) &= \sup_{v_1} \max\{w_1(v_1),\ w_1(v_1) - z_1(v_1) + x_1\} \\ &= \max\{\sup_{v_1}\{w_1(v_1)\},\ x_1 + \sup_{v_1}\{w_1(v_1) - z_1(v_1)\}\} \\ &= \max\{\tilde{w}_1,\ \tilde{w}_1 + x_1 - \tilde{z}_1\} = \tilde{w}_1 + (x_1 - \tilde{z}_1)^+\end{aligned}$$

so that $\tilde{y}_1 \in \tilde{Y}_1$. From (7.38) we conclude that the replacement of $y_1$ by $\tilde{y}_1$ does not disturb the feasibility condition $y \geqq f$. In the same way, $\tilde{y}_k \in \tilde{Y}_k$ can be constructed, successively for $k = 2,\ldots,n$, such that

$$y_k(x_k) \geqq \tilde{y}_k(x_k) \geqq f(x) - \Sigma_{i<k}\tilde{y}_i(x_i) - \Sigma_{i>k}y_i(x_i),\ x \in \mathbb{R}^n.$$

For $k = n$ we have the desired result. □

Proposition 7.7 shows that in ID(T) one may without loss of generality restrict the attention to functions $y \in \tilde{Y}$ rather than $y \in Y$. Doing this, a great simplification is possible: if $y_i(x_i) = w_i + (x_i - z_i)^+$ then

$$\Sigma_{i=1}^n y_i(x_i) \geqq f(x)$$

is true *for all* $x \in \mathbb{R}^n$ *iff it holds for* $x = z$; that is iff $\Sigma_{i=1}^n w_i \geqq f(z)$. Hence ID(T) can be reduced to

$$(\mathrm{I}\tilde{\mathrm{D}}(\mathrm{T})) \qquad \operatorname*{minimize}_{z \in \mathbb{R}^n} \{f(z) + G(z)\},$$

where, as before, $f(z) = (R(z) - T)^+$ and where $G(z)$ is the separable function defined by

$$(7.39)\qquad \begin{aligned} G(z) &:= \Sigma_{i=1}^n G_i(z_i),\ z \in \mathbb{R}^n, \text{ and for } z_i \in \mathbb{R}: \\ G_i(z_i) &:= E(\xi_i - z_i)^+ = \int (x_i - z_i)^+ dF_i(x_i) = \int_{z_i}^{\infty} (1 - F_i(x_i))dx_i. \end{aligned}$$

Clearly, $G_i$ is a finite, nonnegative, nonincreasing, convex function on $\mathbb{R}$. Its asymptotes for $|z_i| \to \infty$ are given by $(\bar{\xi}_i - z_i)^+$; in fact, $G_i(z_i) = \bar{\xi}_i - z_i$ iff $z_i \leqq a_i$ and $G_i(z_i) = 0$ iff $z_i \geqq b_i$ (see Figure 7.1). $G_i(z_i)$ has an obvious interpretation: it is the expected delay of activity i beyond $z_i$.

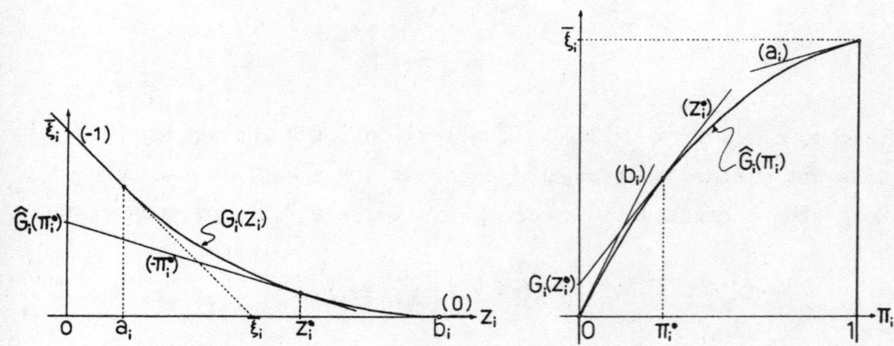


FIGURE 7.1. The function $G_i$ defined in (7.39) and its concave conjugate function $\hat{G}_i$ defined in (7.49). Slopes are indicated by (.); $z_i^o$ and $\pi_i^o$ are related by subgradient duality, see (7.58).

THEOREM 7.8. ID(T) and $\widetilde{\text{ID}}$(T) are equivalent.

*a. For each* $T \in \mathbb{R}$ *the minimum in* $\widetilde{\text{ID}}$(T) *is attained, and*

$$\text{(7.40)} \qquad \min \widetilde{\text{ID}}(T) = \min \text{ID}(T) = \max \text{IP}(T) = h(T), \ t \in \mathbb{R}.$$

*b. If* $z^* \in \mathbb{R}^n$ *is optimal for* $\widetilde{\text{ID}}$(T) *then* $y^* \in \tilde{Y}$, *defined by*

$$y_i^*(x_i) := w_i + (x_i - z_i^*)^+, \ i = 1,\ldots,n, \text{ where the } w_i \text{ are arbitrary reals satisfying } \Sigma_{i=1}^n w_i = f(z^*),$$

*is optimal for* ID(T). *If* $y^* \in Y$ *is optimal for* ID(T), *then a* $\tilde{y} \in \tilde{Y}$ *exists which is also optimal for* ID(T), *with* $P_{H^*}(\tilde{y} = y^*) = 1$ *for any optimal distribution* $H^*$ *for* IP(T).

PROOF. Follows directly from Theorem 7.1 and (the discussion after) Proposition 7.7. □

The reduced dual problem $I\tilde{D}(T)$ is a finite-dimensional convex program. In fact, it can be seen as a simple recourse formulation of the incompletely specified stochastic programming problem

$$\text{minimize}_{z\in\mathbb{R}^n} \{(R(z) - T)^+ : z \geqq \xi\}.$$

We interpret $I\tilde{D}(T)$ therefore as the problem of finding reference values or *promises* z for the durations of the activities, in such a way that the delay (beyond T) of the project completion based on z, is balanced with the sum of the expected delays (beyond z) of the random activity durations. It is easy to prove, that an optimal solution $z^*$ must exist with $a_i \leqq z_i^* \leqq b_i$: the objective is a nonincreasing function of $z_i$ if $z_i \leqq a_i$, and nondecreasing if $z_i \geqq b_i$. Hence, see (7.40),

$$(7.41) \qquad h(T) = \min_{z\in\mathbb{R}^n} \{(R(z) - T)^+ + G(z)\} = \min_{a \leqq z \leqq b}\{(R(z) - T)^+ + G(z)\}.$$

<u>PROPOSITION 7.9</u>. <u>Characterization of h.</u>
*The function* h *is finite, nonnegative, nonincreasing and convex. Its asymptotes are given by* $(T_c - T)^+$, *where* $T_c$ *is the finite number*

$$(7.42) \qquad T_c := \inf_{z\in\mathbb{R}^n} \{R(z) + G(z)\} \geqq R(\bar{\xi}).$$

*In fact, the set of minimizers in (7.42),*

$$Z_c := \arg\min_{z\in\mathbb{R}^n} \{R(z) + G(z)\},$$

*is nonempty, and*

$$(7.43) \qquad \begin{aligned} h(T) &= T_c - T \;\; \textit{iff} \;\; T \leqq T_1 := \sup\{ R(z) : z \in Z_c\}, \\ h(T) &= 0 \qquad\quad \textit{iff} \;\; T \geqq T_0 := R(b). \end{aligned}$$

*Since all* $a_i \geqq 0$ *it follows that* $T_1 \geqq 0$. *Moreover,* $T_1 \leqq T_c \leqq T_0 \leqq \infty$, *and* $T_0$ *is finite iff all* $b_i$ *are finite.*

PROOF. The first statement is trivial. Furthermore, applying Jensen's inequality we get

$$G(z) \geqq \Sigma_{i=1}^{n}(\bar{\xi}_i - z_i)^+, \; z \in \mathbb{R}^n.$$

From (7.21) we know that

$$R(z) \geqq R(\bar{\xi}) - \Sigma_{i=1}^{n}(\bar{\xi}_i - z_i)^+, \; z \in \mathbb{R}^n,$$

so that (7.42) is true. This implies that $T_c$ is finite. The infimum in (7.42) is attained, since, as in (7.41), one might restrict the attention to $z \in a + \mathbb{R}^n_+$, and $R(z) + G(z) \to \infty$ if $\|z\| \to \infty$, $z \in a + \mathbb{R}^n_+$. Hence, $Z_c \neq \emptyset$, even $Z_c \cap (a + \mathbb{R}^n_+) \neq \emptyset$, and $T_1$ is well-defined and $> -\infty$.

Let us now prove (7.43). From (7.41) it follows that

$$h(T) \geqq \min_{z \in \mathbb{R}^n} \{R(z) - T + G(z)\} = T_c - T, \; T \in \mathbb{R}.$$

We have to show that equality holds iff $T \leqq T_1$. For that purpose, rewrite (7.41) as

$$h(T) + T - T_c = \min_{z \in \mathbb{R}^n} \{\max\{R(z) + G(z) - T_c, \; T + G(z) - T_c\}\}.$$

Recalling that the minimum is attained, and that $R(z) + G(z) \geqq T_c$ for all $z \in \mathbb{R}^n$, we conclude that $h(T) = T_c - T$ iff $\exists \tilde{z} \in \mathbb{R}^n$ with $R(\tilde{z}) + G(\tilde{z}) - T_c = 0$ as well as $T + G(\tilde{z}) - T_c \leqq 0$. These conditions are equivalent to

$$\tilde{z} \in Z_c, \; T \leqq R(\tilde{z}).$$

Hence, by definition of $T_1$, $h(T) = T_c - T$ if $T < T_1$ and $h(T) > T_c - T$ if $T > T_1$. Since h is continuous, we must have $h(T) = T_c - T$ if $T \leqq T_1$, actually. This completes the proof of the first statement in (7.43). The second one follows directly from (7.41): since both $(R(z) - T)^+$ and $G(z)$ are nonnegative for all $z \in \mathbb{R}^n$, also $h(T) \geqq 0$ where equality holds iff

$$\exists \tilde{z} \in \mathbb{R}^n, \; R(\tilde{z}) \leqq T, \; G(\tilde{z}) = 0.$$

Since R is nondecreasing, and since $G(\tilde{z}) = 0$ iff $\tilde{z} \geqq b$, this condition is equivalent to $T \geqq T_0 := R(b)$. Of course, $h(T) > 0$ for all $T \in \mathbb{R}$ if $T_0 = \infty$. In that case one has $\lim_{T\to\infty} h(T) = 0$, as can be shown by selecting, for an arbitrary $\varepsilon > 0$, $z_\varepsilon \in \mathbb{R}^n$ with $G(z_\varepsilon) < \varepsilon$: then $h(T) < \varepsilon$ if $T \geqq R(z_\varepsilon)$. Consequently, we did not only show (7.43), but also that the asymptotes of h are given by $(T_c - T)^+$.

Let us prove that $T_1 \geqq 0$. Recall that $Z_c \cap (a + \mathbb{R}^n_+) \neq \emptyset$. Take any $\hat{z} \in Z_c \cap (a + \mathbb{R}^n_+)$. Then $T_1 \geqq R(\hat{z}) \geqq R(a) \geqq 0$; the first inequality follows from $\hat{z} \in Z_c$, the second one is a consequence of $\hat{z} \geqq a$, and the last one holds since $a \geqq 0$.

Finally, $T_1 \leqq T_c \leqq T_0$ follows from $h(T) \geqq (T_c - T)^+$ for all T, together with (7.43). □

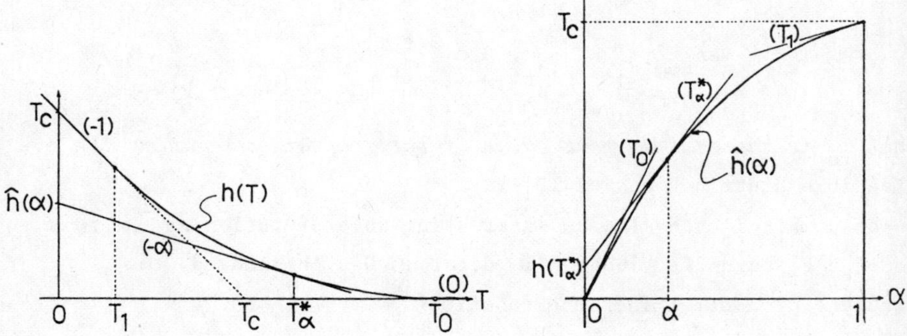


FIGURE 7.2. The function h and its concave conjugate function $\hat{h}$. Slopes are indicated by (.); $T^*_\alpha$ and $\alpha$ are related by subgradient duality, see (7.62).

<u>COROLLARY 7.10. Optimal solutions of $I\tilde{D}(T)$ for $T \notin (T_1,T_0)$.</u>
*Assume* $T \notin (T_1,T_0)$. *Then the following conditions are necessary and sufficient for* $z^* \in \mathbb{R}^n$ *to be optimal for* $I\tilde{D}(T)$.
*If* $T_0 < \infty$, $T \geqq T_0$: $z^* \geqq b$ *and* $R(z^*) \leqq T$,
*If* $T \leqq T_1$ : $z^* \in Z_c$ *and* $R(z^*) \geqq T$.

<u>PROOF</u>. This is a direct consequence of (the proof of) (7.43) and $h(T) = \min I\tilde{D}(T)$. □

The values $T := T_0$ and $T := T_1$ have a direct interpretation in problem IP(T). On the one hand, $T_0$ is the smallest value of T for which

$$\max_{H\in\mathcal{H}(F_1,\ldots,F_n)} E_H(R(\xi) - T)^+ = h(T) = 0,$$

that is, for which there will be *no delay* beyond T *with probability 1*, with respect to each distribution H that is feasible for IP(T). On the other hand, $T_1$ is the largest value of T for which

$$\max_{H\in\mathcal{H}(F_1,\ldots,F_n)} E_H\max(R(\xi),T) = h(T) + T = T_c,$$

that is, for which there will be positive *delay* beyond T *with probability 1*, with respect to each distribution $H^*$ that is optimal for IP(T). The latter formula gives also a direct interpretation of $T_c$ in terms of problem IP(T): for $T \leqq T_1$ we get

$$T_c := \max_{H\in\mathcal{H}(F_1,\ldots,F_n)} E_H R(\xi),$$

so that $T_c$ is the *maximum mean value of the project completion time* over all feasible distributions of IP(T).

The number $T_c$ has also an interesting interpretation as the mean value of a certain distribution $\Phi_c$ determined by h. Indeed, the subgradient of h determines the function

$$\text{(7.44)} \qquad \Phi_c(t) := 1 + \min\{u\colon u \in \partial h(t)\}, \quad t \in \mathbb{R},$$

and from the characterization of h in Proposition 7.9 it follows that $\Phi_c$ is a probability distribution function of a random variable $\eta$ with support in $[T_1,T_0]$. Representing h as the integral of its subgradient [31 Theorem 24.2] we get

$$\text{(7.45)} \qquad h(T) = \int_T^\infty (1 - \Phi_c(t))dt = \int (t-T)^+ d\Phi_c(t), \quad T \in \mathbb{R},$$

so that

$$\int t\,d\Phi_c(t) = \lim_{T\to-\infty} \int \max(t,T)d\Phi_c(t) = \lim_{T\to-\infty} (h(T) + T) = T_c.$$

The distribution function $\Phi_c$ is interesting by itself since for all $T \in \mathbb{R}$

$$h(T) = \max_{H\in H(F_1,\ldots,F_n)} E_H(R(\xi) - T)^+ = E_{\Phi_c}(\tau-T)^+.$$

In general, there is *no* distribution of $\xi$ in $H(F_1,\ldots,F_n)$ for which $\tau = R(\xi)$ is distributed according to $\Phi_c$. The existence of such a distribution is equivalent to having the *same* worst-case distribution for IP(T) *for all* $T \in \mathbb{R}$. $\Phi_c$ can be interpreted as the smallest upper bound for all feasible distributions of $\tau$ in the sense of *convex ordering* of probability distributions [27,29].

Up to now we have shown, that the dual ID(T) of the inner problem IP(T) can be reduced to a finite-dimensional convex programming problem $\tilde{ID}(T)$. As a consequence, we were able to characterize the optimal value h(T) of all these problems as a function of T. Moreover, we gave a complete characterization of the optimal solutions of the reduced dual $\tilde{ID}(T)$ for extreme specifications of T. Later on we will also characterize the optimal solutions of $\tilde{ID}(T)$ in the more interesting case $T \in [T_1,T_0]$. First we combine the (IP(T),ID(T)) duality and the equivalence of ID(T) and $\tilde{ID}(T)$ to get optimality conditions for IP(T) in terms of $\tilde{ID}(T)$.

THEOREM 7.11. Duality of IP(T) and $\tilde{ID}(T)$.
*In order that* $H^* \in H(F_1,\ldots,F_n)$ *is optimal for* IP(T) *it is necessary and sufficient that its support is contained in* $X_T(z^*)$ *for any optimal solution* $z^*$ *of* $\tilde{ID}(T)$. *Here, as in* (7.28), *we define*

$$X_T(z) := z + \cup_j\{K_j: j \in CP_T(z)\},$$

*where* $K_j$, $j = 0,1,\ldots,p$ *and* $CP_T(z)$ *are defined in* (7.20), (7.24) *and* (7.25), *respectively.*

PROOF. From Theorem 7.1, applied to the dual pair (IP(T),ID(T)) and the relations between ID(T) and $\tilde{ID}(T)$ described in Theorem 7.8, it is clear that the support of $H^*$ has to be restricted to the set

$$\{x \in \mathbb{R}^n:\ f(z^*) + \Sigma_{i=1}^n (x_i - z_i^*)^+ = f(x)\}.$$

Since by definition $f(x) = (R(x) - T)^+$, $x \in \mathbb{R}^n$, this set is precisely $X_T(z^*)$, see Lemma 7.6. □

The structure of the support of a worst-case distribution $H^*$ (i.e. a solution of IP(T)) shows the dependence of the random durations $\xi_1,\dots,\xi_n$ under $H^*$. For example, assume that $R(z^*) = T$. (In Theorem 7.13 we will show that this case necessarily occurs if one does not specify T too extremely i.e. if $T \in [T_1,T_0]$). Then with probability 1 there are only two possibilities for the realizations of $\xi$ in relation to the promises $z^*$, at least if we simplify a little bit by assuming that $P_{H^*}(\xi_i = z_i^*) = 0$ for all i:

*either* all actitities are ready in time: $\xi_i < z_i^*$ and $R(\xi) < T$,
*or* the project has a delay due to lateness of *all* activities of *one* critical path $j \in CP(z^*)$ *only*; then $\xi_i < z_i^*$ for all $i \notin B_j$, whereas $\xi_i > z_i^*$ for all $i \in B_j$. Hence $CP(\xi) = \{j\}$.

7.4.3. The dual of the reduced dual problem. Theorem 7.11 will be used to describe explicitly all worst-case distributions in terms of the optimal solutions $z^*$ of $I\tilde{D}(T)$. For this purpose it appears to be very useful to consider a *dual problem of* $I\tilde{D}(T)$ at the same time. This second dualization is related to the $(LP_1,LP_2)$ duality in the previous section: just as in the deterministic case the dual of the project planning problem $I\tilde{D}(T)$ appears to be a *flow problem* in the project network; it will be denoted by $I\tilde{P}(T)$. The dualization of $I\tilde{D}(T)$ is done in a standard way, as follows. By expressing the polyhedral function $(R(z) - T)^+$ as the value of a linear program, one gets the following separable convex program which is equivalent to $I\tilde{D}(T)$:

$$(7.46) \qquad \underset{\substack{y\in\mathbb{R} \\ z\in\mathbb{R}^n}}{\text{minimize}} \left\{ y + G(z) : \begin{pmatrix} -1 & 0 \\ -e & A \end{pmatrix} \begin{pmatrix} y \\ z \end{pmatrix} \leqq \begin{pmatrix} 0 \\ T.e \end{pmatrix} \right\}$$

where, as in Section 7.3, $e := (1,1,\dots,1)^T \in \mathbb{R}^p$ and A is the incidence matrix of the underlying graph. With $\lambda_0$ being the multiplier of the first constraint, and $\lambda := (\lambda_1,\dots,\lambda_p)$ the row vector of the multipliers of the other constraints, the Lagrangian function of (7.46) is given by

$$(7.47) \qquad L(y,z;\lambda_0,\lambda) := y + G(z) - (\lambda_0 + \lambda e)y + \lambda A z - T.\lambda e, \quad \text{if } y \in \mathbb{R},\ z \in \mathbb{R}^n,\ \lambda_0 \in \mathbb{R}_+,\ \lambda \in \mathbb{R}^p_+,$$

$:= -\infty \qquad$ otherwise.

The corresponding dual objective function $\psi$, to be maximized over $(\lambda_0,\lambda)$, is defined by

(7.48) $$\psi(\lambda_0,\lambda) := \inf_{y\in\mathbb{R},z\in\mathbb{R}^n} L(y,z;\lambda_0,\lambda).$$

Due to the separable convex structure of (7.47), which is available since $G(z) = \Sigma_{i=1}^n G_i(z_i)$, the dual objective function can be calculated explicitly: the infimization can be carried out for each $y_i$, $z_i$, $i = 1,\ldots,n$, separately. Hence

$$\begin{aligned}\psi(\lambda_0,\lambda_1) &= -T(1-\lambda_0) + \Sigma_{i=1}^n \hat{G}_i((\lambda A)_i) \\ &\qquad \text{if } \lambda_0 \geqq 0,\ \lambda \geqq 0 \text{ and } \lambda_0 + \lambda e = 1, \\ &= -\infty \qquad \text{otherwise,}\end{aligned}$$

where

(7.49) $$\hat{G}_i(\pi_i) := \inf_{z_i\in\mathbb{R}} \{\pi_i z_i + G_i(z_i)\},\ \pi_i \in \mathbb{R},\ i = 1,\ldots,n.$$

With $\hat{G}$ defined by

(7.50) $$\hat{G}(\pi) := \Sigma_{i=1}^n \hat{G}_i(\pi_i),\ \pi = (\pi_1,\ldots,\pi_n) \in \mathbb{R}^n,$$

we get the following representation of the dual problem of $\widetilde{\text{ID}}(T)$

(7.51) $$\underset{\substack{(\pi_0,\pi)\in\mathbb{R}^{1+n}\\(\lambda_0,\lambda)\in\mathbb{R}^{1+p}}}{\text{maximize}} \left\{ -\pi_0 T + \hat{G}(\pi) : \begin{array}{l}\pi = \lambda A,\ 1-\lambda_0 = \pi_0 = \lambda e \\ \lambda_0 \geqq 0,\ \lambda \geqq 0\end{array}\right\}.$$

The constraints indicate, that the variables of this problem can be interpreted as flows in the network: using the definitions of Section 7.3 we have

$$\begin{aligned}\lambda_j &\equiv \text{flow in path } j,\ j = 1,\ldots,p, \\ \pi_i &\equiv \text{flow in arc } i,\ i = 1,\ldots,n,\end{aligned}$$

$$\pi_0 \equiv 1 - \lambda_0 \equiv \text{total flow in the network.}$$

It is possible to eliminate the variables $(\lambda_0,\lambda)$: it is easily seen that the constraints in (7.51) are equivalent to $\pi_0 \leqq 1$ and

(7.52) $\quad \pi \in \Pi(\pi_0),\ \pi_0 \geqq 0,$

for $\Pi(\alpha)$ defined in (7.15). Vectors $(\pi_0,\pi)$ satisfying (7.52) can be interpreted as feasible *circulations* in the augmented network, defined by adding the arc 0 from sink to source: the flow $\pi_0$ in the arc 0 is precisely the total flow from source to sink in the original network, so that necessarily $\pi_i \leqq \pi_0$, $i = 1,\dots,n$. So we get the following concise formulation of the dual problem of $I\tilde{D}(T)$

$(I\tilde{P}(T))$ $\quad \text{maximize}_{(\pi_0,\pi)}\{-\pi_0 T + \hat{G}(\pi)\colon \pi \in \Pi(\pi_0),\ 0 \leqq \pi_0 \leqq 1\}.$

In spite of the fact that the variables $(\lambda_0,\lambda)$ can easily be eliminated from (7.51) we sometimes need to introduce them again. Notice that they satisfy $\lambda_j \geqq 0$, $j = 0,1,\dots,p$, and $\Sigma_{j=0}^{p}\lambda_j = 1$, or, using a notation introduced in (7.16): $\lambda \in \Lambda(1 - \lambda_0)$, $0 \leqq \lambda_0 \leqq 1$. We say that $(\lambda_0,\lambda)$ satisfying these conditions, and $(\pi_0,\pi)$ satisfying $\pi \in \Pi(\pi_0)$, $0 \leqq \pi_0 \leqq 1$, are *related* to each other if

(7.53) $\quad \pi_0 = 1 - \lambda_0,\ \pi = \lambda A.$

Related vectors $(\lambda_0,\lambda)$ and $(\pi_0,\pi)$ describe the same flow in the network. For every $(\lambda_0,\lambda)$ with $\lambda \in \Lambda(1 - \lambda_0)$, $0 \leqq \lambda_0 \leqq 1$, there exists precisely one $(\pi_0,\pi)$ with $\pi \in \Pi(\pi_0)$, $0 \leqq \pi_0 \leqq 1$, which is related to $(\lambda_0,\lambda)$, but for every such $(\pi_0,\pi)$ there exist in general more than one related $(\lambda_0,\lambda)$ vectors. This is due to the fact that every flow in the original network can be decomposed into flows in the paths, but this decomposition is not unique in general. Of course, $(\pi_0,\pi,\lambda_0,\lambda)$ solves (7.51) iff $(\pi_0,\pi)$ solves $I\tilde{P}(T)$ and $(\lambda_0,\lambda)$, $(\pi_0,\pi)$ are related.

Before we establish the duality relations between $I\tilde{D}(T)$ and $I\tilde{P}(T)$, we will first characterize the functions $\hat{G}_i$, $i = 1,\dots,n$, which occur in the objective function of $I\tilde{P}(T)$. Notice, that the definition in (7.49) implies that $\hat{G}_i$ is the *concave conjugate* function of $G_i$ (see Chapter 2 for the definition). This relation enables us to characterize $\hat{G}_i$ and its subgradients completely. Recall that the subgradient of the convex function $G_i$ at $z_i$ is given by

$$(7.54) \qquad \partial G_i(z_i) := \{\pi_i \in \mathbb{R}: G_i(x) \geqq G_i(z_i) + \pi_i(x - z_i) \ \forall x \in \mathbb{R}\},$$

whereas the definition of the subgradient (or supergradient) of the concave function $\hat{G}_i$ at $\pi_i$ is

$$(7.55) \qquad \partial \hat{G}_i(\pi_i) := \{z_i \in \mathbb{R}: \hat{G}_i(y) \leqq \hat{G}_i(\pi_i) + z_i(y - \pi_i) \ \forall y \in \mathbb{R}\}.$$

PROPOSITION 7.12. The concave conjugate function $\hat{G}_i$ of $G_i$. $\hat{G}_i(\pi_i) = -\infty$ *if* $\pi_i < 0$ *or* $\pi_i > 1$, *and*

$$(7.56) \qquad \hat{G}_i(\pi_i) = \int_0^{\pi_i} F_i^{-1}(1 - t_i)dt_i, \ 0 \leqq \pi_i \leqq 1,$$

*where* $F_i^{-1}$ *is any real function on* (0,1) *satisfying* $F_i^{-1}(t) = x \Rightarrow F_i(x) \leqq t \leqq F_i(x+0)$. *On* [0,1] $\hat{G}_i$ *is a finite nonnegative increasing concave function, with* $\hat{G}_i(0) = 0$, $\hat{G}_i(1) = \bar{\xi}_i$, *and with slope* $b_i$ *in* $\pi_i = 0+$ *and with slope* $a_i$ *in* $\pi_i = 1-$. *Moreover,*

$$(7.57) \qquad G_i(z_i) = \sup_{0 \leqq \pi_i \leqq 1} \{-\pi_i z_i + \hat{G}_i(\pi_i)\}, \ z_i \in \mathbb{R}.$$

*The minimizing* $z_i$ *in* (7.49) *and the maximizing* $\pi_i$ *in* (7.57) *are related via subgradient duality:*

$$(7.58) \qquad z_i \in \partial \hat{G}_i(\pi_i) \Leftrightarrow -\pi_i \in \partial G_i(z_i) \Leftrightarrow \hat{G}_i(\pi_i) = \pi_i z_i + G_i(z_i),$$

*and these conditions are equivalent to* $z_i \in F_i^{-1}(1 - \pi_i)$, *which is, of course, to be interpreted as*

$$(7.59) \qquad F_i(z_i) \leqq 1 - \pi_i \leqq F_i(z_i + 0).$$

PROOF. (7.57) and (7.58) are a direct consequence of the fact, that $-\hat{G}_i(-\pi_i)$ is the convex conjugate function of the finite convex function $G_i$ [31 Theorem 23.5]. Formula (7.56) is true, since it gives the correct supergradients, see (7.39) and (7.58), and the correct value at $\pi_i = 0$, see (7.49) and (7.58). Since $G_i(z_i) \geqq \bar{\xi}_i - z_i$ with equality for $z_i \leqq a_i$ it follows that $\hat{G}_i(1) = \bar{\xi}_i$. Finally, the statement on the directional derivatives in 0+ and 1- follows from (7.58), and (7.59) follows from (7.39) and (7.58). □

The graph of the function $\hat{G}_i$ is given in Figure 7.1, and the relation between $G_i$ and $\hat{G}_i$ is illustrated there.

<u>THEOREM 7.13</u>. <u>Duality of $I\tilde{D}(T)$ and $I\tilde{P}(T)$.</u>
*For every* $T \in \mathbb{R}$,

$$(7.60) \qquad h(T) = \min I\tilde{D}(T) = \max I\tilde{P}(T).$$

*The circulation* $(\pi_0^*, \pi^*)$ *with* $0 \leqq \pi_0^* \leqq 1$, $\pi^* \in \Pi(\pi_0^*)$, *is optimal for* $I\tilde{P}(T)$ *and* $z^*$ *is optimal for* $I\tilde{D}(T)$ *iff the following statements are true.*

a. (7.59) *holds for* $i = 1,\dots,n$, *with* $(\pi_i, z_i) := (\pi_i^*, z_i^*)$.

b. *If* $\pi_i^* > 0$ *then* $i \in CA(z^*)$, $i = 1,\dots,n$.

c. *If* $\pi_0^* > 0$ *then* $R(z^*) \geqq T$; *if* $\pi_0^* < 1$ *then* $R(z^*) \leqq T$.

*If* $(\pi_0^*, \pi^*)$ *is optimal for* $I\tilde{P}(T)$ *then* $-\pi_0^* \in \partial h(T)$. *If* $z^*$ *is optimal for* $I\tilde{D}(T)$, *for a* $T \in [T_1, T_0]$, *then* $R(z^*) = T$.

<u>PROOF</u>. Notice that the convex program (7.46) is stable: its optimal value remains bounded under small perturbations of the right-hand sides, since Slater's constraint qualification holds. Therefore, see Chapter 2,

$$\inf I\tilde{D}(T) = \max I\tilde{P}(T).$$

Since we know already that $h(T) = \min I\tilde{D}(T)$, see (7.40), the proof of (7.60) is complete.

Because of the stability of (7.46) the optimal solutions of $I\tilde{D}(T)$ and $I\tilde{P}(T)$ are precisely the saddle-points of the Lagrangian function L, defined in (7.47), see Chapter 2. The Karush-Kuhn-Tucker conditions are then necessary and sufficient conditions for $w^* := (y^*, z^*; \lambda_0^*, \lambda^*)$ to be a saddle-point of L. These conditions are written down below. Notice that the multi-dimensional subgradient condition $0 \in \partial_{y,z} L(w^*)$

has been replaced by the one-dimensional conditions $0 \in \partial_y L(w^*)$, $0 \in \partial_{z_i} L(w^*)$, $i = 1,\dots,n$. This is justified since L is separable as a function of (y,z).

KKT conditions for (7.46).

(i) Primal feasibility (and optimality): $y^* = (R(z^*) - T)^+$.

(ii) Dual feasibility (and $0 \in \partial_y L(w^*)$): $\lambda^* \in \Lambda(1-\lambda_0^*)$, $\lambda_0^* \in [0,1]$.

(iii) $0 \in \partial_{z_i} L(w^*)$: $-(\lambda^* A)_i \in \partial G_i(z_i^*)$, $i = 1,\dots,n$.

(iv) Complementary slackness: $\lambda_0^* y^* = 0$, and for $j = 1,\dots,p$: $\lambda_j^*((Az^*)_j - T - y^*) = 0$. That is,

$$\text{if } \lambda_0^* > 0 \text{ then } R(z^*) \leqq T, \text{ and}$$

$$\text{if } \lambda_j^* > 0,\ j \geqq 1, \text{ then } R(z^*) \geqq T \text{ and } j \in CP(z^*),$$

so that the complementary slackness conditions can simply be described, with $CP_T(z)$ defined in (7.25), as

$$\text{if } \lambda_j^* > 0 \text{ then } j \in CP_T(z^*),\ j = 0,1,\dots,p.$$

It is an elementary exercise to show, that these conditions are equivalent to (a), (b) and (c) in Theorem 7.13, for $(\pi_0^*, \pi^*)$ and $(\lambda_0^*, \lambda^*)$ related to each other in the sense of (7.53). Use has to be made of (7.13) and the equivalence of (7.58) and (7.59).

The last statements of Theorem 7.13 are consequences of the next proposition, which deals with the relations between the function h(T) and its concave conjugate function $\hat{h}$, defined in (7.33).

PROPOSITION 7.14. The concave conjugate function $\hat{h}$ of h.
$\hat{h}(\alpha) = -\infty$ *if* $\alpha \notin [0,1]$, *and*

$$(7.61) \qquad \hat{h}(\alpha) = \int_0^\alpha \Phi_c^{-1}(1-t)\,dt,\ 0 \leqq \alpha \leqq 1,$$

*where* $\Phi_c^{-1}$ *is any real function on* (0,1) *satisfying* $\Phi_c^{-1}(t) = x \Rightarrow \Phi_c(x) \leqq t \leqq \Phi_c(x+0)$; $\Phi_c$ *is the convex upper bound of the distribution of* $R(\xi)$ *defined in* (7.44). *On* [0,1] $\hat{h}$ *is a finite nonnegative increasing concave function, with* $\hat{h}(0) = 0$, $\hat{h}(1) = T_c$, *and with slope* $T_0$ *in* $\alpha = 0+$ *and with slope* $T_1$ *in* $\alpha = 1-$. *Moreover, the inversion formula* (7.34) *holds, where, of course, one may restrict the supremization to* $\alpha \in [0,1]$. *The minimizing* T *in* (7.33) *and the maximizing* $\alpha$ *in* (7.34) *are related via subgradient duality*

(7.62) $T \in \partial\hat{h}(\alpha) \Leftrightarrow -\alpha \in \partial h(T) \Leftrightarrow \hat{h}(\alpha) = \alpha T + h(T)$,

*and these conditions are equivalent to* $T \in \Phi_c^{-1}(1-\alpha)$, *to be interpreted as*

(7.63) $\Phi_c(T) \leqq 1-\alpha \leqq \Phi_c(T+0)$.

*Finally,*

(7.64) $\hat{h}(\alpha) = \max_{\pi\in\Pi(\alpha)}\hat{G}(\pi),\ 0 \leqq \alpha \leqq 1$.

PROOF. Apart from (7.64) every statement follows completely similar as the corresponding statement on $\hat{G}_i$ in Proposition 7.12: just replace $F_i$ by $\Phi_c$, see (7.45).

We sketch the proof of (7.64). Combining (7.33) and (7.41) we get for $0 \leqq \alpha \leqq 1$

$$\hat{h}(\alpha) = \inf_{T\in\mathbb{R}}\{\alpha T + \inf_{z\in\mathbb{R}^n}\{(R(z) - T)^+ + G(z)\}\}$$

$$= \inf_{z\in\mathbb{R}^n}\{\inf_{T\in\mathbb{R}}\{\alpha T + (R(z) - T)^+ + G(z)\}.$$

The inner infimum is attained by $T = R(z)$, since $\alpha \in [0,1]$, so that

(7.65) $\hat{h}(\alpha) = \inf_{z\in\mathbb{R}^n}\{\alpha R(z) + G(z)\}$.

This convex programming problem can be dualized in the same way as we derived the dual $I\tilde{P}(T)$ from $I\tilde{D}(T)$. As a result one gets the flow problem (7.64) as the dual of the stable problem (7.65), so that the optimal values coincide. The details of this dualization are left to the reader.□

PROOF OF THEOREM 7.13. (continued)
The last two statements have yet to be proved. Assume that $(\pi_0^*, \pi^*)$ is optimal for $I\tilde{P}(T)$ for any $T \in \mathbb{R}$. Then $\alpha := \pi_0^*$ solves

$$\text{maximize}_{0\leqq\alpha\leqq 1}\{-\alpha T + \sup_{\pi\in\Pi(\alpha)}\hat{G}(\pi)\}.$$

Because of (7.64) this is nothing other than

$$\text{maximize}_{0 \leqq \alpha \leqq 1} \{-\alpha T + \hat{h}(\alpha)\} =$$

$$\text{maximize}_{\alpha \in \mathbb{R}} \{-\alpha T + \hat{h}(\alpha)\}$$

so that $T \in \partial\hat{h}(\pi_0^*)$; that is, $-\pi_0^* \in \partial h(T)$, see (7.62).

Finally we have to show that if $T \in [T_1, T_0]$ any optimal solution $z^*$ of $\widetilde{ID}(T)$ satisfies $R(z^*) = T$. For $T = T_1$ and for $T = T_0$ (if finite) this statement follows from Corollary 7.10. If $T_1 < T < T_0$ we employ a dualization of $\widetilde{ID}(T)$ which differs slightly from $\widetilde{IP}(T)$. If $z^*$ solves $\widetilde{ID}(T)$ then $y := (R(z^*) - T)^+$ and $z := z^*$ solve the stable convex program

$$\text{minimize}_{y,z} \{y + G(z) \colon R(z) - y \leqq T,\ -y \leqq 0\}.$$

Let $\alpha$ and $\beta$ be the multipliers for the constraints, then the dual problem is

$$\text{maximize}_{\alpha,\beta \geqq 0} \inf_{y,z} \{-\alpha T + y(1-\alpha-\beta) + G(z) + \alpha R(z)\}$$

so that $\beta = 1 - \alpha$ necessarily, and with (7.65) the dual is simplified to

$$\text{maximize}_{0 \leqq \alpha \leqq 1} \{-\alpha T + \min_z \{\alpha R(z) + G(z)\}\} =$$

$$\text{maximize}_{0 \leqq \alpha \leqq 1} \{-\alpha T + \hat{h}(\alpha)\}.$$

Optimal values for $\alpha$ and $\beta$ are therefore determined by $T \in \partial\hat{h}(\alpha^*)$, that is $-\alpha^* \in \partial h(T)$, see (7.62), and $\beta^* = 1 - \alpha^*$. Consequently $\alpha^* > 0$ and $\beta^* > 0$ if $T_1 < T < T_0$, and the complementary slackness conditions imply $R(z^*) = T$. □

7.4.4. The optimal solutions of the inner problem. In Theorem 7.11 we characterized the optimal solutions of IP(T), the least favorable or worst-case distributions, in terms of the optimal solutions of its reduced dual $\widetilde{ID}(T)$. Theorem 7.13 relates the latter solutions in their turn to the optimal solutions of the "double-dual" problem $\widetilde{IP}(T)$. So it is only natural to expect that the optimal flows of $\widetilde{IP}(T)$ are closely related to the worst-case distributions of IP(T), and that these relations can be revealed by combining both theorems. To give these relations is the main purpose of the present subsection. The optimal

flows appear to be *criticality numbers* of the worst-case distributions.

Theorem 7.13 indicates that the worst-case distributions $H^*$ might be constructed by conditioning upon $z^* + K_j$, $j \in CP_T(z^*)$, as follows. Any probability distribution with support in $z^* + \cup_{j=0}^{p} K_j$ may be decomposed into

$$(7.66) \qquad H(x) = \Sigma_{j=0}^{p} \lambda_j . H_j(x), \ x \in \mathbb{R}^n,$$

where $H_j$ is a probability distribution with support in $z^* + K_j$, and where the coefficients satisfy $\lambda_j \geqq 0$ and $\Sigma_{j=0}^{p} \lambda_j = 1$, that is $\lambda_0 \in [0,1]$ and $\lambda \in \Lambda(1-\lambda_0)$. The terminology introduced around (7.53) appears to be useful to describe the relation between IP(T) and $\widetilde{\mathrm{IP}}$(T). We say that the distribution H defined in (7.66) and the feasible solution $(\pi_0,\pi)$ of $\widetilde{\mathrm{IP}}$(T) are *related* to each other if the coefficients $(\lambda_0,\lambda)$ in (7.66) and $(\pi_0,\pi)$ are related in the sense of (7.53).

The question arises: how to define $\lambda_j$ and $H_j$ in (7.66) in order that H solves IP(T)? We will show that the condition on the coefficients is that H must be related to an optimal solution of $\widetilde{\mathrm{IP}}$(T). The conditions on the $H_j$ appear to be only conditions on their marginals. For the formulation of these conditions we need to introduce two conditional distributions in terms of $F_i$, $(\pi_0^*,\pi^*) = (\pi_0^*,\pi_1^*,\ldots,\pi_n^*)$ and $z^* = (z_1^*,\ldots z_n^*)$, where $(\pi_0^*,\pi^*)$ is optimal for $\widetilde{\mathrm{IP}}$(T) and $z^*$ is optimal for $\widetilde{\mathrm{ID}}$(T). If $z_i^* \leqq a_i$ then we define

$$(7.67) \qquad \begin{aligned} G_i^-(x_i) &:= 0 \text{ if } x_i \leqq z_i^*, \\ &:= 1 \text{ if } x_i > z_i^*. \end{aligned}$$

If $z_i^* > a_i$ (hence $0 < F_i(z_i^*) \leqq 1 - \pi_i^* \leqq F_i(z_i^* + 0)$) we define for $x_i \in \mathbb{R}$

$$(7.68) \qquad \begin{aligned} G_i^-(x_i) := \beta_i . P_{F_i}(\xi_i < x_i | \xi_i < z_i^*) &+ \\ &+ (1 - \beta_i).P_{F_i}(\xi_i < x_i | \xi_i \leqq z_i^*) \end{aligned}$$

where $\beta_i \in [0,1]$ satisfies

$$(7.69) \qquad \frac{1}{1 - \pi_i^*} = \frac{\beta_i}{F_i(z_i^*)} + \frac{1 - \beta_i}{F_i(z_i^* + 0)} .$$

Clearly, if $F_i(\{z_i^*\}) = 0$ then the value of $\beta_i$ does not matter, but if $F_i(\{z_i^*\}) > 0$ we must have

$$\beta_i = \frac{F_i(z_i^*)}{1 - \pi_i^*} \cdot \frac{F_i(z_i^* + 0) - (1 - \pi_i^*)}{F_i(\{z_i^*\})} .$$

It is easy to show that in all cases $G_i^-$ is a probability distribution function with support in $(-\infty, z_i^*]$. If $z_i^* \geqq b_i$ we define

$$(7.70) \qquad \begin{aligned} G_i^+(x_i) &:= 0 \text{ if } x_i \leqq z_i^*, \\ &:= 1 \text{ if } x_i > z_i^*. \end{aligned}$$

If $z_i^* < b_i$ (hence $F_i(z_i^*) \leqq 1 - \pi_i^* \leqq F_i(z_i^* + 0) < 1$) we define for $x_i \in \mathbb{R}$

$$(7.71) \qquad \begin{aligned} G_i^+(x_i) := \gamma_i . P_{F_i}(\xi_i < x_i \mid \xi_i \geqq z_i^*) &+ \\ + (1 - \gamma_i) . P_{F_i}(\xi_i < x_i \mid \xi_i > z_i^*) \end{aligned}$$

where $\gamma_i \in [0,1]$ satisfies

$$(7.72) \qquad \frac{1}{\pi_i^*} = \frac{\gamma_i}{1 - F_i(z_i^*)} + \frac{1 - \gamma_i}{1 - F_i(z_i^* + 0)} .$$

Again, if $F_i(\{z_i^*\}) = 0$ the value of $\gamma_i$ does not matter, but if $F_i(\{z_i^*\}) > 0$ we must have

$$\gamma_i = \frac{1 - F_i(z_i^*)}{F_i(\{z_i^*\})} \cdot \frac{F_i(z_i^* + 0) - (1 - \pi_i^*)}{\pi_i^*} .$$

In all cases, $G_i^+$ is a probability distribution function with support in $[z_i^*, \infty)$. Finally, recall that $B_j$ is the index set of all activities of path j, and that $D_i$ is the index set of all paths through activity i in the project network.

Now we are ready to state a major result on the worst-case distributions.

THEOREM 7.15. Characterization of optimal solutions of IP(T).
*Let* $z^*$ *be any optimal solution of* $\widetilde{ID}(T)$, *and let* H *be defined by* (7.66) *for certain coefficients* $(\lambda_0,\lambda)$ *and certain distributions* $H_j$ *with support in* $z^* + K_j$, $j = 0,1,\ldots,p$.

a. *If* H *is optimal for* IP(T), *then it is related to an optimal solution of* $\widetilde{IP}(T)$.

b. *Let* H *be related to an optimal solution of* $\widetilde{IP}(T)$. *Then it is optimal for* IP(T), *if* $H_j \in \mathcal{H}(G_1^j,\ldots,G_n^j)$, $j = 0,1,\ldots,p$, *where*

$$G_i^j := G_i^- \text{ if } i \notin B_j,$$
$$G_i^j := G_i^+ \text{ if } i \in B_j.$$

PROOF. (a) Let H solve IP(T). From Theorem 7.11 it follows that without loss of generality we may take $\lambda_j = 0$ if $j \notin CP_T(z^*)$. We shall show that $(\pi_0,\pi)$ determined by $(\lambda_0,\lambda)$ via (7.53) satisfies the conditions of Theorem 7.13. It is obvious that $(\pi_0,\pi)$ is feasible for $\widetilde{ID}(T)$. If $\pi_0 < 1$ then $\lambda_0 > 0$ hence $0 \in CP_T(z^*)$ so that $R(z^*) \leqq T$. If $\pi_0 > 0$ then $\exists j \geqq 1$ with $\lambda_j > 0$, so that $j \in CP_T(z^*)$ and $R(z^*) \geqq T$ follows. In the same way, if $\pi_i > 0$ for an $i \geqq 1$ then $\exists j \geqq 1$ with $i \in B_j$, $\lambda_j > 0$ so that $j \in CP(z^*)$ and $i \in CA(z^*)$. (7.59) follows from the fact that $F_i$ must be the i-th marginal of H. Indeed,

$$F_i(z_i^*) = P_H(\xi_i < z_i^*) = \Sigma_j \lambda_j . P_{H_j}(\xi_i < z_i^*) \leqq \Sigma_{j \notin D_i} \lambda_j = 1 - \pi_i \tag{7.73}$$

where the inequality is a consequence of $P_{H_j}(z^* + K_j) = 1$ so that

$$P_{H_j}(\xi_i < z_i^*) \begin{cases} = 0 \text{ if } i \in B_j, \\ \leqq 1 \text{ if } i \notin B_j. \end{cases} \tag{7.74}$$

Similarly

$$P_{H_j}(\xi_i \leqq z_i^*) \begin{cases} \geqq 0 \text{ if } i \in B_j, \\ = 1 \text{ if } i \notin B_j, \end{cases} \tag{7.75}$$

so that

$$(7.76) \qquad F_i(z_i^* + 0) = P_H(\xi_i \leqq z_i^*) = \Sigma_j \lambda_j . P_{H_j}(\xi_i \leqq z_i^*) \geqq \Sigma_{j \notin D_i} \lambda_j = 1 - \pi_i .$$

From (7.73) and (7.76) (7.59) follows, which completes the proof.
(b) Wishing to apply Theorem 7.11 we have to show that H has the correct support and the correct marginals. Its support is contained in $X_T(z^*)$, see Theorem 7.13, since $\lambda_j > 0$ for a $j \geqq 1$ implies $R(z^*) \leqq T$ and $\pi_i^* > 0$ $\forall i \in B_j$. Hence $i \in CA(z^*)$ for all $i \in B_j$, so that $j \in CP(z^*)$, and $\lambda_0 > 0$ imply $\pi_0^* < 1$ so that $R(z^*) \geqq T$. In order to show that $F_i$ is the i-th marginal of H, it is sufficient to prove

$$(7.77) \qquad P_H(\xi_i < x_i) = F_i(x_i) \qquad \text{for all } x_i < z_i^*,$$

$$(7.78) \qquad P_H(\xi_i > x_i) = 1 - F_i(x_i + 0) \text{ for all } x_i > z_i^*.$$

Since $P_{H_j}(\xi_i < x_i) = 0$ if $x_i < z_i^*$ and $i \in B_j$, we have for $x_i < z_i^*$

$$P_H(\xi_i < x_i) = \Sigma_{j \notin D_i} \lambda_j . P_{H_j}(\xi_i < x_i) = (1 - \pi_i^*) . P_{G_i^-}(\xi_i < x_i).$$

If $z_i^* \leqq a_i$ the right-hand side vanishes, so that (7.77) holds in that case. Suppose now that $z_i^* > a_i$. Then from (7.68) and (7.69)

$$P_{G_i^-}(\xi_i < x_i) = \left(\frac{\beta_i}{F_i(z_i^*)} + \frac{1 - \beta_i}{F_i(z_i^* + 0)}\right) F_i(x_i) = \frac{1}{1 - \pi_i^*} F_i(x_i)$$

so that (7.77) is also true if $z_i^* > a_i$. Similarly, (7.78) follows from

$$P_H(\xi_i > x_i) = \Sigma_{j \in D_i} \lambda_j . P_{H_j}(\xi_i > x_i) = \pi_i^* . P_{G_i^+}(\xi_i > x_i), \; x_i > z_i^*,$$

using (7.70), (7.71) and (7.72). □

In general many worst-case distributions exist for IP(T), T fixed. Even if $I\widetilde{D}(T)$ and $I\widetilde{P}(T)$ have unique solutions (as occurs e.g. for $T_1 < T < T_0$ if $\hat{G}$ is strictly concave and differentiable, in particular if each marginal distribution has a positive density on $(a_i,b_i)$) the related vector of coefficients $(\lambda_0,\lambda)$ is not unique, generally. Depending on the structure of the network one has no or little or ample freedom in decomposing a feasible circulation into flows through paths. Moreover, the freedom of choice of $H_j$ in (7.66) is enormous. Although no attempt has been made to characterize all $H_j$, it is clear from Theorem 7.15 that it is sufficient to prescribe only their marginals. In fact, Theorem 7.15 part b specifies for each relevant j a distribution problem on $z^* + K_j$, comparable with P' or P in Section 7.2, with the distinction that *each* feasible solution is good: one may take the independent solution, the NW Solution, etc. The distinction between different worst-case distributions does not seem to be very important, since the relevant information is contained in $\pi$ and $\lambda$ which have an interpretation as criticality numbers. This is most easily seen for continuous distributions $F_i$ as shown in the next theorem.

THEOREM 7.16. Criticality numbers of worst-case distributions.
*Suppose that* $F_i$ *has a positive density function on* $(a_i,b_i)$, *for each* $i = 1,\ldots,n$. *Then, for each* $T \in [T_1,T_0]$, $\widetilde{IP}(T)$ *has a unique solution* $(\pi_0^*,\pi^*) = (\pi_0^*,\pi_1^*,\ldots,\pi_n^*)$, *and each optimal solution* $H^*$ *for* IP(T) *satisfies then:*

$$P_{H^*}(R(\xi) > T) = \pi_0^* = 1 - \lambda_0^*, \tag{7.79}$$

$$P_{H^*}(R(\xi) > T,\ i \in CA(\xi)) = \pi_i^*,\ i = 1,\ldots,n, \tag{7.80}$$

$$P_{H^*}(R(\xi) > T,\ j \in CP(\xi)) = \lambda_j^*,\ j = 1,\ldots,p, \tag{7.81}$$

*where* $(\lambda_0^*,\lambda^*) = (\lambda_0^*,\lambda_1^*,\ldots,\lambda_p^*)$ *is related to* $(\pi_0^*,\pi^*)$ *by* (7.53). *Such* $(\lambda_0^*,\lambda^*)$ *is not unique, generally; for each* $(\lambda_0^*,\lambda^*)$ *that is related to* $(\pi_0^*,\pi^*)$ *there exists corresponding worst-case distributions.*

PROOF. Fix $T \in [T_1,T_o]$ and consider $\widetilde{ID}(T)$. Since necessarily $R(z^*) \leqq T$ (even $R(z^*) = T$) for any optimal solution $z^*$ (Theorem 7.13), the set of optimal solutions of $\widetilde{D}(T)$ can be written as

$$Z^* := \arg\min_{z\in\mathbb{R}^n} \{G(z): R(z) \leqq T\}.$$

Obviously, $Z_i^* := \{u \in \mathbb{R}: \exists z^* \in Z^* \text{ with } z_i^* = u\}$ is a closed convex set, $i = 1,\dots,n$. Since $F_i$ is strictly increasing on $[a_i,b_i]$, $Z_i^*$ can only be more than a singleton if it is contained in either $(-\infty,a_i]$ or $[b_i,\infty)$. In all cases, $\pi_i^*$ is unique, since $F_i$ is continuous, see (7.59). Of course, uniqueness of $\pi_i^*$, $i = 1,\dots,n$, implies uniqueness of $\pi_0^*$. In order to show (7.79)-(7.81), represent any worst-case distribution $H^*$ by (7.66), with $(\lambda_0,\lambda)$ related to $(\pi_0^*,\pi^*)$ by (7.53). Since $R(z^*) = T$, we have

$$\begin{aligned} z^* + \operatorname{int} K_0 &\subset \{x \in X_T(z^*): R(x) < T\} \subset z^* + K_0, \\ z^* + \operatorname{int} K_j &\subset \{x \in X_T(z^*): R(x) > T, \{j\} = CP(x)\} \subset z^* + K_j,\ j \geqq 1, \end{aligned}$$

where $X_T(z^*)$ is defined in Theorem 7.11. Hence, the events $\{R(x) = T\}$ and $\{|CP(x)| \geqq 2\}$ are contained in the (partly common) boundaries of the translated orthants, which have zero probability under $H^*$ since $F_i(\{z_i^*\}) = 0\ \forall i$. Therefore

$$\begin{aligned} P_{H^*}(R(\xi) < T) &= P_{H^*}(R(\xi) \leqq T) = P_{H^*}(z^* + K_0) = \lambda_0^*, \\ P_{H^*}(R(\xi) > T,\ j \in CP(\xi)) &= P_{H^*}(R(\xi) > T,\ \{j\} = CP(\xi)) = \\ &= P_{H^*}(z^* + K_j) = \lambda_j^*,\ j \geqq 1, \\ P_{H^*}(R(\xi) > T,\ i \in CA(\xi)) &= \Sigma_{j\in D_i} P_{H^*}(R(\xi) > T,\ j \in CP(\xi)) = \\ &= \Sigma_{j\in D_i} \lambda_j = \pi_i^*. \qquad \square \end{aligned}$$

From the proof it is clear that (7.79)-(7.81) also hold for discrete marginals $F_i$, if $z^*$ can be chosen such that $F_i(\{z_i^*\}) = 0\ \forall i$; but this is not possible in general. If $F_i(\{z_i^*\}) > 0$ for some i the analysis is more difficult since the common boundaries of the translated orthants may have positive probability. If in (7.66) the representation is changed to

$$H := \Sigma_j\{\hat\lambda_j.\hat H_j: j \in CP_T(z^*)\}$$

with $P_{\hat H_j}(z^* + \hat K_j) = 1$, where $\hat K_j \subset K_j$, $\cup_j \hat K_j = \cup_j K_j$, and $\hat K_j \cap \hat K_h = \emptyset$ for $j \neq h$, then it is not difficult to show, that (7.79)-(7.81) generalize to

$$P_{H^*}(R(\xi) > T) \leqq \hat{\pi}_0 = 1 - \hat{\lambda}_0 \leqq P_{H^*}(R(\xi) \geqq T),$$

$$P_{H^*}(R(\xi) > T, \{j\} = CP(\xi)) \leqq \hat{\lambda}_j \leqq P_{H^*}(R(\xi) \geqq T, j \in CP(\xi)), \quad j \geqq 1,$$

$$P_{H^*}(R(\xi) > T, i \in CA(\xi), |CP(\xi)| = 1) \leqq \hat{\pi}_i \leqq P_{H^*}(R(\xi) \geqq T, i \in CA(\xi)), \; i \geqq 1.$$

7.4.5. The outer problem. We now return to the outer problem introduced in the beginning of this section:

$$(OP(\alpha)) \qquad \text{minimize}_{T \in \mathbb{R}} \{\alpha T + h(T)\}, \; \alpha \in [0,1].$$

Recall that its decision variable T represents the promise to be made with respect to the unknown project completion time; h(T) is the maximum expected delay beyond T. As before, h(T) is the optimal value of the inner problem IP(T). By analyzing IP(T) we got several representations of the function h in the previous part of this section. These will be applied now to get corresponding reformulations of OP($\alpha$).

First of all, by dualizing IP(T) and by reducing its dual problem we got h(T) = min $\widetilde{ID}$(T), that is

$$h(T) = \min_{z \in \mathbb{R}^n} \{(R(z) - T)^+ + G(z)\}, \; T \in \mathbb{R},$$

see (7.41). Substitution in OP($\alpha$) gives the following reformulation of OP($\alpha$):

$$(\widetilde{OD}(\alpha)) \qquad \text{minimize}_{z \in \mathbb{R}^n} \{\alpha R(z) + G(z)\}, \; 0 \leqq \alpha \leqq 1,$$

see (7.65). Its interpretation is clear: the problem to determine an optimal promise T for the project completion time is transformed to a problem to determine optimal promises z for the random durations of the activities separately. In $\widetilde{OD}(\alpha)$ a balance is aimed at between the promised project completion time T = R(z) and the sum G(z) of the expected delays of the activities beyond z. Just as $\widetilde{ID}$(T), $\widetilde{OP}(\alpha)$ is a finite-dimensional convex programming problem. It can be seen as a simple recourse formulation of the incompletely specified stochastic programming problem

$$\text{minimize}_{z \in \mathbb{R}^n} \{R(z) : z \geqq \xi\}.$$

Just as $\widetilde{ID}(T)$ also $\widetilde{OD}(\alpha)$ can be dualized, with a flow problem in the project network as the result:

$$(\widetilde{OP}(\alpha)) \qquad \text{maximize}_{\pi \in \Pi(\alpha)} \hat{G}(\pi), \ 0 \leqq \alpha \leqq 1,$$

see (7.64). In $\widetilde{OP}(\alpha)$ one is looking for a flow in the network with total flow value $\alpha$, in such a way that "risky activities" are preferred: recall that $\hat{G}(\pi) = \Sigma_{i=1}^n \hat{G}_i(\pi_i)$ with $\partial\hat{G}_i(\pi_i) = F_i^{-1}(1 - \pi_i)$, see (7.58) and (7.59).

The following theorem summarizes the results on the various versions of the outer problem.

THEOREM 7.17. Relations between $OP(\alpha)$, $\widetilde{OD}(\alpha)$ and $\widetilde{OP}(\alpha)$, $\alpha \in [0,1]$.
*a. The problems* $OP(\alpha)$, $\widetilde{OD}(\alpha)$ *and* $\widetilde{OP}(\alpha)$ *have the same finite optimal value for each* $\alpha \in [0,1]$*:*

$$\hat{h}(\alpha) = \sup OP(\alpha) = \inf \widetilde{OD}(\alpha) = \sup \widetilde{OP}(\alpha), \ 0 \leqq \alpha \leqq 1,$$

*where* $\hat{h}$ *is the concave conjugate function of* h, *defined in* (7.33) *and characterized in Proposition* 7.14.
*b. If* $\alpha = 0$ *and* $T_0 := R(b) = \infty$ *then* $OP(\alpha)$ *and* $\widetilde{OD}(\alpha)$ *do not have an optimal solution. In all other cases optimal solutions exist for each of the three problems.*
*c.* $T^*$ *is optimal for* $OP(\alpha)$ *iff* $T^* \in \partial\hat{h}(\alpha)$. *This condition is equivalent to* $T^* \in \Phi_c^{-1}(1-\alpha)$, *where* $\Phi_c$ *is the distribution function defined in* (7.44). *Hence* $T^* \in (T_1,T_0)$ *if* $\alpha \in (0,1)$. *Values for* $T^*$ *outside* $[T_1,T_0]$ *can only occur if* $\alpha = 1$ *(but then also* $T^* = T_1$ *itself is good) or if* $\alpha = 0$ *(but then also* $T^* = T_0$ *is good).*
*d.* $T^*$ *is an optimal solution for* $OP(\alpha)$ *iff an optimal solution* $z^*$ *of* $\widetilde{OD}(\alpha)$ *exists with* $T^* = R(z^*)$.
*e.* $z^*$ *is optimal for* $\widetilde{OD}(\alpha)$ *and* $\pi^* \in \Pi(\alpha)$ *is optimal for* $\widetilde{OP}(\alpha)$ *iff* $z_i^* \in F_i^{-1}(1 - \pi_i^*)$ *and* $\pi_i^* > 0 \Rightarrow i \in CA(z^*)$, *both for* $i = 1,\ldots,n$.

PROOF. (a) See (7.65) and (7.64).
(d) Follows from the remark before (7.65).

(c) Since h is finite and convex, $T^*$ is optimal for OP($\alpha$) iff $-\alpha \in \partial h(T^*)$, that is $T^* \in \partial\hat{h}(\alpha)$, or equivalently, $T^* \in \Phi_c^{-1}(1-\alpha)$, see (7.62) and (7.63). The remaining statements follow then from Proposition 7.14.

(b) From (c) and (d) it follows that OP($\alpha$) and $O\tilde{D}(\alpha)$ have optimal solutions iff $\partial\hat{h}(\alpha) \neq \emptyset$, and this condition holds unless $\alpha = 0$ and $T_0 = \infty$, see Proposition 7.14. Since for $\alpha \in [0,1]$ $\Pi(\alpha)$ is a nonempty bounded polyhedron and $\hat{G}$ is continuous on it, $O\tilde{P}(\alpha)$ has optimal solutions for all $\alpha \in [0,1]$.

(e) As indicated in the proof of Proposition 7.14 $O\tilde{P}(\alpha)$ is the dual problem of $O\tilde{D}(\alpha)$. By reformulating the corresponding Karush-Kuhn-Tucker conditions one gets the optimality conditions to be proved. The analysis is completely similar to that of Theorem 7.13. We omit the details. □

Of course, the worst-case distributions corresponding to an optimal solution $T^*$ of the outer problem OP($\alpha$) are the optimal solutions of the inner problem IP($T^*$). They are characterized in Theorem 7.15, and their criticality numbers are the solution of $I\tilde{P}(T^*)$, see Theorem 7.16. In fact, if one solves OP($\alpha$) by means of solving $O\tilde{P}(\alpha)$, one gets these criticality numbers right away, as can be verified by a detailed comparison of the flow problems $O\tilde{P}(\alpha)$ and $I\tilde{P}(T^*)$.

We conclude the discussion on the outer problem with a remark on sensitivity analysis. Instead of fixing $\alpha$ one might be interested in the optimal solutions of OP($\alpha$) for various values of $\alpha$, even for the whole range $\alpha \in [0,1]$. In the latter case the family of outer problems $\{OP(\alpha),\ \alpha \in [0,1]\}$ is completely equivalent to the family $\{IP(T),\ T \in \mathbb{R}\}$ of inner problems, in the sense that the optimal values as functions of the parameters are the conjugate functions h and $\hat{h}$, which determine each other completely. For instance, one might solve IP(T) for a number of values of T, say $T := T_k$, $k = 1,\ldots,K$, by solving $I\tilde{P}(T_k)$. Let $(\pi_0^{(k)}, \pi^{(k)})$ be the optimal solution of $I\tilde{P}(T_k)$. Then $T_k$ must be an optimal solution for OP($\alpha_k$) for $\alpha_k := \pi_0^{(k)}$, since $-\alpha_k \in \partial h(T_k)$, see Theorem 7.13, and this is equivalent to $T_k \in \partial\hat{h}(\alpha_k)$. Moreover, since $\hat{h}$ is concave it follows that optimal solutions $T_\alpha^*$ for OP($\alpha$) are decreasing functions of $\alpha$, so that interpolation is possible:

if $\alpha_k \leqq \alpha \leqq \alpha_1$ then $T_1 \leqq T^*_\alpha \leqq T_k$.

7.4.6. Worst-case marginals. Up to now we assumed that the marginal distribution $F_i$ of the various activities of the project are known completely. From a practical point of view this assumption is not realistic. We will now relax this assumption. It appears that the results on worst-case joint distributions can be combined in a natural way with a minimax approach to the marginals, too. Then problem IP(T) is replaced by

$$\text{maximize}_{F_i} \{\sup_{H\in\mathcal{H}(F_1,\ldots,F_n)} E_H f_T(\xi) : F_i \in \mathcal{F}_i, \; i = 1,\ldots,n\}$$

where $\mathcal{F}_i$ is the class of all distribution functions for $\xi_i$ compatible with the available information. The equivalence of IP(T) and $\widetilde{\text{ID}}$(T) shows, that we may also write

$$\text{maximize}_{\mathcal{F}_i} \{\min_{z\in\mathbb{R}^n} \{f_T(z) + \Sigma_{i=1}^n E_{F_i}(\xi_i - z_i)^+\}\}.$$

If for i = 1,...,n the class $\mathcal{F}_i$ contains a "*uniformly*" worst-case element, that is, if an $F_i^* \in \mathcal{F}_i$ exists with

$$\text{maximum}_{F_i\in\mathcal{F}_i} E_{F_i}(\xi_i - z_i)^+ = E_{F_i^*}(\xi_i - z_i)^+$$

$$\textit{for all } z_i \in \mathbb{R},$$

then $\max_{\mathcal{F}_i}$ and $\min_z$ may be interchanged. In that case the analysis of IP(T) is directly applicable, by taking for $F_i$ the worst-case marginal distribution $F_i^*$.

EXAMPLES 7.18. If $\mathcal{F}_i$ is the class of all distributions with given bounded range $[a_i,b_i]$, then $F_i^*(\{b_i\}) = 1$, since $r(x) := (x-z)^+$ is nondecreasing. Less trivial is the situation if also the mean value $\bar{\xi}_i$ is known; then the convexity of r implies that now the support of $F_i^*$ is $\{a_i,b_i\}$, see [43,5]:

$$\text{(7.82)} \qquad F_i^*(\{a_i\}) := \frac{b_i - \bar{\xi}_i}{b_i - a_i}, \quad F_i^*(\{b_i\}) := \frac{\bar{\xi}_i - a_i}{b_i - a_i}.$$

In [43] it is also shown, that if $\mathcal{F}_i$ is the class of all symmetrical unimodular distributions on $[a_i, b_i]$, then $F_i^*$ is the uniform distribution on $[a_i, b_i]$. If $\mathcal{F}_i$ is the class of all unimodular distributions on $[a_i, b_i]$ with a given mode $m_i$ then $F_i^*$ is the uniform distribution on $[m_i, b_i]$, see [2]. This last specification of $\mathcal{F}_i$ is the most interesting, since it contains precisely the information which usually is supposed to be known in PERT networks.

7.4.7. Numerical calculations. The convex, polyhedral function $f_T(z) = (R(z) - T)^+$ can be represented as the solution of an LP program (as in (7.46), or in a more common dynamic formulation in terms of earliest times for all nodes). Therefore, $\widetilde{\mathrm{ID}}(T)$ and $\widetilde{\mathrm{OD}}(\alpha)$ are simple recourse models for which good algorithms exist. For example, if all $F_i$ are discrete then also each $G_i$ can be put into an LP form, so that $\widetilde{\mathrm{ID}}(T)$ and $\widetilde{\mathrm{OD}}(\alpha)$ are LP programs. Of course, also the original programs IP(T) and OP($\alpha$) are then finite-dimensional LP problems but their size is tremendously larger. If the $F_i$ are uniform distributions $\widetilde{\mathrm{ID}}(T)$ and $\widetilde{\mathrm{OD}}(\alpha)$ are quadratic with linear constraints; in the general case the objective G(z) is separable convex. Similar statements hold for the flow formulations $\widetilde{\mathrm{IP}}(T)$ and $\widetilde{\mathrm{OP}}(\alpha)$; e.g.

$$\hat{G}_i(\pi_i) = \max_y \{\Sigma_{j=1}^m t_j y_j : \Sigma_{j=1}^m y_j = \pi_i,\ 0 \leq y_j \leq p_j \quad \forall_j\}$$

if $F_i$ is the discrete distribution $F_i(\{t_j\}) = p_j$, $j = 1,\dots,m$, with $t_1 > t_2 > \dots > t_m$ and $\Sigma_{j=1}^m p_j = 1$, $p_j > 0$. Moreover, the flow structure may be exploited to devise special algorithms, see [3,8]. In fact, in [3] a simple recourse model, which is slightly more general than $\widetilde{\mathrm{ID}}(T)$, is taken as a starting point, and a well-known network flow algorithm is adjusted to solve the problem with discrete distributions.

In order to get an idea about the influence of alternative specifications problem $\widetilde{\mathrm{OP}}(\alpha)$ has been solved for the project network of Figure 7.3. This network is a slight modification of the Electronic Module Development Project as described in [28 p. 216]. Apart from the network structure, the only information on the duration of activity i

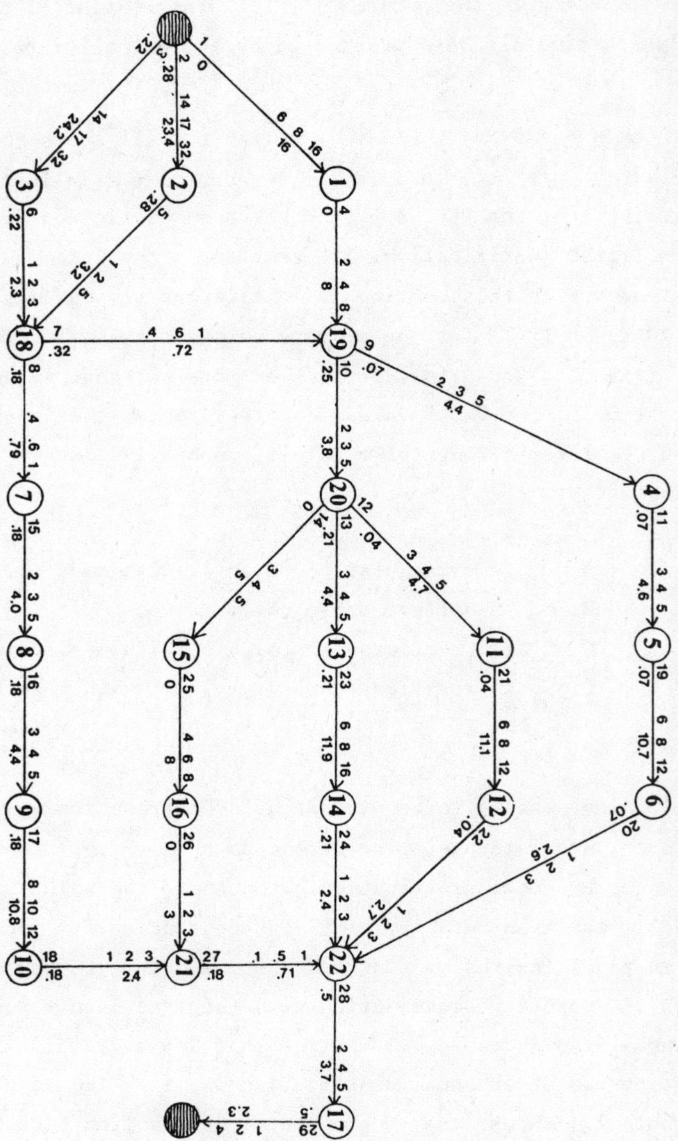

FIGURE 7.3. The Electronic Module Development Project. Above the arrow of each activity i the data i, $a_i$, $m_i$ and $b_i$ are given; underneath the solutions $\pi_i^*$ and $z_i^*$ for the triangular distribution A and $\alpha = 0.5$ are shown.

is given by $a_i$, $m_i$ and $b_i$ (see Figure 7.3), where $[a_i,b_i]$ is the support and $m_i$ the mode of $F_i$. This is the "standard" information needed for PERT. According to this heuristic, the distribution of the project completion time can be approximated by the normal distribution $N(\mu,\sigma^2)$, where $\mu := \Sigma_{i\in B_j} \bar{\xi}_i$, $\sigma^2 := \Sigma_{i\in B_j} \sigma_i^2$ for a $j \in CP(\bar{\xi})$, with $\bar{\xi}_i :=$ $(a_i + 4m_i + b_i)/6$ and $\sigma_i := (b_i - a_i)/6$ for all i. In the example we have $\mu = 47.7$ and $\sigma \approx 4.4$. In spite of its charming simplicity it is not surprising that the PERT approach has been criticized for its dubious probabilistic justification. It seems to us that the minimax approach as worked out in this section is an alternative. One might solve IP(T) (via $I\tilde{D}(T)$ or $I\tilde{P}(T)$) for one or more relevant values of T, or solve $OP(\alpha)$ (via $O\tilde{D}(\alpha)$ or $O\tilde{P}(\alpha)$) for one or more relevant values of $\alpha$, or solve IP(T) or $OP(\alpha)$ for "all" values in order to get the upper bound $\Phi_c$. We solved $OP(\alpha)$ for a range of $\alpha$-values, under four assumptions on the marginals:

A. $F_i$ is triangular on $[a_i,b_i]$, with mode $m_i$.
B. $F_i$ is uniform on $[m_i,b_i]$.
C. $F_i$ is restricted to $\{a_i,b_i\}$, with mean value $m_i$, see (7.82).
D. $F_i$ is restricted to $\{a_i,m_i,b_i\}$, with mean value $m_i$ and $F_i(\{m_i\}) = \frac{1}{2}$.

Distribution A seems to be a reasonable distribution compatible with the underlying interpretation of the data $(a_i,m_i,b_i)$, whereas B is the worst-case unimodal distribution. In C and D the value for $m_i$ is interpreted as the mean value $\bar{\xi}_i$ of $\xi_i$ rather than as its mode; C is the worst-case marginal for that situation, whereas D is a rather arbitrary "unimodal" discrete distribution with the same mean value. The results are summarized in Table 7.1. Notice that Table 7.1 gives also quantiles of the convex upper bound distribution $\Phi_c$, i.e. $1-\alpha$ as a function of $T^*_\alpha$. Table 7.1 shows, that the uniform distribution B is indeed much less favorable than the triangular distribution A; the two-point distribution C gives much higher values for $T^*_\alpha$ than the three-point distribution D for not too large values for $\alpha$. Of course, in any case the solutions coincide (approximately) for $\alpha$ = (approximately) 0. For distribution A and $\alpha = 0.5$ more details about the solutions are given in Figure 7.3: for each activity i the values for $\pi^*_i$ and $z^*_i$ are in-

$T^*_\alpha$

| | α=1.0 | 0.9 | 0.8 | 0.7 | 0.6 | 0.5 | 0.4 | 0.3 | 0.2 | 0.1 | 0.0 |
|---|---|---|---|---|---|---|---|---|---|---|---|
| A | 46.8 | 49.3 | 50.9 | 52.4 | 54.0 | 55.6 | 57.4 | 59.4 | 61.9 | 65.3 | 76.0 |
| B | 58.0 | 59.4 | 60.7 | 62.2 | 63.6 | 65.1 | 66.5 | 68.2 | 70.2 | 72.4 | 76.0 |
| C | 41.4 | 45.4 | 46.0 | 46.0 | 51.0 | 56.0 | 56.0 | 71.0 | 71.0 | 76.0 | 76.0 |
| D | 42.1 | 42.1 | 44.1 | 44.1 | 48.6 | 49.6 | 52.0 | 55.0 | 57.0 | 71.0 | 76.0 |

$\hat{h}(\alpha)$

| | α=1.0 | 0.9 | 0.8 | 0.7 | 0.6 | 0.5 | 0.4 | 0.3 | 0.2 | 0.1 | 0.0 |
|---|---|---|---|---|---|---|---|---|---|---|---|
| A | 56.5 | 51.7 | 46.7 | 41.5 | 36.2 | 30.7 | 25.1 | 19.3 | 13.2 | 6.9 | 0.0 |
| B | 66.0 | 59.7 | 53.7 | 47.5 | 41.3 | 34.8 | 28.2 | 21.5 | 14.6 | 7.4 | 0.0 |
| C | 57.0 | 53.3 | 48.7 | 44.1 | 39.3 | 34.0 | 28.4 | 22.2 | 15.1 | 7.6 | 0.0 |
| D | 52.0 | 48.6 | 44.3 | 39.9 | 35.2 | 30.3 | 25.0 | 19.8 | 14.2 | 7.6 | 0.0 |

TABLE 7.1. The optimal solutions $T^*_\alpha$ and the corresponding optimal values $\hat{h}(\alpha)$ of OP($\alpha$) for different distributions and $\alpha$-values; $T_1 = T^*_1$, $T_c = \hat{h}(1)$, $T_0 = T^*_0$.

dicated. If, loosely speaking, we interpret $\alpha$ as the maximum acceptable "risk" which we are ready to take under unfavorable conditions, $\pi^*_i$ indicates the (uniquely) determined contribution of activity i to that risk. If $\pi^*_i = 0$ then activity i is not important in that respect, and the promise $z^*_i$ can be taken maximal $= b_i$, but the more $\pi^*_i/\alpha$ is close to one, the more risk we have to accept in activity i. By the way, this example also illustrates that the criticality numbers of the paths are far from unique: every optimal flow passes node 18, and the flows in each subpath from source to node 18 and in each subpath from node 18 to the sink are uniquely determined by $\pi^*$, but there are many possibilities to combine the flows in subpaths to flows in paths.

REFERENCES

1. S. CAMBANIS, G. SIMONS, W. STOUT (1976). Inequalities for Ek(X,Y) when the marginals are fixed. *Z. Wahrsch. Verw. Gebiete 36*, 285-294.
2. T. CIPRA (1978). Class of unimodal distributions and its transformations. *Časopis Pěst. Mat. 103*, 17-26.

3. H.J. CLEEF, W. GAUL (1982). Project scheduling via stochastic programming. *Math. Operationsforsch. Statist. Ser. Optim. 13*, 449-469.
4. B. DODIN (1985). Bounding the project completion time distribution in PERT networks. *Oper. Res. 33*, 862-881.
5. J. DUPACOVA (1976). Stochastic and dynamic programming: experience in stochastic programming models. A. PREKOPA (ed.). *Survey of Mathematical Programming*, Proceedings of the IXth International Mathematical Programming Symposium, Hungarian Academy of Sciences, Budapest, part 5, 99-105.
6. S.E. ELMAGHRABY (1967). On the expected duration of PERT type networks. *Management Sci. 5*, 229-306.
7. S.E. ELMAGHRABY (1977). *Activity Networks*, Wiley, New York-London-Sydney-Toronto.
8. D.R. FULKERSON (1961). A network flow computation for project cost curves. *Management Sci. 7*, 167-178.
9. D.R. FULKERSON (1962). Expected critical path length in PERT type networks. *Oper. Res. 10*, 808-817.
10. N. GAFFKE, L. RÜSCHENDORF (1981). On a class of extremal problems in statistics. *Math. Operationsforsch. Statist. Ser. Optim. 12*, 123-135.
11. N. GAFFKE, L. RÜSCHENDORF. On the existence of probability measures with given marginals. *Statist. Decisions*, to appear.
12. U. GARCIA-PALOMARES, E. GINE (1977). On the linear programming approach to the optimality of Prokhorov's distance. *J. Math. Anal. Appl. 60*, 596-600.
13. W. GAUL (1981). Bounds for the expected duration of a stochastic project planning model. *J. Inform. Optim. Sci. 2*, 45-63.
14. W. GAUL (1983). Bounding distributions for random project-completion-times. M.J. BECKMANN, W. EICHHORN, W. KRELLE (eds.). *Mathematische Systemen in der Oekonomie*, Königstein/Ts, 169-179.
15. W.-R. HEILMANN (1981). A mathematical programming approach to Strassen's theorem on distributions with given marginals. *Math. Operationsforsch. Statist. Ser. Optim. 12*, 593-596.
16. H.G. KELLERER (1961). Functionen auf Produkträumen mit vorgegebenen Marginalfunktionen. *Math. Ann. 144*, 323-344.
17. H.G. KELLERER (1964). Verteilungsfunktionen mit gegebenen Marginalverteilungen. *Z. Wahrsch. Verw. Gebiete 3*, 247-270.

18. H.G. KELLERER (1964). Masstheoretische Marginalprobleme. *Math. Ann. 153*, 168-198.

19. H.G. KELLERER (1984). Duality theorems and probability metrics. M. IOSIFESCU (ed.). *Proceedings of the Seventh Conference on Probability Theory*, Rumanian Academy of Sciences, Bucureşti, 211-220.

20. H.G. KELLERER (1984). Duality theorems for marginal problems. *Z. Wahrsch. Verw. Gebiete 67*, 339-432.

21. G.B. KLEINDORFER (1971). Bounding distributions for a stochastic acyclic network. *Oper. Res. 19*, 1586-1601.

22. W.K. KLEIN HANEVELD (1984). Distributions with known marginals and duality of mathematical programming with application to PERT. M. IOSIFESCU (ed.). *Proceedings of the Seventh Conference on Probability Theory*, Rumanian Academy of Sciences, Bucureşti, 221-235.

23. T.L. LAI, H. ROBBINS (1978). A class of dependent random variables and their maxima. *Z. Wahrsch. Verw. Gebiete 42*, 89-111.

24. E.L. LEHMAN (1966). Some concepts of dependence. *Ann. Math. Statist. 37*, 1137-1152.

25. A.W. MARSHALL, I. OLKIN (1979). *Inequalities: Theory of Majorization and its Applications*, Academic Press, New York-London-Toronto-Sydney-San Francisco.

26. J.J. MARTIN (1965). Distribution of the time through a directed, acyclic network. *Oper. Res. 13*, 46-66.

27. I. MEILIJSON, A. NADAS (1979). Convex majorization with an application to the length of critical paths. *J. Appl. Probab. 16*, 671-677.

28. J.J. MODER, C.R. PHILLIPS (1964). *Project Management with CPM and PERT*, Reinhold, New York.

29. A. NADAS (1979). Probabilistic PERT. *IBM J. Res. Develop. 23*, 339-347.

30. P. ROBILLARD, M. TRAHAM (1977). The completion time of PERT networks. *Oper. Res. 25*, 15-29.

31. R.T. ROCKAFELLAR (1970). *Convex Analysis*, Princeton University Press, Princeton NJ.

32. L. RÜSCHENDORF (1980). Inequalities for the expectation of Δ-monotone functions. *Z. Wahrsch. Verw. gebiete 54*, 341-349.

33. L. RÜSCHENDORF (1981). Sharpness of Fréchet-bounds. *Z. Wahrsch. Verw. Gebiete 57*, 293-302.

34. L. RÜSCHENDORF (1982). Random variables with maximum sums. *Adv. in Appl. Probab. 14*, 623-632.

35. L. RÜSCHENDORF (1983). Solution of a statistical optimization problem by rearrangement methods. *Metrika 30*, 55-61.

36. L. RÜSCHENDORF (1981). *Robustness against Dependence*, Report Technical University Aachen, Aachen.

37. G. SCHAY (1974). Nearest random variables with given distributions. *Ann. Probab. 2*, 163-166.

38. A.W. SHOGAN (1977). Bounding distributions for a stochastic PERT network. *Networks 7*, 359-381.

39. V. STRASSEN (1965). The existence of probability measures with given marginals. *Ann. Math. Statist. 36*, 423-439.

40. A. SZULGA (1978). On the Wasserstein metric. J. KOZESNIK (ed.). *Transactions of the 8th Prague Conference on Information Theory, Statistical Decision Functions, Random Processes*, Czechoslovak Academy of Sciences, Prague, Volume B, 267-273.

41. A. SZULGA (1982). On minimal metrics in the space of random variables. *Theory Probab. 2*, 401-405.

42. W. WHITT (1976). Bivariate distributions with given marginals. *Ann. Statist. 4*, 1280-1289.

43. J. ZACKOVA (1966). On minimax solutions of stochastic linear programming problems. *Časopis Pěst. Mat. 91*, 423-429.

# CHAPTER 8

# A DUAL OF A DYNAMIC INVENTORY CONTROL MODEL: THE DETERMINISTIC AND THE STOCHASTIC CASE

## 8.1 INTRODUCTION

In this chapter we study a simple multistage production-inventory control model. Two cases are considered: first it is assumed that the demands are deterministic, and after that the demands are supposed to be random variables.

Usually, such a problem is solved by means of the dynamic programming algorithm (DPA). Here it is also formulated as a linear programming problem. The reason is that we want to analyze the dual problem arising from perturbations of the demands. In particular, we are interested in the economic interpretation of the dual problem and its optimal solution as well as in the relation between the dual problem and the dynamic programming algorithm.

Already in Section 4.6 we derived a dual problem to a general stochastic dynamic programming (SDP) problem, which includes the (stochastic) inventory control model of this chapter as a very special case. We showed there that an arbitrary SDP problem can be formulated in a natural way as an abstract linear programming problem. Its decision variables are the state-action probability measures, and its constraints express the dynamic structure of the model: they describe the relations between the state-action probabilities of subsequent stages. The corresponding dual problem appeared to be a linear programming representation of the backward recursion DPA. The dual pair of linear programming problems to be studied in the present chapter is completely different. The variables of the primal linear program are, at least in the stochastic case, *functions* rather than probability measures: they represent productions and inventories as functions of the demands in the past. The constraints describe, as in Section 4.6, the dynamic structure of the problem, but the dynamics are now represented by the system function rather than by a transition probability. It will appear that the abstract linear programming duality theory of Section 4.2 is also applicable to problems in function spaces, as is the case in the present chapter. The dual problem is easily formulated; its variables form a "price system" related to the optimal way of satisfying the demands.

In order that the approach by linear programming in function spaces works it is necessary to assume that all costs (production costs, holding costs and shortage costs) are linear. Of course, duality of convex programming could be applied if the cost function are convex rather than linear. But the duality setting of this chapter would fail if e.g. set-up costs would be present. We will restrict ourselves to the linear case, since we want to get a dual problem in its most explicit form. For simplicity, we will also assume that there is only one commodity, and that there are no production, shortage, labor, liquidity, ... constraints. It is also assumed that backlogging of demand is possible; for analogous results in the no-backlogging case we refer to [8]. Finally, we consider only models with a finite number of stages. Within these restrictions the model of this chapter is general: the cost functions are nonstationary, and, in the stochastic case, we allow for the possibility that the demands in the various periods are stochastically *dependent*. In this respect the model is more general that the Markovian inventory control models discussed in Section 6.4.

The contents of this chapter is summarized as follows. In Section 8.2 we study the case that all demands are deterministic. Actually, the problem is then a transshipment problem, and the optimal dual solution is easily characterized completely. It is intimately related to the Dijkstra tree of the underlying graph. It is also possible to solve the dual problem iteratively. Two such methods, to be called *dual dynamic programming methods*, are described. The first one is equivalent to the primal dynamic programming method in (sub-)differentiable form; it can also be described as an iterative proces in which a Dijkstra tree of a subgraph is adjusted to get a Dijkstra tree of a larger subgraph. The second one is obtained by application of conjugacy to the primal dynamic programming method.

In Section 8.3 it is assumed that the future demands are unknown, except for their probability distribution. Then the problem is a multistage stochastic linear program as described in Chapter 3. Applying the abstract linear programming duality theory developed in Section 4.2 we derive analogous results as in the deterministic case for the primal as well as for the dual problem, except for the fact that the Dijkstra trees do not play a dominant role. If the demand distribution has a density the dual problem can be solved iteratively by the differentiable form of the dynamic programming recursion. As in the deterministic case a second dual dynamic programming method is developed by application of conjugacy. Furthermore, explicit expressions for the optimal dual solution are given in terms of probabilities of certain events. These expressions seem to be appropriate to study the sensitivity of the model with respect to the distribution of the demands.

In Section 8.4 we give an economic interpretation of the dual problems and their optimal solutions.

## 8.2 DETERMINISTIC DEMANDS

This section deals with the inventory control model under the assumption that all parameters of the problem, the future demands being included, are known. This assumption reduces the model to a very simple one. The main reason for the analysis of the deterministic model is that we want to have the results at our disposal when the stochastic case is studied in the next section.

The contents of this section is divided into six subsections. In 8.2.1 the optimization problem is introduced. The dynamic programming approach to this problem is summarized in 8.2.2. In the next subsection the dynamic programming recursion is transformed into a backward recursion for the derivatives of the cost-to-go functions. The linear programming formulation of the optimization problem is given in 8.2.4. Special attention is given to the dual problem. Finally, in 8.2.5 and 8.2.6 two dual dynamic programming methods are described.

8.2.1. Formulation of the problem. Consider the following production-inventory control model. A shop deals with a single commodity. During N successive time periods customers demand the amounts $\xi_n \in \mathbb{R}^+$, $n = 1,\ldots,N$. In the beginning of period n it is possible to produce an amout $x_n \in \mathbb{R}^+$, at costs $k_n.x_n$. Let $y_n$ be the stock after that production. If $y_n \geqq \xi_n$ then the demand is going to be satisfied, the initial stock in the next period is $y_n - \xi_n$ and the holding cost is $h_n.(y_n - \xi_n)$. If, however, $y_n < \xi_n$, then the shortage has to be satisfied in a later period (so that there is *backlogging*) and the shortage cost $q_n.(\xi_n - y_n)$ is incurred. It is convenient to represent shortages by negative inventories. Let $s_0 \in \mathbb{R}$ be the initial inventory, and $s_n \in \mathbb{R}$ the inventory at the end of period n, then $s_n = s_{n-1} + x_n - \xi_n$, and the holding-and-shortage cost in period n is $h_n.(s_n)^+ + q_n.(s_n)^-$, $n = 1,\ldots,N$, where $x^\pm := \max(0,\pm x)$. The terminal stock is $s_N$. If $s_N < 0$ the shortage $-s_N$ has to be replenished by an additional purchase elsewhere, at the unit price $k_{N+1}$. If $s_N > 0$ this remaining stock is sold at the salvage unit price r. So the terminal costs are $k_{N+1}.(s_N)^- - r.(s_N)^+$. More generally, we suppose that $\bar{s}$ is a nonnegative target value of the terminal stock and therefore the terminal costs are $k_{N+1}.(s_N - \bar{s})^- - r.(s_N - \bar{s})^+$. The inventory level as a function of time is illustrated in Figure 8.1.

We assume that $s_0$, $\bar{s}$ and all unit costs are known to the shopkeeper together with the demands $\xi_1,\ldots,\xi_N$. The optimization problem is then:

($P_o$) find productions $x_1,\ldots,x_N \geqq 0$ such that the total cost is minimized.

We assume that problem $P_o$ is *proper*. By this we mean that the unit costs satisfy the following conditions:

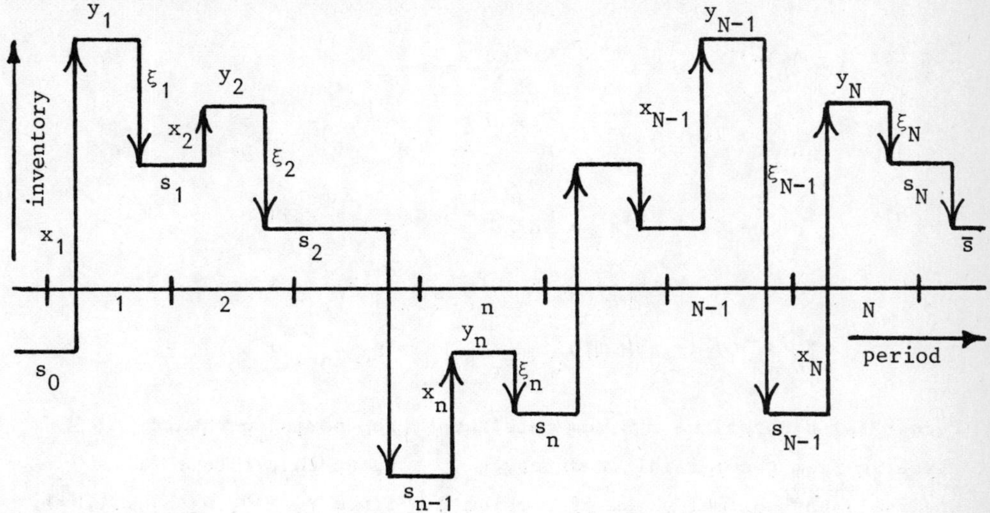


FIGURE 8.1. Diagram of the inventory level as a function of time.

$$
(8.1) \qquad \begin{aligned} &\gamma_n := q_n + h_n \geqq 0, && n = 1,\dots,N, \\ &\delta_n := k_n + \Sigma_{i=n}^{N} h_i - r \geqq 0, && n = 1,\dots,N+1. \end{aligned}
$$

If for some n we would have that $\gamma_n < 0$, then at stage n it were profitable not to deliver even if the stock at hand were positive, since the combined unit holding-and-shortage cost would be negative. If for some n we would have that $\delta_n < 0$, then in period n it were profitable to produce unbounded amounts to get unbounded low costs.

In order to simplify certain formulae we will sometimes use the following additional notations:

$$
h_0 := q_0 := \gamma_0 := 0;\ q_{N+1} := k_{N+1},\ h_{N+1} := -r,\ \gamma_{N+1} := \delta_{N+1};
$$

$$
\xi_0 := -s_0,\ \xi_{N+1} := \bar{s},\ \xi_{-1} := -\Sigma_{i=0}^{N+1}\xi_i;\ s_{N+1} := s_N - \xi_{N+1}.
$$

8.2.2. <u>The dynamic programming recursion</u>. It is well-known that problem $P_o$ is easily solved by dynamic programming. Then one calculates the value functions (or cost-to-go functions) $f_{N+1}, f_N, \dots, f_1$ successively as follows:

(8.2a) $f_{N+1}(s_N) := k_{N+1}.(s_N - \bar{s})^- - r.(s_N - \bar{s})^+,\ s_N \in \mathbb{R},$

and for $n = N, N-1, \ldots, 1$, $y_n \in \mathbb{R}$ and $s_{n-1} \in \mathbb{R}$,

$$g_n(y_n) := h_n.(y_n - \xi_n)^+ + q_n.(y_n - \xi_n)^- + f_{n+1}(y_n - \xi_n),$$

(8.2b) $$f_n(s_{n-1}) := \inf_{x_n \geqq 0}\{k_n.x_n + g_n(s_{n-1} + x_n)\} =$$

$$= -k_n.s_{n-1} + \inf_{y_n \geqq s_{n-1}}\{k_n.y_n + g_n(y_n)\},$$

$$X_n^*(s_{n-1}) := \arg\min_{x_n \geqq 0}\{k_n.x_n + g_n(s_{n-1} + x_n)\}.$$

Here $f_n(s)$ denotes the minimum total cost from period n onward starting from the initial inventory $s_{n-1} = s$, so that $f_1(s_0)$ is precisely the optimal value of problem $P_o$. Since $\gamma_n \geqq 0$, $n = 1,\ldots,N+1$, all functions $f_n$ and $g_n$ are convex, as can be shown by an inductive proof. They are even polyhedral. The conditions $\delta_n \geqq 0$, $n = 1,\ldots,N$, guarantee that $f_n$ and $g_n$ are finite, since

$$k_n.y + g_n(y) \geqq \delta_n.y + \text{constant},\ y \in \mathbb{R},$$

which can be proved by induction. Together with the polyhedrality of $g_n$ this inequality implies that $X_n^*(s) \neq \emptyset$ for all $s \in \mathbb{R}$, $n = 1,\ldots,N$. Therefore, $P_o$ has optimal solutions, and $(x_1^*, x_2^*, \ldots, x_N^*)$ solves $P_o$ iff $x_n^* \in X_n^*(s_{n-1}^*)$ where $s_0^* := s_0$ and $s_n^* := s_{n-1}^* + x_n^* - \xi_n$, $n = 1,\ldots,N$.

We call any element $y_n^*$ of

(8.2c) $$Y_n^* := \arg\min_{y_n \in [-\infty,\infty]}\{k_n.y_n + g_n(y_n)\}$$

an *optimal inventory level* for stage n, $n = 1,\ldots,N$. This definition is self-evident since optimal productions are completely characterized by optimal inventory levels:

$$x_n^* \in X_n^*(s) \Leftrightarrow \exists y_n^* \in Y_n^*,\ y_n^* < \infty:\ x_n^* = (y_n^* - s)^+,$$

as follows from the convexity of the functions $g_n$. Therefore, we may and shall interpret $P_o$ as the problem to find all optimal level sets

$Y_n^*$, n = 1,...,N.

In the next subsection we will give a more explicit characterization of the optimal level sets $Y_n^*$ by a more detailed analysis of the functions $f_n$ and $g_n$, see Figure 8.2. The analysis is based on the (sub-)differentiable form of the dynamic programming algorithm, and it exploits the polyhedrality of the functions $f_n$ and $g_n$.

8.2.3. The differentiable form of the dynamic programming recursion. Since the functions $f_n$ and $g_n$ defined in (8.2) are finite and polyhedral, they are differentiable except in a finite number of points, and between these points the derivatives are constant. We will show that the dynamic programming recursion (8.2) can be transformed into a similar backward recursion for the derivatives, and that the optimal inventory level sets can also be found by this differentiable form of the dynamic programming algorithm.

We introduce the following notations. For n = 1,...,N+1

$$\begin{aligned} &\eta_{n,n-1} := -\infty, \\ (8.3) \qquad &\eta_{n,i} := \Sigma_{j=n}^{i} \xi_j, \; i = n,n+1,\dots,N+1, \\ &\eta_{n,N+2} := +\infty. \end{aligned}$$

Inspection of (8.2) shows, that for each n the functions $f_n$ and $g_n$ are differentiable everywhere except possibly at the points $\eta_{ni}$, i = n,n+1,...,N+1, and that their derivatives satisfy the recursion, with $\chi_A(s) := 1$ if $s \in A$ and $\chi_A(s) := 0$ if not:

$$f'_{N+1}(s_N) = -k_{N+1} + \gamma_{N+1} \cdot \chi_{(\xi_{N+1},\infty)}(s_N), \; s_N \neq \eta_{N+1,N+1},$$

and for n = N,N-1,...,1

$$\begin{aligned} g'_n(y_n) &= -q_n + \gamma_n \cdot \chi_{(\xi_n,\infty)}(y_n) + f'_{n+1}(y_n - \xi_n) \\ &\qquad\qquad y_n \neq \eta_{ni}, \; i = n,n+1,\dots,N+1, \\ f'_n(s_{n-1}) &= \max\{-k_n, g'_n(s_{n-1})\}, \\ &\qquad\qquad s_{n-1} \neq \eta_{ni}, \; i = n,n+1,\dots,N+1. \end{aligned}$$

The expression for $f_n'(s_{n-1})$ holds since $g_n$ is convex. Moreover, since $f_n$ and $g_n$ are polyhedral functions their derivatives are constant on intervals. Therefore, we introduce for all relevant values of n and i

$$\sigma_{ni} := g_n'(y) \text{ if } \eta_{n,i-1} < y < \eta_{ni},$$

$$\rho_{ni} := f_n'(s) \text{ if } \eta_{n,i-1} < s < \eta_{ni}.$$

Then we have the following recursion for $\rho_{ni}$ and $\sigma_{ni}$:

$$\text{(8.4a)} \qquad \rho_{N+1,N+1} = -k_{N+1}, \ \rho_{N+1,N+2} = -r,$$

and for $n = N, N-1, \ldots, 1$

$$\sigma_{nn} = -q_n + \rho_{n+1,n+1},$$

$$\text{(8.4b)} \qquad \sigma_{ni} = h_n + \rho_{n+1,i} \quad , \ i = n+1, \ldots, N+2,$$

$$\rho_{ni} = \max\{-k_n, \sigma_{ni}\} \quad , \ i = n, \ldots, N+2.$$

We call (8.4a,b) the (sub-)*differentiable form of the dynamic programming recursion* (8.2a,b). It is not difficult to derive, that the optimal inventory levels are completely determined by (8.4b):

$$\text{(8.4c)} \qquad \eta_{ni} \in Y_n^* \text{ iff } -\sigma_{n,i+1} \leq k_n \leq -\sigma_{ni}, \ i = n-1, \ldots, N+2,$$

where we set $\sigma_{n,n-1} := -\infty$ and $\sigma_{n,N+3} := +\infty$. Of course, $Y_n^*$ is the convex hull of all elements which are found in (8.4c). In Figure 8.2 we sketched the graphs of $f_n$ and $g_n$.

<u>8.2.4. The linear program and its dual</u>. We are interested in the sensitivity of $P_o$ with respect to the values of $\xi_0 = -s_0$, $\xi_1, \ldots, \xi_N, \xi_{N+1} = \bar{s}$. This sensitivity can easily be studied by putting $P_o$ in a linear programming format. Define the following linear programming problem in the real variables $u_0$, $v_0$, $x_1$, $u_1, v_1, x_2, \ldots, u_N, v_N, x_{N+1}$ and w, where $\xi_{-1} := -\Sigma_{i=0}^{N+1}\xi_i$ and $h_0 := q_0 := 0$:

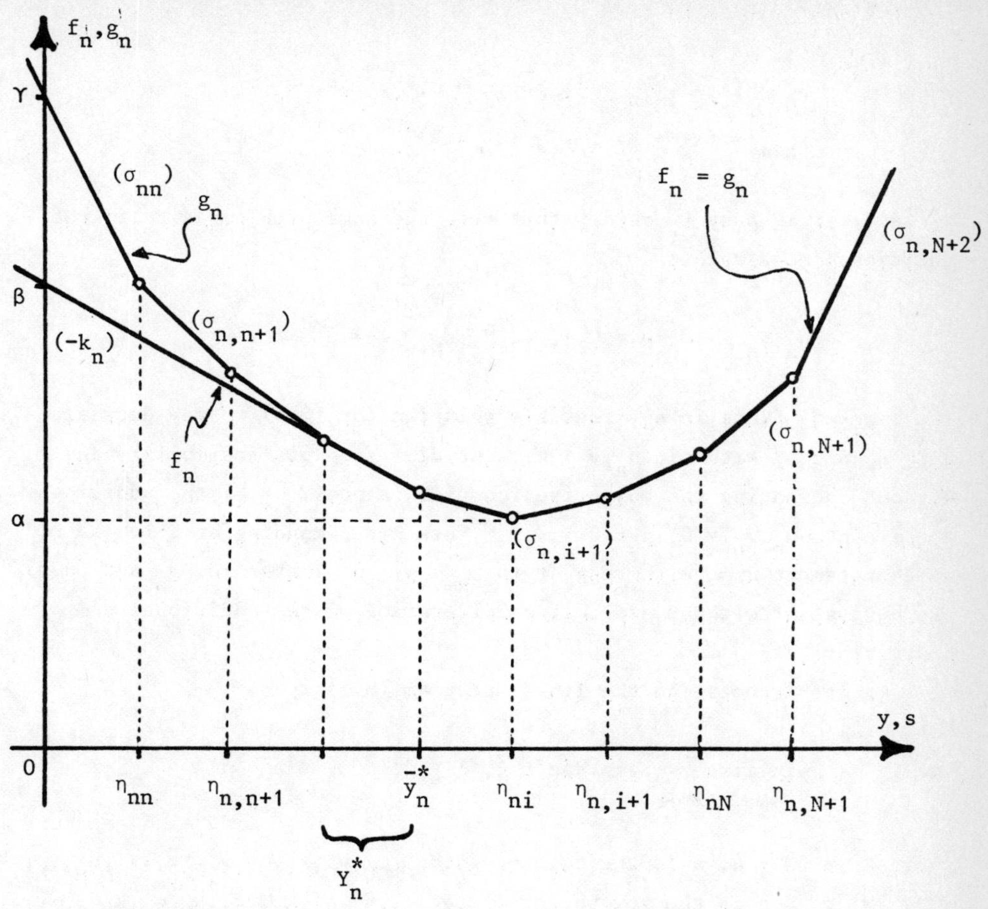


FIGURE 8.2. The functions $g_n$ and $f_n$ as defined in (8.2). Slopes are indicated by (.). The numbers $\alpha := \inf g_n$, $\beta := f_n(0)$, $\gamma := g_n(0)$ are the same as in Figure 8.5.

$$
(8.5)\quad \begin{cases}
\text{minimize } \Sigma_{n=0}^{N}(h_n.u_n + q_n.v_n + k_{n+1}.x_{n+1}) - r.w \\
\text{subject to} \\
-u_0 + v_0 = \xi_0 \\
u_{n-1} - v_{n-1} + x_n - u_n + v_n = \xi_n \qquad n = 1,\ldots,N \\
u_N - v_N + x_{N+1} - w = \xi_{N+1} \\
-x_1 - x_2 \ldots. - x_{N+1} + w = \xi_{-1} \\
\text{and all } u_n, v_n, x_n, w \geqq 0.
\end{cases}
$$

Of course the condition $-\Sigma_{i=1}^{N+1} x_i + w = \xi_{-1}$ is superflous since it is equal to minus the sum of the other equalities. The relation of (8.5) to problem $P_o$ is given by

$$u_n = (s_n)^+,\ v_n = (s_n)^-,\ n = 0,1,\dots,N,$$
$$x_{N+1} = (s_N - \bar{s})^-,\ w = (s_N - \bar{s})^+.$$

In fact, it is easy to verify that P is the same problem as $P_o$ if the additional constraints

$$u_n \cdot v_n = 0,\ n = 0,1,\dots,N,\ x_{N+1} \cdot w = 0$$

are imposed. Since in any feasible solution for (8.5) we may decrease both $u_n$ and $v_n$ with $\min(u_n,v_n)$ without violating its feasibility and without increasing the objective function (since $\gamma_n \geqq 0$) the additional constraints $u_n \cdot v_n = 0$, $n = 0,1,\dots,N$, are not demanding at all. The same statement holds with respect to $x_{N+1} \cdot w = 0$. Therefore, $P_o$ and (8.5) are equivalent without explicit consideration of the additional constraints.

In matrix notation the linear program (8.5) reads

$$\text{(P)} \qquad \underset{z \in \mathbb{R}^M}{\text{minimize}} \{ez: \bar{A}z = \bar{\xi},\ z \geqq 0\}$$

where $M := 3N + 4$, z is the column vector $(u_0,v_0,x_1,u_1,v_1,x_2,\dots,u_N,v_N,x_{N+1},w)^T$, e is the row vector $(h_0,q_0,k_1,h_1,q_1,k_2,\dots,h_N,q_N,k_{N+1},-r)$, $\bar{\xi}$ is the column vector $(\xi_0,\xi_1,\dots,\xi_N,\xi_{N+1},\xi_{-1})^T$ and $\bar{A}$ is the constraint matrix implicitly defined in (8.5). A fundamental property of P is that it is a *transshipment* problem, since the constraint matrix $\bar{A}$ is the incidence matrix of a finite directed connected graph G, see Figure 8.3. The vertex set of G is $V := \{V_{-1},V_0,V_1,\dots,V_N,V_{N+1}\}$ and its edge set is $E := \{H_n,Q_n,K_{n+1}$ for $n = 0,\dots,N$, and R$\}$. The unit transportation cost of $H_n$, $Q_n$, $K_{n+1}$ and R is $h_n$, $q_n$, $k_{n+1}$ and $-r$, respectively. For any n, if $\xi_n < 0$ then $V_n$ represents a source with supply $-\xi_n$, and if $\xi_n > 0$ then $V_n$ represents a sink with demand $\xi_n$. If $\xi_n = 0$ then $V_n$ is an intermediate point. Notice that our assumptions imply that the only possible sources are $V_0$ and $V_{-1}$. The variables $u_n$, $v_n$, $x_n$ and w describe the flows in $H_n$, $Q_n$, $K_n$ and R, respectively. The constraints $\bar{A}z = \bar{\xi}$ assure the conservation of flow at each vertex of G, and the nonnegativity of the flows means that they must be in the direction of

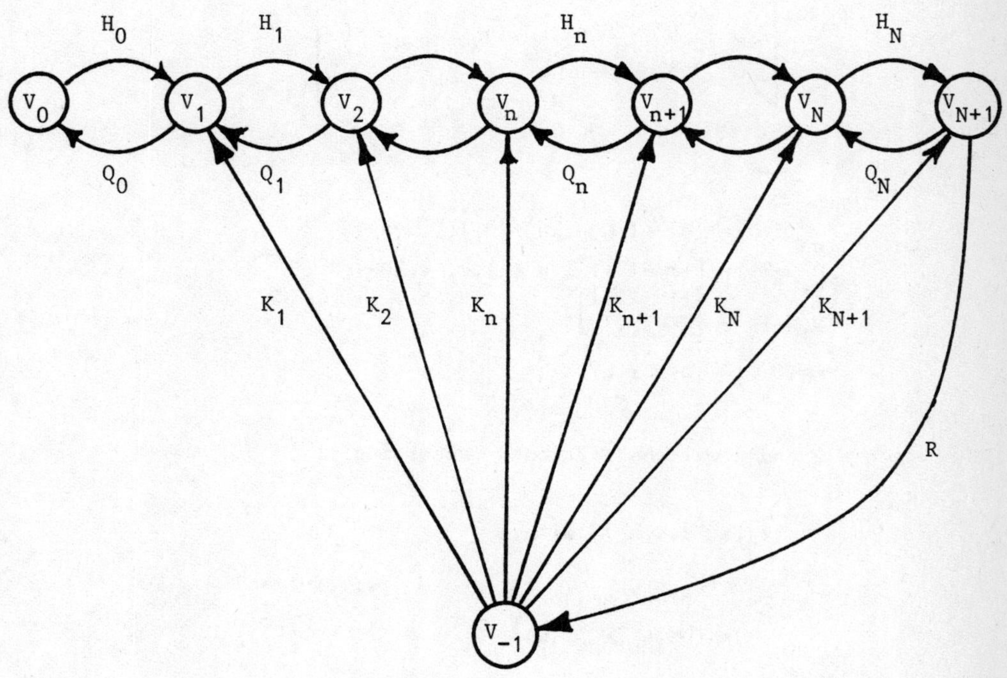


FIGURE 8.3. The graph G.

the edges. Hence P is the problem to find feasible flows with minimum total cost.

Since G is connected and $\xi_{-1} = -\Sigma_{i=0}^{N+1}\xi_i$, P is feasible for all values of $(\xi_0,\xi_1,\ldots,\xi_{N+1})$. Its optimal value is finite if and only if it is not profitable to transport around some directed cycle in G. This is true precisely if the unit costs satisfy (8.1), i.e. iff $P_o$ is proper.

Omitting the superfluous constraint $-\Sigma_{i=1}^{N+1}x_i + w = \xi_{-1}$ we get the equivalent formulation

$$\text{(P)} \qquad \underset{z\in\mathbb{R}^M}{\text{minimize}} \{ez\colon Az = \xi,\ z \geqq 0\}$$

where $\xi := (\xi_0,\xi_1,\ldots,\xi_{N+1})^T$ and where A is a lower block stair-case matrix

$$A = \begin{bmatrix} A_{00} & 0 & 0 & \cdots & 0 & 0 \\ A_{10} & A_{11} & 0 & \cdots & 0 & 0 \\ 0 & A_{21} & A_{22} & \cdots & 0 & 0 \\ \vdots & \vdots & \vdots & & \vdots & \vdots \\ 0 & 0 & 0 & \cdots & A_{N-1,N-1} & 0 \\ 0 & 0 & 0 & \cdots & A_{N,N-1} & A_{NN} \end{bmatrix}$$

with

$$\begin{aligned} A_{nn} &:= [-1\ \ +1\ \ 0]\ ,\ n = 0,1,\dots,N-1, \\ A_{n,n-1} &:= [+1\ \ -1\ \ +1],\ n = 1,2,\dots,N-1, \\ A_{N,N-1} &:= \begin{bmatrix} +1 & -1 & +1 \\ 0 & 0 & 0 \end{bmatrix}, \\ A_{NN} &:= \begin{bmatrix} -1 & +1 & 0 & 0 \\ +1 & -1 & +1 & -1 \end{bmatrix}. \end{aligned}$$

The vectors z and e will be partioned accordingly:

$$z = (z_0, z_1, \dots, z_N)^T \text{ with}$$

$$z_n := (u_n, v_n, x_{n+1})^T \text{ for } n = 0,1,\dots,N-1 \text{ and}$$

$$z_N := (u_N, v_N, x_{N+1}, w)^T,$$

$$e = (e_0, e_1, \dots, e_N) \text{ with}$$

$$e_n := (h_n, q_n, k_{n+1}) \text{ for } n = 0,1,\dots,N-1 \text{ and}$$

$$e_N := (h_N, q_N, k_{N+1}, -r).$$

The usual dual problem of P, related to the perturbation of the right-hand side $\xi$, is

(D) $$\text{maximize}_{\pi \in \mathbb{R}^{N+2}} \{\pi\xi : \pi A \leqq e\}$$

where $\pi := (\pi_0, \dots, \pi_{N+1})$ is the row vector of dual variables. In more detail

(D)
$$\begin{aligned} &\text{maximize} \quad \textstyle\sum_{n=0}^{N+1} \pi_n \cdot \xi_n \\ &\text{subject to } -h_n \leqq \pi_n - \pi_{n+1} \leqq q_n,\ n = 0,1,\dots,N, \\ &\qquad\qquad\qquad\qquad \pi_n \leqq k_n,\ n = 1,\dots,N, \\ &\qquad\qquad r \leqq \qquad \pi_{N+1} \leqq k_{N+1}. \end{aligned}$$

Not surprisingly, the optimal values of the dual variables can be interpreted as equilibrium prices; more precisely as price differences $\pi_n - \pi_{-1}$ with $\pi_{-1} := 0$. The constraints in D express the fact that in an equilibrium the price $\pi_k$ of the commodity at a vertex $V_k$ may not exceed the price $\pi_n$ at an adjacent vertex $V_n$ plus the cost of transportation from $V_n$ to $V_k$. We will work out this interpretation in Section 8.4. Another useful way of looking at problem P is to interpret the unit cost of each edge as its "length". Then D can be seen as a generalized shortest path problem: for example, if we set all $\xi_n$ equal to 0 except $\xi_k := 1$ then the optimal value $\pi_k^*$ of D is precisely the distance (= length of the shortest path in G) from $V_{-1}$ to $V_k$.

The complementary slackness conditions of P and D are $(\pi A - e)z = 0$, that is

$$\begin{aligned}
u_n &.(h_n + \pi_n - \pi_{n+1}) = 0, \; n = 0,1,\dots,N,\\
v_n &.(q_n - \pi_n + \pi_{n+1}) = 0, \; n = 0,1,\dots,N,\\
x_{n+1} &.(k_{n+1} - \pi_{n+1}) = 0, \; n = 0,1,\dots,N,\\
w &.(\pi_{N+1} - r) = 0.
\end{aligned}$$

These conditions stipulate that optimal flows use only those edges for which equality holds in the dual constraints.

Before characterizing the optimal solutions of P and D, let us first recall some well-known results about *basic* solutions of $Az = \xi$, see for instance [4,3]. The set of columns of A indexed by $T \subset \mathcal{E}$ is a column-basis for A if and only if T is the edge set of a spanning tree of G. By slight abuse of language we will identify a tree with its edge set. If we denote the flow in an arbitrary edge $E \in \mathcal{E}$ by $z_E$ and the unit transportation cost of E by $e_E$, then the basic solutions of P and D, respectively, corresponding to an arbitrary spanning tree T can be written as

$$\text{(8.6)} \quad \begin{aligned}
z_E &= z_E(T) := \Sigma_{n=0}^{N+1} \alpha_{En}(T).\xi_n, \; E \in \mathcal{E},\\
\pi_n &= \pi_n(T) := \Sigma_{E \in \mathcal{E}} e_E.\alpha_{En}(T), \; n = 0,\dots,N+1,
\end{aligned}$$

where $\alpha_{En}(T) = +1$ if the unique path from $V_{-1}$ to $V_n$ in T contains E, and E is directed away from $V_{-1}$,

$\alpha_{En}(T) = -1$ if the unique path from $V_{-1}$ to $V_n$ in T contains E, and E is directed towards $V_{-1}$,

$\alpha_{En}(T) = 0$ otherwise.

Denote these basic solutions by $z(T) := \{z_E(T)\}$ and $\pi(T) := \{\pi_E(T)\}$, respectively. Of course, (8.6) satisfies the complementary slackness conditions automatically, and the corresponding values of the objective function of P and D coincide. Therefore, (8.6) describes optimal solutions for P and D, respectively, if and only if z(T) is feasible for P and $\pi(T)$ is feasible for D. In that case we call the spanning tree T *optimal*.

As indicated before, the nonnegativity of the demands $\xi_1,\ldots,\xi_N$ as well as of the target value $\xi_{N+1}$ of the terminal inventory implies that only $V_0$ and $V_{-1}$ can be sources in the transshipment problem. Apart from the trivial case $\xi_0 = \xi_1 = \ldots = \xi_N = \xi_{N+1} = 0$ there are three possibilities:

(i) if $-\xi_0 \geqq \Sigma_{i=1}^{N+1}\xi_i$ then only $V_0$ is a source,

(ii) if $\xi_0 \geqq 0$ then only $V_{-1}$ is a source,

(iii) if $0 < -\xi_0 < \Sigma_{i=1}^{N+1}\xi_i$ then both $V_{-1}$ and $V_0$ are sources.

Suppose for a moment that there is only one source. Then it is clear that a spanning tree T is primal feasible (i.e. z(T) is feasible for P) if T is a *branching* of G rooted in the source, i.e. if all edges of T are directed away from the source (see Figure 8.4).This condition is also necessary for primal feasibility if all other vertices have strictly positive demands. Moreover, it is obvious that a braching is primal optimal (dual feasible) if it is a *Dijkstra tree* rooted in the source. By definition, a Dijkstra tree of G with root $V_k$ is a branching T of G with root $V_k$ such that for each vertex $V_n$ the unique path in T from $V_k$ to $V_n$ is a shortest path in G. This characterization of optimal branches enables us to describe optimal spanning trees for each $\xi$.

Case (i). If $-\xi_0 \geqq \Sigma_{i=1}^{N+1}\xi_i$ then only $V_0$ is a source. Moreover, since there is only one simple directed path from $V_0$ to each other vertex of G, there is only one branching rooted in $V_0$, namely $T_0 := \{H_0,H_1,\ldots,H_N,R\}$. Therefore $T_0$ must be optimal. That is, if the initial stock is sufficient for the total demand, the optimal solution is simply to produce never anything. The corresponding optimal solution for

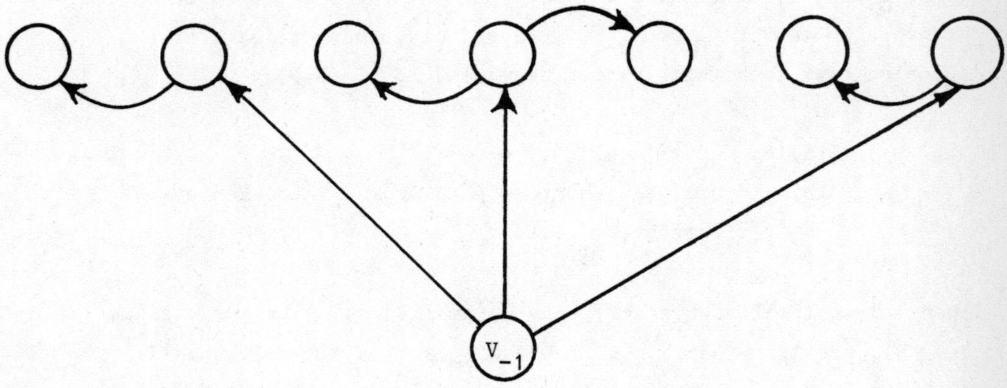


FIGURE 8.4. A branching of G rooted in $V_{-1}$.

D is $\pi_n = r - \Sigma_{j=n}^{N} h_j$, n = 0,1,...,N+1, see (8.6).

Case (ii). If $\xi_0 \geqq 0$ then only $V_{-1}$ is a source. We conclude that *if* $\xi_n \geqq 0$, $n = 0,\ldots,N+1$ *then in order that the spanning tree* $T^*$ *determines an optimal solution for* P *and* D *it is sufficient that* $T^*$ *is a Dijkstra tree rooted in* $V_{-1}$. *If each* $\xi_n > 0$ *then this condition is necessary, too.* A simple but important observation is that the actual values of $\xi_n$, $n = 0,\ldots,N+1$ are more or less irrelevant. In terms of the dual problem this fact can be expressed as follows. *There is a unique solution* $\pi^* = (\pi_0^*,\pi_1^*,\ldots,\pi_{N+1}^*)$ *of* D *which is feasible and optimal for all* $(\xi_0,\xi_1,\ldots,\xi_{N+1}) \geqq 0$. So the optimal cost $\Sigma_{n=0}^{N+1}\pi_n^*.\xi_n$ is linear on $\xi \geqq 0$, and the optimal price $\pi_n^*$ is equal to the distance in G from $V_{-1}$ to $V_n$. The distances can be easily calculated with efficient shortest path algorithms. In formula

$$(8.7) \qquad \pi_n^* = \min_{m=1,2,\ldots,N+1}\{\Sigma_{j=n}^{m-1} q_j + k_m + \Sigma_{j=m}^{n-1} h_j\},\ n = 0,1,\ldots,N+1,$$

where an empty sum is interpreted to be equal to zero. One calculates all optimal solutions of P by looking at the Dijkstra trees $T^*$ determined by $\pi^*$ and by using (8.6) i.e. by using complementary slackness. We get $x_n^* = 0$ if $K_n \notin T^*$ and $x_n^* = \Sigma_{i=a}^{b}\xi_i$ if $K_n \in T^*$ for some a and b, $0 \leqq a \leqq n \leqq b \leqq N+1$, depending on n and $T^*$ *but not on*

$\xi_0, \xi_1, \ldots, \xi_{N+1}$.

Case (iii). If $-\Sigma_{j=1}^{N+1}\xi_j < \xi_0 < 0$ then there are two sources, $V_0$ and $V_{-1}$. However, since there is only one directed path from $V_0$ to each sink, it is easy to characterize optimal spanning trees in terms of optimal branches rooted in $V_0$ and $V_{-1}$, respectively. Indeed, let T be a spanning tree such that, with $m := \min\{n: \Sigma_{i=0}^{n}\xi_i > 0\}$,

(a) $H_i \in T$, $i = 0,1,\ldots,m$,

(b) T contains a shortest path from $V_{-1}$ to each of the vertices $V_m, V_{m+1}, \ldots, V_{N+1}$,

then T is optimal. Conversely, any optimal spanning tree satisfies (a) and (b), provided that $\Sigma_{i=0}^{m-1}\xi_i < 0$ and $\xi_n > 0$, $n = m, m+1, \ldots, N+1$. By (8.6) we find as the optimal solution for D corresponding to T:

$$\pi_n = \pi_m^* - \Sigma_{j=n}^{m-1} h_j \quad \text{if } n < m,$$
$$\pi_n = \pi_n^* \quad \text{if } n \geqq m.$$

Finally, let us write down the value of $\pi_1$ in the optimal solution of D as a function $\pi^{(1)}(s)$ of $s = -\xi_0$, letting $\xi_1, \xi_2, \ldots, \xi_{N+1}$ be non-negative constants. This value is uniquely defined if $s \neq \Sigma_{j=1}^{i}\xi_j$, $i = 1,2,\ldots,N+1$, and the analysis above shows that

$$(8.8)\qquad \pi^{(1)}(s) = \begin{cases} \pi_1^* & \text{if } s < \xi_1, \\ -\Sigma_{j=1}^{m-1}h_j + \pi_m^* & \text{if } \Sigma_{j=1}^{m-1}\xi_j < s < \Sigma_{j=1}^{m}\xi_j,\ m = 2,\ldots,N+1, \\ -\Sigma_{j=1}^{N}h_j + r & \text{if } \Sigma_{j=1}^{N+1}\xi_j < s. \end{cases}$$

We can interpret $\pi^{(1)}(s)$ as the "to-the-initial-stock-s-adjusted-distance" from $V_{-1}$ to $V_1$ in the graph G. Using (8.1) and the definition of $\pi_i^*$ one easily shows that $\pi^{(1)}(s)$ is nonincreasing with s.

8.2.5. A dual dynamic programming method. We now will investigate, how the optimal inventory levels $y_1^*, \ldots, y_N^*$ can be derived from optimal spanning trees. Consider first $y_1^*$. Let $T^*$ be any Dijkstra tree with root $V_{-1}$. If $\{K_1, H_0, \ldots, H_{b-1}\} \subset T^*$, but, if $b \leqq N$, $H_b \notin T^*$ then one derives from the above discussion that there exists an optimal solution for P with $x_1 = (\xi_0 + \xi_1 + \ldots + \xi_b)^+$ for all $\xi_0 \in \mathbb{R}$. Hence $y_1^* = \Sigma_{i=1}^b \xi_i$ is optimal in that case. If $K_1 \notin T^*$ then we find the optimal level $y_1^* = -\infty$ analogously. Of course, if $T^*$ is not unique, b is not necessarily unique either and one gets all optimal inventory levels $y_1^*$ by taking convex combinations.

In the same way, the optimal inventory levels $y_n^*$ of an arbitrary period n are easily found by constructing the Dijkstra trees $T_n^*$ of the subgraph $G_n$ of G, where $G_n$ is defined by omitting from G the vertices $V_0, V_1, \ldots, V_{n-1}$ and the corresponding edges. In fact, one way of determining the sets $Y_n^*$ of optimal inventory levels is the following recursion scheme:

$$\pi_{N+1}^{(N+1)} := k_{N+1} \tag{8.9a}$$

and for n = N,N-1,...,1

$$\begin{aligned} \pi_n^{(n)} &:= \min\{k_n, q_n + \pi_{n+1}^{(n+1)}\}, \\ \pi_{i+1}^{(n)} &:= \min\{k_n + \Sigma_{j=n}^i h_j, \pi_{i+1}^{(n+1)}\}, \quad i = n, n+1, \ldots, N. \end{aligned} \tag{8.9b}$$

Actually, if in (8.9b) the second number is the minimum for some $i = i_0$ then the same holds for all $i \geq i_0$. Clearly, $\pi_i^{(n)}$ *is the distance from* $V_{-1}$ *to* $V_i$ *in* $G_n$, so the recursion (8.9b) can be seen as the construction of $T_n^*$ from $T_{n+1}^*$. In order to find the optimal inventory levels one has to keep track of which number between the curly brackets determines the minimum in (8.9b); for $\eta_{ni}$ defined in (8.3) we have:

$$\begin{aligned} \eta_{n,n-1} &\in Y_n^* \text{ iff } k_n \geq q_n + \pi_{n+1}^{(n+1)}, \\ \eta_{nn} &\in Y_n^* \text{ iff } k_n \leq q_n + \pi_{n+1}^{(n+1)} \text{ and } k_n + h_n \geq \pi_{n+1}^{(n+1)}, \\ \eta_{nb} &\in Y_n^* \text{ iff } k_n + \Sigma_{j=n}^{b-1} h_j \leq \pi_b^{(n+1)} \text{ and} \\ &\qquad\quad k_n + \Sigma_{j=n}^{b} h_j \geq \pi_{b+1}^{(n+1)}, \quad b = n+1, \ldots, N, \\ \eta_{n,N+1} &\in Y_n^* \text{ iff } k_n + \Sigma_{j=n}^{N} h_j \leq \pi_{N+1}^{(n+1)}, \\ \eta_{n,N+2} &\in Y_n^* \text{ iff } k_n + \Sigma_{j=n}^{N} h_j = r. \end{aligned} \tag{8.9c}$$

We remark that $Y_n^*$ is the convex hull of those elements which are discovered by (8.9c). Of course, it is possible to give formulae for $\pi_i^{(n)}$ analogous to (8.7): for n = 1,2,...,N+1 and i = n,...,N+1 we have that

$$\pi_i^{(n)} = \min_{m=n,n+1,\dots,N+1}(\Sigma_{j=i}^{m-1}q_j + k_m + \Sigma_{j=m}^{i-1}h_j).$$

We call the solution method for $P_o$ described by (8.9) a *dual dynamic programming* method, since it finds the optimal inventory level sets $Y_n^*$ by solving the *dual* problem by recursion. For the same reason (8.2) is then called the *primal* dynamic programming method. In both methods the recursion is backward.

The dual value functions $\pi^{(n)}$, n = N+1,...,1, of (8.9) are closely related to the derivatives of the primal cost-to-go functions. Recall from Subsection 8.2.3 that the derivatives of $f_n$ and $g_n$ are constant on intervals, and that their values $\rho_{ni}$ and $\sigma_{ni}$ satisfy the recursion (8.4). By comparison of (8.4) and (8.9) one derives easily by induction that the $\rho_{ni}$ in fact are simple transformations of the distances $\pi_i^{(n)}$ in the subgraph $G_n$; for n = N+1,...,1 we have

(8.9e)
$$\rho_{ni} = \Sigma_{j=n}^{i-1}h_j - \pi_i^{(n)}, \quad i = n,\dots,N+1,$$
$$\rho_{n,N+2} = \Sigma_{j=n}^{N}h_j - r.$$

The formula for $\rho_{n,N+2}$ holds since $\delta_n = k_n + \Sigma_{j=n}^{N}h_j - r \geqq 0$. Similarly, for n = N,N-1,...,1

(8.9f)
$$\sigma_{nn} = -q_n - \pi_{n+1}^{(n+1)},$$
$$\sigma_{ni} = \Sigma_{j=n}^{i-1}h_j - \pi_i^{(n+1)}, \quad i = n+1,\dots,N+1,$$
$$\sigma_{n,N+2} = \Sigma_{j=n}^{N}h_j - r.$$

One may interpret $\sigma_{ni}$ in the following way. Suppose one has to decide in period n about the production destined for the demand in period $i \geqq n$. If $k_n < -\sigma_{ni}$ then $k_n + \Sigma_{j=n}^{i-1}h_j < \pi_i^{(n+1)}$ so immediate production is less expensive than a delay; if $k_n > -\sigma_{ni}$ it is the other way around. So $-\sigma_{ni}$ is a critical value of $k_n$ as far as the decision about the production in period n for period $i \geqq n$ is concerned. This interpretation of $\sigma_{ni}$ agrees completely with (8.4c). Also $\rho_{ni}$ has an interpretation in $G_n$: $k_n + \rho_{ni} = (k_n + \Sigma_{j=n}^{i-1} h_j) - \pi_i^{(n)} \geqq 0$ is the extra unit cost arising from production in period n for demand in period $i \geqq n$ instead of producing for period i at the best time $\geqq n$. Another interpretation of $\rho_{ni}$ is anologous to (8.8). Consider the inventory control model from period n onward, and denote the initial stock $s_{n-1}$ by s, and suppose $s \neq \eta_{ni}$, $i = n,n+1,\ldots,N+1$. Then the "to-s-adjusted-distance" from $V_{-1}$ to $V_n$ in the graph $G_n$, defined similarly to (8.8) and denoted by $\pi^{(n)}(s)$, is equal to

$$\pi^{(n)}(s) := \begin{cases} -\Sigma_{j=n}^{i-1} h_j + \pi_i^{(n)} & \text{if } \eta_{n,i-1} < s < \eta_{ni}, \\ & \qquad i = n,n+1,\ldots,N+1, \\ -\Sigma_{j=n}^{N} h_j + r & \text{if } \eta_{n,N+1} < s. \end{cases}$$

Hence

$$(8.10) \qquad \rho_{ni} = -\pi^{(n)}(s) \text{ for i and s related as above.}$$

<u>8.2.6. The conjugate dual dynamic programming method</u>. The recursion (8.9) is not the only candidate to be called a dual dynamic programming method. We will now propose a dual dynamic programming method which is more "dual" than (8.9) in the sense that its state variables are "prices" (dual variables) rather than "inventories" (primal variables). Recall that the dual problem can be written as

$$\text{maximize}_{\pi \in \mathbb{R}^{N+2}} \{\pi\xi : \pi A \leqq e\}$$

where A is the lower block stair-case matrix specified before. By using the structure of A, and by calculating the supremum iteratively,

$$\sup_{\pi \in \mathbb{R}^{N+2}} \{\ldots\} = \sup_{\pi_0 \in \mathbb{R}} \{\sup_{\pi_1 \in \mathbb{R}} \{\ldots\{\sup_{\pi_{N+1} \in \mathbb{R}} \{\ldots\}\}\ldots\}\},$$

one finds the following recursion scheme involving functions $p_n$: $\mathbb{R} \to [-\infty,\infty)$, $n = N,N-1,\ldots,0$, which solves D.

$$p_N(\pi_N) := \sup_{\pi_{N+1}\in\mathbb{R}} \{\pi_{N+1}.\xi_{N+1}: (\pi_N,\pi_{N+1})A_{NN} \leqq e_N\},$$

(8.11a)

$$P^*_{N+1}(\pi_N) := \arg\max_{\pi_{N+1}\in\mathbb{R}} \{\ldots:\ldots\},$$

and for $n = N-1,\ldots,0$,

$$p_n(\pi_n) := \sup_{\pi_{n+1}\in\mathbb{R}} \{\pi_{n+1}.\xi_{n+1} + p_{n+1}(\pi_{n+1}):$$

(8.11b)

$$\pi_n A_{nn} + \pi_{n+1}A_{n+1,n} \leqq e_n\},$$

$$P^*_{n+1}(\pi_n) := \arg\max_{\pi_{n+1}\in\mathbb{R}} \{\ldots:\ldots\},$$

and finally

$$\sup D = \sup_{\pi_0\in\mathbb{R}} \{\pi_0.\xi_0 + p_0(\pi_0)\},$$

(8.11c)

$$P^*_0 := \arg\max_{\pi_0\in\mathbb{R}} \{\ldots\}.$$

Obviously, $\pi^* = (\pi^*_0,\pi^*_1,\ldots,\pi^*_{N+1})$ solves D if and only if $\pi^*_0 \in P^*_0$, $\pi^*_{n+1} \in P^*_{n+1}(\pi^*_n)$, $n = 0,1,\ldots,N$. An elementary inductive proof shows that for $n \geqq 1$, $p_n(\pi_n)$ is finite iff

(8.12) $\sigma_{nn} \leqq -\pi_n \leqq \sigma_{n,N+2}$

for $\sigma_{ni}$ defined as in (8.4), and that each $p_n$ is a nondecreasing concave polyhedral function on its effective domain. Both statements are also direct consequences of the properties of the primal cost-to-go functions $g_n$ as we will show. The relation between $p_n$ and $g_n$ can be described nicely by conjugacy. Recall that the *convex conjugate function* $g^*$ of any function $g: \mathbb{R} \to (-\infty,\infty]$ is defined by

$$g^*(\pi) := \sup_{y\in\mathbb{R}} \{\pi.y - g(y)\},\ \pi \in \mathbb{R},$$

and that $(g^*)^* = g$ if g is finite and convex [13 Theorem 12.2]. For our purposes the *concave conjugate function* $\hat{g}$ of g is suitable; it is defined by $\hat{g}(\pi) := -g^*(-\pi)$, $\pi \in \mathbb{R}$. Since each function $g_n$, defined in (8.2), is finite and convex, we have for $n = 1,\ldots,N$

$$(8.13)\qquad \hat{g}_n(\pi_n) := \inf_{y_n \in \mathbb{R}} \{\pi_n . y_n + g_n(y_n)\},\ \pi_n \in \mathbb{R},$$

$$(8.14)\qquad g_n(y_n) = \sup_{\pi_n \in \mathbb{R}} \{-\pi_n . y_n + \hat{g}_n(\pi_n)\},\ y_n \in \mathbb{R}.$$

Now we claim that $p_n$ *is the concave conjugate function of* $g_n(. + \xi_n)$, $n = 1,2,\dots,N$; that is:

$$(8.15)\qquad \begin{aligned} p_n(\pi_n) &= \inf_{y_n \in \mathbb{R}} \{\pi_n . y_n + g_n(y_n + \xi_n)\} \\ &= -\pi_n . \xi_n + \hat{g}_n(\pi_n),\ \pi_n \in \mathbb{R}, \end{aligned}$$

or, equivalently,

$$(8.16)\qquad g_n(y_n + \xi_n) = \sup_{\pi_n \in \mathbb{R}} \{-\pi_n . y_n + p_n(\pi_n)\},\ y_n \in \mathbb{R}.$$

Let us prove (8.16). Its right-hand side can be written as

$$\sup_{(\pi_n,\pi_{n+1},\dots,\pi_{N+1})} \{\pi_n . (-y_n) + \Sigma_{i=n+1}^{N+1} \pi_i . \xi_i :$$
$$\pi_j A_{jj} + \pi_{j+1} A_{j+1,j} \leqq e_j,\ j = n,\dots,N-1,\ (\pi_N,\pi_{N+1}) A_{NN} \leqq e_N\},$$

and by linear programming duality this is equal to the optimal value of the feasible problem

$$\inf_{(z_n,z_{n+1},\dots,z_N)} \{\Sigma_{j=n}^{N} e_j z_j :$$
$$\begin{aligned} A_{nn} z_n &= -y_n, \\ A_{i,i-1} z_{i-1} + A_{ii} z_i &= \xi_i,\ i = n+1,\dots,N-1, \\ A_{N,N-1} z_{N-1} + A_{NN} z_N &= (\xi_N,\xi_{N+1})^T\}. \end{aligned}$$

This infimum is precisely the minimal total cost from period n+1 onward, if the initial stock $s_n$ is equal to $y_n$. That is, it is equal to $g_n(y_n + \xi_n)$. This completes the proof of (8.16), and of (8.15).

The basic relation between a pair of conjugate functions is that their subgradients are each other's inverse [13 Theorem 23.5]. Therefore,

$$(8.17) \qquad -\pi_n \in \partial g_n(y_n) \;\Leftrightarrow\; y_n \in \partial \hat{g}_n(\pi_n) \;\Leftrightarrow\; \hat{g}_n(\pi_n) = \pi_n . y_n + g_n(y_n).$$

Recall that the subgradient of a scalar *convex* function g at a point y is defined by

$$\partial g(y) := \{\pi \in \mathbb{R}:\; g(z) \geqq g(y) + \pi.(z-y) \;\forall z \in \mathbb{R}\},$$

whereas the subgradient (or supergradient) of a scalar *concave* function p at a point $\pi$ is defined by

$$\partial p(\pi) := \{y \in \mathbb{R}:\; p(z) \leqq p(\pi) + y.(z-\pi) \;\forall z \in \mathbb{R}\}.$$

The subgradient duality expressed in (8.17) gives us an opportunity to describe $\hat{g}_n$ (and $p_n$) in terms of $g_n$; see Figure 8.5 and compare it with Figure 8.2. The same duality shows that the recursion (8.11) solves the primal problem P. In order to see this, notice from (8.2c) that

$$Y_n^* \cap \mathbb{R} = \{y_n \in \mathbb{R}:\; -k_n \in \partial g_n(y_n)\},\; n = 1,\ldots,N.$$

Therefore, with (8.17) and (8.15),

$$(8.18) \qquad Y_n^* \cap \mathbb{R} = \partial \hat{g}_n(k_n) = \xi_n + \partial p_n(k_n),\; n = 1,\ldots,N.$$

This agrees completely with (8.4c) as far as the finite optimal levels are concerned; simple adjustments can be made to describe the conditions for $-\infty \in Y_n^*$ or $+\infty \in Y_n^*$.

It is tempting to transform the whole of the primal dynamic programming recursion (8.2) to a recursion for the concave conjugates $\hat{f}_n$ and $\hat{g}_n$ of $f_n$ and $g_n$, respectively. This transformation can be done by elementary convex analysis. As a result we get

$$(8.19a) \qquad \hat{f}_{N+1}(\pi_{N+1}) = \begin{cases} \pi_{N+1} . \xi_{N+1} & \text{if } r \leqq \pi_{N+1} \leqq k_{N+1}, \\ -\infty & \text{otherwise,} \end{cases}$$

and for $n = N, N-1, \ldots, 1$

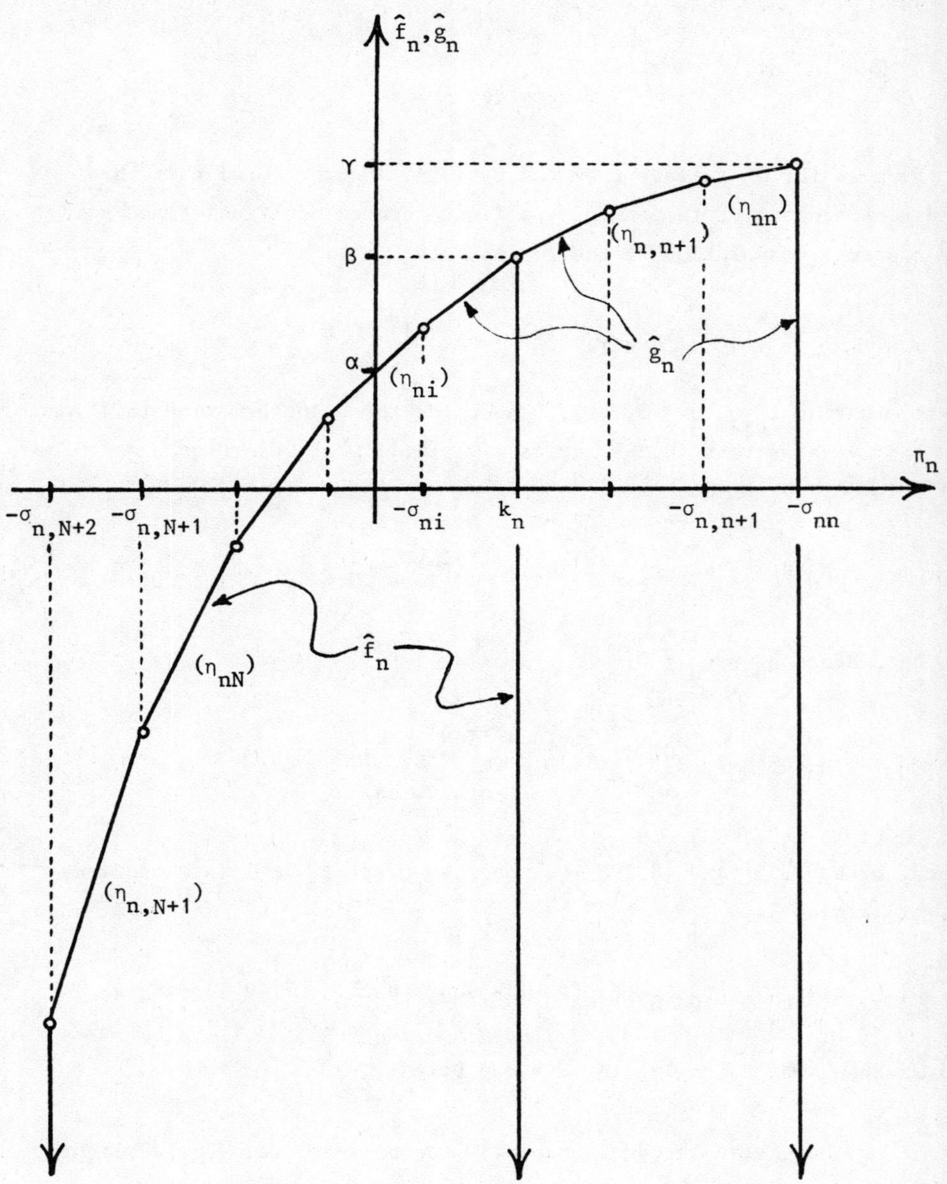


FIGURE 8.5. The concave conjugate functions $\hat{g}_n$ and $\hat{f}_n$, see (8.13) and (8.19b). Slopes are indicated by (.). The numbers $\alpha := \hat{g}_n(0)$, $\beta := \sup \hat{f}_n$, $\gamma := \sup \hat{g}_n$ are the same as in Figure 8.2. $\hat{f}_n(\pi_n) = -\infty$ if $\pi_n \notin [-\sigma_{n,N+2}, k_n]$ and $\hat{g}_n(\pi_n) = -\infty$ if $\pi_n \notin [-\sigma_{n,N+2}, -\sigma_{nn}]$.

$$\hat{g}_n(\pi_n) = \pi_n.\xi_n + \sup_{\pi_{n+1}} \{\hat{f}_{n+1}(\pi_{n+1}): -h_n \leqq \pi_n - \pi_{n+1} \leqq q_n\},$$

(8.19b)

$$\hat{f}_n(\pi_n) = \begin{cases} \hat{g}_n(\pi_n) & \text{if } \pi_n \leqq k_n, \\ -\infty & \text{otherwise.} \end{cases}$$

The formula for $\hat{g}_n$ follows from the fact [13 Theorem 16.4] that the concave conjugate of the sum of two finite convex functions f and g is the supremal convolution of their conjugates

$$(f+g)^\wedge(\pi) = \sup_{t\in\mathbb{R}} \{\hat{f}(t) + \hat{g}(\pi-t)\},\ \pi \in \mathbb{R}.$$

By eliminating $\hat{f}_{n+1}$, $n = N,\ldots,1$, and using the transformation (8.15) one easily shows that (8.19) agrees with (8.11), at least for $n \geqq 1$. For $n = 0$ in (8.11b) we have

$$p_0(\pi_0) = \sup_{\pi_1\in\mathbb{R}} \{\pi_1.\xi_1 + p_1(\pi_1): -h_0 \leqq \pi_0 - \pi_1 \leqq q_0,\ \pi_1 \leqq k_1\},$$

so that, since $h_0 = q_0 = 0$,

$$p_0(\pi_0) = \begin{cases} \pi_0.\xi_1 + p_1(\pi_0) & \text{if } \pi_0 \leqq k_1, \\ -\infty & \text{otherwise.} \end{cases}$$

Hence, with (8.15) and (8.19b) we conclude that $p_0 = \hat{f}_1$. This confirms (8.11c) since

(8.20) $$\sup_{\pi_0\in\mathbb{R}} \{\pi_0.\xi_0 + p_0(\pi_0)\} = \sup_{\pi_0} \{\pi_0.\xi_0 + \hat{f}_1(\pi_0)\} = f_1(-\xi_0)$$

and we know that $f_1(-\xi_0) = \inf P = \sup D$.

As we have seen the dual problem D can be interpreted as a slightly generalized shortest path problem. Its solution represents the to-the-initial-stock-adjusted-distances from $V_{-1}$ to the other vertices of G. The dual dynamic programming method (8.9a,b,c) solves D by constructing, iteratively for $n = N,N-1,\ldots,0$, the Dijkstra tree of the subgraph $G_n$ from that of the smaller subgraph $G_{n+1}$. The conjugate dual method has also an interpretation as a shortest path problem. In order to see this, we introduce a modification $\hat{G}_n(\pi_n)$ of the subgraph $G_n$. By definition, $\hat{G}_n(\pi_n)$ is the subgraph of G that arises if all vertices $V_0,V_1,\ldots,V_{n-1}$ and the corresponding edges are omitted, if an edge is added from $V_n$ to $V_{-1}$ with length $-\pi_n$ and if the length $k_n$ of $K_n$ is replaced by $\pi_n$. From (8.9f) it follows directly that $\hat{G}_n(\pi_n)$ has no negative cycles iff (8.12) holds. The distance from $V_{-1}$ to $V_n$ in $\hat{G}_n(\pi_n)$ is then precisely $\pi_n$. One may verify that $\pi_{n+1} \in P^*_{n+1}(\pi_n)$ *if and only if* $\pi_{n+1}$ *is the distance from* $V_{-1}$ *to* $V_{n+1}$ *in* $\hat{G}_n(\pi_n)$, at least if $\xi_j > 0$, $j \geqq n+1$.

We conclude that both dual dynamic programming methods (8.9) and (8.11) are transformations of the primal dynamic programming method (8.2). The former is equivalent to a recursion for the *derivatives* of the primal cost-to-go functions, whereas the latter is equivalent to a recursion for the concave *conjugates* of these functions.

## 8.3 STOCHASTIC DEMANDS

In this section we analyze the same inventory control model as in the previous section, but now under the assumption that the demands are random variables. More precisely, it is assumed that the realizations of the previous demands are known to the shopkeeper, but that he has to decide on the productions without knowing the values of the future demands; only their probability distribution is available. It is *not* assumed that demands in different periods are independent random variables. Therefore, in each period precisely the conditional distribution of the future demands given the realizations of the previous demands is known.

This section contains 7 subsections. We summarize their contents. After a formal statement of the problem in 8.3.1, which includes the introduction of policies, we deal with some technicalities on conditional expectations in 8.3.2. In 8.3.3 it is shown, that the dynamic programming recursion is well-defined, and that it generates optimal policies. These optimal policies are characterized by optimal inventory levels, which are functions of the previous demands due to the possible dependence between the demands. In 8.3.4 we formulate the problem as a multistage stochastic linear program in function spaces. It appears that the abstract linear programming duality theory, developed in Section 4.2, is useful here: the dual problem is easily derived, and optimality is characterized by feasibility and complementary slackness. The constraints of the dual problem have a form, that suggests a recursive solution method. In order to be able to formulate such a "dual dynamic programming method" in detail, we first analyze in 8.3.5 the derivatives of the (primal) cost-to-go functions. Assuming that the demands have *continuous* conditional distributions, it is shown not only that these derivatives exist, but also that the optimal dual solution can be expressed in terms of them. Therefore, the derivatives of the primal cost-to-go functions are studied in order to derive results on the dual problem. It appears, that the derivatives satisfy a backward recursion, to be called the differentiable form of the primal dynamic programming algorithm. Moreover, by solving it we get explicit expressions for the derivatives and, consequently, for the optimal dual solution too. In 8.3.6. the differentiable form of the primal dynamic programming method is transformed to a dual dynamic programming method. Finally, in 8.3.7 a second dual dynamic programming method is proposed by applying conjugacy to the primal method.

8.3.1. Formulation of the problem. We now suppose that the shopkeeper has no complete information about the future demands. In period n-1 the actual demand $\xi_{n-1}$ is observed, but the decision about the production $x_n$ has to be made before the demands $\xi_n, \xi_{n+1}, \ldots$ can be observed. The joint probability distribution of $(\xi_0, \ldots, \xi_N)$, where as before $-\xi_0$ stands for the initial inventory $s_0$ (which now will be interpreted as a random variable) is supposed to be known, so $x_n$ may depend on $(\xi_0, \xi_1, \ldots, \xi_{n-1})$, and also on the conditional probability distribution of $(\xi_n, \xi_{n+1}, \ldots, \xi_N)$ given the observed values of $(\xi_0, \xi_1, \ldots, \xi_{n-1})$ but not on their realizations. We assume that the target value of the terminal inventory $\bar{s} = \xi_{N+1} \geqq 0$ as well as the unit costs $k_n$, $h_n$, $q_n$, $-r$ are known constants.

A sequence $x = (x_1, x_2, \ldots, x_N)$ of nonnegative functions $x_n : \mathbb{R} \times \mathbb{R}_+^{n-1} \to \mathbb{R}_+$ is called a production *policy* if each function is Borel measurable. The set of policies is denoted by X. Clearly, $X \neq \emptyset$; e.g. $0 \in X$. The optimization problem is then

$(SP_o)$ find an $x \in X$ such that the expected total cost is minimum.

If such an $x \in X$ exists it is called optimal. The following assumptions assure that $\inf SP_o$ is finite.

(i) $E|\xi_i| < \infty$, $i = 0,1,\dots,N$.

(ii) $SP_o$ is proper, i.e. (8.1) holds.

In order to see this, remark first that the properness condition $\gamma_n \geq 0$ implies that the combined holding-and-shortage cost in period n, $1 \leq n \leq N$, is minorized by the linear function $h_n.s_n$ of the (positive of negative) inventory $s_n$ at the end of period n; in a similar way the terminal costs are minorized by $h_{N+1}.s_{N+1}$ where $s_{N+1} := s_N - \xi_{N+1}$. Therefore, the total cost is minorized by

$$\Sigma_{n=1}^{N+1} h_n.s_n + \Sigma_{n=1}^{N} k_n.x_n.$$

Substitution of $s_n = -\Sigma_{i=0}^{n}\xi_i + \Sigma_{i=1}^{n}x_i$, $n = 0,1,\dots,N$, gives

$$\Sigma_{n=1}^{N+1} h_n.(-\Sigma_{i=0}^{n}\xi_i) + \Sigma_{n=1}^{N}\delta_n.x_n.$$

Because of the properness assumption we know that each $\delta_n$ is nonnegative, so that the second term is nonnegative for all policies. Hence, the total cost is minorized by the first term, which has a finite expectation because of assumption (i). Hence $\inf SP_o > -\infty$. Also the inequality $\inf SP_o < +\infty$ can easily be established, namely by calculating the total cost $C_0(\xi)$ of the policy "never to produce anything", i.e. $x = 0$. One has

$$\begin{aligned} C_0(\xi) &= \Sigma_{n=1}^{N+1}\{h_n.(\Sigma_{i=0}^{n}\xi_i)^- + q_n.(\Sigma_{i=0}^{n}\xi_i)^+\} \\ &\leq c_0.\Sigma_{n=0}^{N+1}|\xi_n| \end{aligned}$$

for a suitable constant $c_0$. Assumption (i) implies that

$$\inf SP_o \leq E\, C_0(\xi) < \infty$$

so that $\inf SP_o$ is indeed finite.

In the sequel we need slightly stronger conditions on the unit costs than those given in (8.1). Let us call $SP_o$ *strictly proper* if (8.1) holds with the addition that $\delta_n > 0$, $n = 1,\dots,N$. This additional requirement will appear to be useful for the proof that optimal policies exist. From now on we assume therefore that $SP_o$ is strictly proper.

8.3.2. Some notations and remarks on conditional expectations. We now make a digression in order to build up the probabilistic framework for the dynamic programming approach to $SP_o$ as well as for its formulation as a multistage stochastic linear program.

Denote the range space of the random variable $\xi_n$ by $\Xi_n$, n = 0,1,...,N; that is $\Xi_0 := \mathbb{R}$, and $\Xi_n := \mathbb{R}_+$ since all demands are supposed to be nonnegative. Since $\xi^N := (\xi_0,\xi_1,\ldots,\xi_N)$ is the basic random vector in $SP_o$, we choose its range space $(\Xi^N,\mathcal{B}^N,\lambda^N)$ as sample space: $\Xi^N := \prod_{n=0}^N \Xi_n$, $\mathcal{B}^N$ is the σ-algebra of Borel sets in $\Xi^N$ and $\lambda^N$ is the given distribution of $\xi^N$. Then an arbitrary random variable is any real Borel measurable function on the sample space. We will not make a notational distinction between a random variable and its realization.

For $0 \leqq n \leqq m \leqq N$ we use the abbreviations:

$\xi_n^m := (\xi_n,\xi_{n+1},\ldots,\xi_m)$,

$\Xi_n^m := \prod_{i=n}^m \Xi_i$,

$\mathcal{B}_n^m :=$ the Borel σ-algebra of $\Xi_n^m$,

$\lambda_n^m :=$ the marginal distribution of $\xi_n^m$ with respect to $\lambda^N$.

In these notations, the lower index n is suppressed if n = 0, and the upper index m is suppressed if m = n. That is, $(\Xi^n,\mathcal{B}^n,\lambda^n)$ is the marginal probability space of $\xi^n := (\xi_0,\ldots,\xi_n)$, and $(\Xi_n,\mathcal{B}_n,\lambda_n)$ is that of $\xi_n$, n = 0,1,...,N.

Because of the sequential nature of $SP_o$ we will often use conditional expectations. In particular we need in period n conditioning upon the past realizations $\xi^{n-1}$. By definition, the *conditional expectation* of any random variable $\varphi(\xi^N)$ given $\xi^{n-1}$, n = 1,...,N(,N+1), is any $\mathcal{B}^{n-1}$-measurable function $E[\varphi|\xi^{n-1}]$ of $\xi^{n-1}$ which satisfies for all $D \in \mathcal{B}^{n-1}$

$$\int_D E[\varphi|\xi^{n-1}]\lambda^{n-1}(d\xi^{n-1}) = \int_{\{\xi^{n-1}\in D\}} \varphi(\xi^N)\lambda^N(d\xi^N).$$

Such a function, which again is a random variable, is also denoted by $E[\varphi(\xi^N)|\xi^{n-1}]$, or simply by $E^{n-1}\varphi(\xi^N)$ or $E^{n-1}\varphi$. Conditional expectations exist, although not uniquely; however, different versions are equal almost surely. Conditional expectations can and will be calculated by means of regular conditional distributions. A function $\mu_n^N(d\xi^N|\xi^{n-1})$: $B^N \times \Xi^{n-1} \to [0,1]$ is called a *regular conditional distribution* for $\xi^N$ given $\xi^{n-1}$ if

(i) for any fixed $B \in B^N$, $\mu_n^N(B|.)$ is a version of $E^{n-1}\chi_B$ where $\chi_B(\xi^N) := 1$ if $\xi^N \in B$ and $\chi_B(\xi^N) := 0$ otherwise;

(ii) for any fixed $\xi^{n-1} \in \Xi^{n-1}$, $\mu_n^N(.|\xi^{n-1})$ is a probability on $(\Xi^N, B^N)$;

or equivalently, if it is a *transition probability* from $(\Xi^{n-1}, B^{n-1})$ to $(\Xi^N, B^N)$ such that $\lambda^{n-1}\mu_n^N = \lambda^N$. (We refer to Section 4.3 for the definition of transition probabilities, and in particular to (4.21) for the definition of the "composition" $\lambda^{n-1}\mu_n^N$; here we use the trivial bounding functions $e_{n-1}(\xi^{n-1}) \equiv 1$ and $e_N(\xi^N) \equiv 1$.)

Since $\Xi^N$ is a Borel space, regular conditional distributions exist [2 Theorem 4.34]. They are useful, since for *all* random variables $\varphi$

$$\int\varphi(\xi^N)\mu_n^N(d\xi^N|\xi^{n-1})$$

is a version of $E^{n-1}\varphi$. But just as conditional expectations regular conditional distributions are not unique. For instance, if $\mu_n^N$ satisfies the conditions (i) and (ii) then so does $\tilde{\mu}_n^N$ defined by

$$\begin{aligned}\tilde{\mu}_n^N(B|\xi^{n-1}) &:= \mu_n^N(B|\xi^{n-1}), \quad B \in B^N,\ \xi^{n-1} \in D,\\ &:= \nu(B) \qquad\quad\ , \quad B \in B^N,\ \xi^{n-1} \notin D,\end{aligned}$$

for any probability $\nu$ on $(\Xi^N, B^N)$ and for any $D \in B^{n-1}$ with $\lambda^{n-1}(D) = 0$. Although the difference between two versions of the same regular conditional distribution is not very important, we will still be using modifications as the above in order to ascertain that some statements are true for *all* $\xi^{n-1} \in \Xi^{n-1}$ rather than only for $\lambda^{n-1}$-almost every $\xi^{n-1}$. For example, we want that for any random variable $\varphi$ depending on $\xi^{n-1}$ only the equality $E^{n-1}\varphi(\xi^{n-1}) = \varphi(\xi^{n-1})$ holds all over $\Xi^{n-1}$ rather than almost surely. This can be arranged as follows. Let $\nu_n^N$ be the marginal distribution of $\xi_n^N$ given $\xi^{n-1}$ with respect to $\mu_n^N$. Then the following modification of $\mu_n^N$ has the described property:

$$\tilde{\mu}_n^N(D \times B|\xi^{n-1}) := \begin{cases} \nu_n^N(B|\xi^{n-1}) & \text{if } \xi^{n-1} \in D, \\ 0 & \text{if } \xi^{n-1} \notin D, \end{cases}$$

for any $D \in \mathcal{B}^{n-1}$, $B \in \mathcal{B}_n^N$. Indeed, for each $\xi^{n-1}$ the measure $\tilde{\mu}_n^N(.|\xi^{n-1})$ is determined uniquely on $\mathcal{B}^N$ if it is defined on $\mathcal{B}^{n-1} \times \mathcal{B}_n^N$. A standard proof then shows, that the use of $\tilde{\mu}_n^N$ in the definition of $E^{n-1}$ implies that

$$E^{n-1}\varphi(\xi^{n-1}) = \varphi(\xi^{n-1}) \text{ for } all \ \xi^{n-1} \in \Xi^{n-1}$$

for any random variable $\varphi$ depending on $\xi^{n-1}$ only. Notice that $\tilde{\mu}_n^N$ is determined completely by its marginal distribution $\nu_n^N$ of $\xi_n^N$. In the sequel we will use only regular conditional distributions of the type $\tilde{\mu}_n^N$. By a slight abuse of notation we will use the notation $\mu_n^N$ for both $\tilde{\mu}_n^N$ and its marginal $\nu_n^N$ of $\xi_n^N$. Also, $\mu_n(d\xi_n|\xi^{n-1})$ and $\mu_n^m(d\xi_n^m|\xi^{n-1})$, $1 \leqq n \leqq m \leqq N$ will be notations of marginals of $\mu_n^N$. Of course, these are regular conditional distributions for $\xi_n$ and $\xi_n^m$, respectively, given $\xi^{n-1}$.

Up to now we were looking for a nice representation of $E^{n-1}$ for fixed n. Of course, the regular conditional distributions $\mu_n^N$, $n = 1,\ldots,N$, are closely related. For instance, just as $\mu_n^N$ also $\mu_n * \mu_{n+1}^N$ is a regular conditional distribution of $\xi_n^N$ given $\xi^{n-1}$, where the composition $*$ of transition probabilities is characterized by

$$(\mu_n * \mu_{n+1}^N)(B_1 \times B_2|\xi^{n-1}) := \int_{B_1} \mu_{n+1}^N(B_2|(\xi^{n-1},t))\mu_n(dt|\xi^{n-1}),$$
$$B_1 \in \mathcal{B}_n,\ B_2 \in \mathcal{B}_{n+1}^N,\ \xi^{n-1} \in \Xi^{n-1},$$

so that for any $B \in \mathcal{B}_{n+1}^N$

$$\mu_n^N(B|\xi^{n-1}) = (\mu_n * \mu_{n+1}^N)(B|\xi^{n-1}) \text{ for almost every } \xi^{n-1}.$$

There is a direct way to improve this to an equality for *all* $\xi^{n-1}$, namely, by considering it as a *definition* for $\mu_n^N$. That is, if $\mu_{n+1}^N$ is defined already, one only has to come up with a version of $\mu_n(d\xi_n|\xi^{n-1})$ in order to specify $\mu_n^N(d\xi_n^N|\xi^{n-1})$ completely. Therefore, from now on we will *assume that regular conditional distributions* $\mu_n(d\xi_n|\xi^{n-1})$, $n = 1,2,\ldots,N$ *have been singled out, and that* $\mu_n^N$ *is defined as*

$$\mu_n^N := \mu_n * \mu_{n+1} * \dots * \mu_N.$$

Consequently, for any random variable $\varphi$ and any $n \in \{1,2,\dots,N\}$ *we define*

$$E^{n-1}\varphi(\xi^N) := \int_{\Xi^N} \varphi(t^N)\mu_n^N(dt^N|\xi^{n-1}), \ \forall \xi^{n-1} \in \Xi^{n-1}$$

*if this integral exists*. A standard proof (see e.g. [1 Proposition 7.28]) shows that whenever $E^{n-1}\varphi(\xi^N)$ exists, it can be calculated as the iterated integral

$$\int_{\Xi_n}\{\int_{\Xi_{n+1}}\{\dots\{\int_{\Xi_{N-1}}\{\int_{\Xi_N}\varphi(\xi^N)\ \mu_N(d\xi_N|\xi^{N-1})\}\ \mu_{N-1}(d\xi_{N-1}|\xi^{N-2})\} \\ \dots\}\ \mu_{n+1}(d\xi_{n+1}|\xi^n)\}\ \mu_n(d\xi_n|\xi^{n-1}).$$

Recall that by definition $\varphi: \Xi^N \to \mathbb{R}$ is a Borel measurable function. It will appear that it is appropriate to extend the class of functions $\varphi$ for which $E^{n-1}\varphi$ is defined. A natural extension is the class of *universally measurable* functions $\varphi: \Xi^N \to \mathbb{R}$. By definition, a real function $\varphi$ on a Borel space $\Omega$ is called universally measurable if for every probability distribution $\lambda$ on the Borel sets of $\Omega$ there is a Borel measurable function $\varphi_\lambda: \Omega \to \mathbb{R}$ such that $\varphi(\omega) = \varphi_\lambda(\omega)$ for $\lambda$-almost every $\omega \in \Omega$, see e.g. [1 Lemma 7.27]. Not surprisingly, $\int_\Omega \varphi(\omega)\lambda(d\omega) := \int_\Omega \varphi_\lambda(\omega)\lambda(d\omega)$ is well-defined, at least unless $\int_\Omega (\varphi_\lambda(\omega))^+\lambda(d\omega) = \int_\Omega (\varphi_\lambda(\omega))^-\lambda(d\omega) = +\infty$. It can be shown [1 Proposition 7.45] that whenever $\int\varphi(\xi^N)\mu_n^N(d\xi^N|\xi^{n-1})$ exists for $\varphi: \Xi^N \to \mathbb{R}$ universally measurable, this conditional expectation can be calculated by iterative integration with respect to $\mu_N, \mu_{N-1}, \dots, \mu_{n+1}$ and $\mu_n$, precisely as in the case of Borel measurable functions $\varphi$.

Summing up, starting with arbitrary regular conditional distributions $\mu_n(d\xi_n|\xi^{n-1})$, $n = 1,\dots,N$, we gave precise definitions for $E^{n-1}\varphi$ for any $n = 1,2,\dots,N$ and any universally measurable function $\varphi: \Xi^N \to \mathbb{R}$, at least if $\varphi$ is sufficiently bounded. We continue our probabilistic exercise by proposing special choices for $\mu_1,\dots,\mu_N$. Recall that it is assumed that $E|\xi_k| < \infty$ for $k = 0,1,\dots,N$. Therefore

$$E^{n-1}|\xi_k| < \infty \text{ for almost every } \xi^{n-1}.$$

We know already that if $k \leqq n-1$ this is even true for *all* $\xi^{n-1}$ since then $E^{n-1}|\xi_k| = |\xi_k|$, because of our definition of $E^{n-1}$. We want to specify $\mu_1,\ldots,\mu_N$ such that it is also true for *all* $\xi^{n-1}$ if $k \geqq n$. That this is possible can be shown as follows. Start with arbitrary regular conditional distributions $\mu_1,\ldots,\mu_N$. Then, recursively for $n = N,N-1,\ldots,1$, define a modification $\tilde{\mu}_n$ of $\mu_n$ which provides the desired property, as follows.

<u>n = N</u>. Define $\tilde{\varphi}_N(\xi^N) := \tilde{\varphi}_N(\xi_N) := |\xi_N|$. Since $E|\xi_N| < \infty$ we know that $\lambda^{N-1}(D^{N-1}) = 0$, where

$$D^{N-1} := \{\xi^{N-1} \in \Xi^{N-1}: \int\tilde{\varphi}_N(\xi^N)\mu_N(d\xi_N|\xi^{N-1}) = +\infty\}.$$

Hence, also $\tilde{\mu}_N(d\xi_N|\xi^{N-1})$, defined by

$$\tilde{\mu}_N(B|\xi^{N-1}) = \begin{cases} \mu_N(B|\xi^{N-1}), & B \in \mathcal{B}_N, \quad \xi^{N-1} \notin D^{N-1}, \\ 1 & , 0 \in B \in \mathcal{B}_N, \ \xi^{N-1} \in D^{N-1}, \\ 0 & , 0 \notin B \in \mathcal{B}_N, \ \xi^{N-1} \in D^{N-1}, \end{cases}$$

is a regular conditional distribution of $\xi_N$ given $\xi^{N-1}$, and

$$\int\tilde{\varphi}_N(\xi^N)\tilde{\mu}_N(d\xi_N|\xi^{N-1}) < +\infty \text{ for } all\ \xi^{N-1} \in \Xi^{N-1}.$$

<u>n = N-1</u>. Define

$$\tilde{\varphi}_{N-1}(\xi^{N-1}) := \max\{|\xi_{N-1}|, \int\tilde{\varphi}_N(\xi^N)\tilde{\mu}_N(d\xi_N|\xi^{N-1})\}, \ \xi^{N-1} \in \Xi^{N-1},$$

and

$$D^{N-2} := \{\xi^{N-2} \in \Xi^{N-2}: \int\tilde{\varphi}_{N-1}(\xi^{N-1})\mu_{N-1}(d\xi_{N-1}|\xi^{N-2}) = +\infty\}.$$

Then $\lambda^{N-2}(D^{N-2}) = 0$, so that similarly as in the case $n = N$ a modification $\tilde{\mu}_{N-1}$ of $\mu_{N-1}$ can be constructed such that

$$\int\tilde{\varphi}_{N-1}(\xi^{N-1})\tilde{\mu}_{N-1}(d\xi_{N-1}|\xi^{N-2}) < \infty, \ \forall\xi^{N-2} \in \Xi^{N-2}.$$

For $n = N-2, N-3, \dots, 1$ the definitions are similar. Finally, $\tilde{\varphi}_0(\xi^0) := \int\tilde{\varphi}_1(\xi^1)\tilde{\mu}_1(d\xi_1|\xi^0)$, $\xi^0 \in \Xi^0$.

Since $\tilde{\mu}_n$ is only different from $\mu_n$ on sets with probability zero, it is also a regular conditional distribution of $\xi_n$ given $\xi^{n-1}$, $n = 1,\dots,N$. Moreover, it is easy to verify that if these modifications are used in the definition of $E^{n-1}$, we have, for each $k \geqq n$,

$$E^{n-1}|\xi_k| < \infty \text{ for all } \xi^{n-1} \in \Xi^{n-1}.$$

Indeed, $E^{n-1}|\xi_k|$ is equal to

$$\int_{\Xi_n}\{\dots\{\int_{\Xi_k}|\xi_k|\tilde{\mu}_k(d\xi_k|\xi^{k-1})\}\dots\}\tilde{\mu}_n(d\xi_n|\xi^{n-1}).$$

Because of $|\xi_k| \leqq \tilde{\varphi}_k(\xi^k)$ the inner integral is bounded by $\tilde{\varphi}_{k-1}(\xi^{k-1})$, and since for $j = k-1,\dots,n$

$$\int\tilde{\varphi}_j(\xi^j)\tilde{\mu}_j(d\xi_j|\xi^{j-1}) \leqq \tilde{\varphi}_{j-1}(\xi^{j-1}), \ \xi^{j-1} \in \Xi^{j-1},$$

it follows that $E^{n-1}|\xi_k| \leqq \tilde{\varphi}_{n-1}(\xi^{n-1}) < \infty$ for all $\xi^{n-1} \in \Xi^{n-1}$.

In the remainder of this section we will use "*nice*" regular conditional distributions $\mu_1,\dots,\mu_N$; that is, *it is assumed that* $E^{n-1}|\xi_k| < \infty$ *all over* $\Xi^{n-1}$ for all $n = 1,\dots,N$ and all $k = 0,1,\dots,N$.

Notice, that in order to specify the distribution $\lambda^N$ of $\xi^N$ completely, one only has to specify $\mu_1,\dots,\mu_N$ and the initial distribution $\lambda_0$ of $\xi_0$, since $\lambda^N = \lambda_0 * \mu_0 * \dots * \mu_N$. Using the same notation, the marginal distribution $\lambda^n$ of $\xi^n$ is given by $\lambda^n = \lambda_0 * \mu_0 * \dots * \mu_n$, $n = 0,1,\dots,N$. Similarly, the marginal distribution $\lambda_n$ of $\xi_n$ is $\lambda_n = \lambda_0\mu_0\dots\mu_n$.

For convenience we will use the notation $E^{N+1}\varphi(\xi^N)$ for $\varphi(\xi^N)$, occasionally.

8.3.3. The dynamic programming recursion. Consider again problem $SP_o$ introduced in 8.3.1. Its solution by means of dynamic programming is based on the following iteration scheme. Define for $s_N \in \mathbb{R}$ and $\xi^N \in \Xi^N$

(8.21a) $\quad f_{N+1}(s_N,\xi^N) := f_{N+1}(s_N) := k_{N+1}.(s_N - \xi_{N+1})^- - r.(s_N - \xi_{N+1})^+,$

and for $n = N,N-1,\ldots,1$, $y_n \in \mathbb{R}$, $s_{n-1} \in \mathbb{R}$, $\xi^{n-1} \in \Xi^{n-1}$

(8.21b)

$$g_n(y_n,\xi^{n-1}) := E^{n-1}[h_n.(y_n-\xi_n)^+ + q_n.(y_n-\xi_n)^- + f_{n+1}(y_n-\xi_n,\xi^n)]$$
$$= \int\{h_n.(y_n-\xi_n)^+ + q_n.(y_n-\xi_n)^- + $$
$$+ f_{n+1}(y_n-\xi_n,\xi^{n-1},\xi_n)\}\mu_n(d\xi_n|\xi^{n-1}),$$

$$f_n(s_{n-1},\xi^{n-1}) := \inf_{x_n \geqq 0}\{k_n.x_n + g_n(s_{n-1}+x_n,\xi^{n-1})\}$$
$$= -k_n.s_{n-1} + \inf_{y_n \geqq s_{n-1}}\{k_n.y_n + g_n(y_n,\xi^{n-1})\},$$

$$X_n^*(s_{n-1},\xi^{n-1}) := \arg\min_{x_n \geqq 0}\{k_n.x_n + g_n(s_{n-1}+x_n,\xi^{n-1})\},$$

and finally:

(8.21c) $\quad f_0 := E\, f_1 = \int f_1(-\xi_0)\lambda_0(d\xi_0).$

Here $\lambda_0$ is the distribution of $\xi_0$, and $\mu_n(d\xi_n|\xi^{n-1})$, $n = 1,\ldots,N$, is an arbitrary fixed nice regular conditional distribution of $\xi_n$ given $\xi^{n-1}$, $n = 1,2,\ldots,N$; see 8.3.2 for the definition. For $n = 1$ we write $f_1(s_0,\xi^0) = f_1(s_0,-s_0)$ as $f_1(s_0)$ or as $f_1(-\xi_0)$, and $X_1^*(s_0,\xi^0) = X_1^*(s_0)$. We will show that $f_0 = \inf SP_0$, see Corollary 8.2.

As has been indicated previously in similar situations (see Sections 4.6 and 6.4) the recursion (8.21) needs further explanation since questions of measurability and finiteness arise. The main problem is that if a bivariate Borel measurable function is infimized over one of its variables, the result is not necessarily Borel measurable with respect to the other variable. However, as shown by Shreve [18 Theorems 2.4 and 2.3; 1 Propositions 7.48 and 7.47] each $g_n$ and $f_n$, $n = N,\ldots,1$ is a well-defined extended real *lower semianalytic* function on $\mathbb{R}\times\Xi^{n-1}$. For the definition of lower semianalytic functions and their properties we refer to [18 Section 2.3] or [1 Section 7.7]. Shreve's results imply that all definitions in (8.21) have a precise meaning. For example, the integrand in the definition of $g_n(y_n,\xi^{n-1})$ is lower semianalytic hence universally measurable, so that the integral exists if the provision $\infty - \infty = \infty$ is accepted.

We will now exploit the special properties of $SP_o$ in order to derive much more detailed information on (8.21). Let us recall the assumptions

(8.22)
a. $SP_o$ is strictly proper.
b. $E|\xi_n| < \infty$, $n = 0,1,\ldots,N$.
c. $\mu_n(d\xi_n | \xi^{n-1})$ is a nice regular conditional distribution of $\xi_n$ given $\xi^{n-1}$, $n = 1,2,\ldots,N$.

THEOREM 8.1. *Assume* (8.22). *Then each function* $f_n$ *and* $g_n$ *defined in* (8.21) *is everywhere finite. Moreover, a lower bound for both* $g_n(y,\xi^{n-1})$ *and* $f_n(y,\xi^{n-1})$ *is given by*

$$\underline{g}_n(y,\xi^{n-1}) := (\Sigma_{j=n}^{N+1} h_j).y - \Sigma_{i=n}^{N+1}(\Sigma_{j=i}^{N+1} h_j).E^{n-1}\xi_i,$$

*and an upper bound for both function values is given by*

$$\bar{g}_n(y,\xi^{n-1}) := (\Sigma_{j=n}^{N+1}\tilde{\gamma}_j).|y| + \Sigma_{i=n}^{N+1}(\Sigma_{j=i}^{N+1}\tilde{\gamma}_j).E^{n-1}|\xi_i|$$

*where* $\tilde{\gamma}_j := |h_j| + |q_j|$, $j = 1,2,\ldots,N+1$.

*For any* $\xi^{n-1} \in \Xi^{n-1}$, *the functions* $f_n(.,\xi^{n-1})$ *and* $g_n(.,\xi^{n-1})$ *are finite and convex hence continuous. They are even Lipschitz continuous.*

*For all* $n = 1,\ldots,N$

$$X_n^*(s_{n-1},\xi^{n-1}) \neq \emptyset \quad \forall (s_{n-1},\xi^{n-1}) \in \mathbb{R} \times \Xi^{n-1}.$$

PROOF. The proof of the theorem is by backward induction.

*(lower bound)* $n = N$. Fix $(y_N,\xi^{N-1}) \in \mathbb{R} \times \Xi^{N-1}$ and consider the definition of $g_N(y_N,\xi^{N-1})$. Since $SP_o$ is proper, we know that $q_N \geqq -h_N$ and $k_{N+1} \geqq r = h_{N+1}$, so that

$$h_N.(y_N - \xi_N)^+ + q_N.(y_N - \xi_N)^- + f_{N+1}(y_N - \xi_N) \geqq$$
$$\geqq (h_N + h_{N+1}).y_N - h_N.\xi_N - (h_N + h_{N+1}).\xi_{N+1}.$$

Integration with respect to $\mu_N(d\xi_N|\xi^{N-1})$ gives

$$g_N(y_N,\xi^{N-1}) \geqq \underline{g}_N(y_N,\xi^{N-1}) \ \forall (y_N,\xi^{N-1}) \in \mathbb{R}\times \Xi^{N-1}.$$

Consequently,

$$k_N.x_N + g_N(s_{N-1} + x_N,\xi^{N-1}) \geqq$$
$$\geqq (k_N + h_N + h_{N+1}).x_N + \underline{g}_N(s_{N-1},\xi^{N-1}).$$

We know that $\delta_N := k_N + h_N + h_{N+1} \geqq 0$ since $SP_o$ is proper, so that the first term is nonnegative for all $x_N \geqq 0$. Therefore, $f_N(s_{N-1},\xi^{N-1}) \geqq \underline{g}_N(s_{N-1},\xi^{N-1})$.

$\underline{n = m}$. Suppose that for any $m \in \{2,3,...,N-1\}$ it has been shown that $f_{m+1}(s_m,\xi^m) \geqq \underline{g}_{m+1}(s_m,\xi^m)$ for all $(s_m,\xi^m) \in \mathbb{R}\times\Xi^m$. Then, since $q_m \geqq -h_m$

$$h_m.(y_m - \xi_m)^+ + q_m.(y_m - \xi_m)^- + f_{m+1}(y_m - \xi_m,\xi^m) \geqq$$
$$\geqq h_m.(y_m - \xi_m) + \underline{g}_{m+1}(y_m - \xi_m,\xi^m)$$
$$= (\Sigma_{j=m}^{N+1} h_j).y_m - \Sigma_{i=m}^{N+1}(\Sigma_{j=i}^{N+1} h_j).E^m\xi_i$$

for all $(y_m,\xi^m) \in \mathbb{R}\times\Xi^m$. Because of $E^{m-1}(E^m\xi_i) = E^{m-1}\xi_i$, $1 \leqq m \leqq i \leqq N+1$, by integration with respect to $\mu_m(d\xi_m|\xi^{m-1})$ we get that

$$g_m(y_m,\xi^{m-1}) \geqq \underline{g}_m(y_m,\xi^{m-1}) \ \forall (y_m,\xi^{m-1}) \in \mathbb{R}\times\Xi^{m-1}.$$

Consequently, for all $(s_{m-1},\xi^{m-1}) \in \mathbb{R}\times\Xi^{m-1}$ and $x_m \in \mathbb{R}^+$

$$k_m.x_m + g_m(s_{m-1} + x_m,\xi^{m-1}) \geqq \delta_m.x_m + \underline{g}_m(s_{m-1},\xi^{m-1}).$$

Since $\delta_m \geqq 0$ and $x_m \geqq 0$ it follows that

$$f_m(s_{m-1},\xi^{m-1}) \geqq \underline{g}_m(s_{m-1},\xi^{m-1}).$$

This completes the inductive step in the proof of the lower bounds.

*(upper bound)* The proof that $\bar{g}_n(y,\xi^{n-1})$ is an upper bound for both $g_n(y,\xi^{n-1})$ and $f_n(y,\xi^{n-1})$ is completely similar. The details of the inductive proof, which uses the inequalities

$$h_n.(y_n - \xi_n)^+ + q_n.(y_n - \xi_n)^- \leqq \tilde{\gamma}_n.(|y_n| + |\xi_n|),$$

$$f_n(s_{n-1},\xi^{n-1}) \leqq g_n(s_{n-1},\xi^{n-1})$$

are left to the reader.

*(finiteness)* Both $\underline{g}_n(y,\xi^{n-1})$ and $\bar{g}_n(y,\xi^{n-1})$ are finite all over $\mathbb{R} \times \Xi^{n-1}$ since $E^{n-1}|\xi_i| < \infty$ $\forall i$ $\forall n$ due to the choice of *nice* regular conditional distributions. Therefore, $f_n$ and $g_n$, $n \geqq 1$, are finite everywhere. Moreover, since $\underline{g}_1(-\xi_0,\xi_0) \leqq f_1(-\xi_0) \leqq \bar{g}_1(-\xi_0,\xi_0)$ it follows that $f_0$ is finite too:

$$-\infty < -\Sigma_{i=0}^{N+1}(\Sigma_{j=i}^{N+1}h_j).E\xi_i \leqq f_0 \leqq \Sigma_{i=0}^{N+1}(\Sigma_{j=i}^{N+1}\tilde{\gamma}_j).E|\xi_i| < \infty.$$

*(convexity)* The convexity of $f_n(.,\xi^{n-1})$ and $g_n(.,\xi^{n-1})$ is also proved by backward induction to n. Indeed, $f_{N+1}$ is convex since $k_{N+1} \geqq r$. If $f_{m+1}(.,\xi^m)$ is convex for all $\xi^m$, then the integrand in the definition of $g_m(.,\xi^{m-1})$ is convex for all $(\xi^{m-1},\xi_m)$ because of $h_m + q_m \geqq 0$. Since integration preserves convexity it follows that $g_m(.,\xi^{m-1})$ is convex for all $\xi^{m-1} \in \Xi^{m-1}$. As a consequence, $f_m(.,\xi^{m-1})$ is convex too, since the infimand in its definition is convex jointly in $s_{m-1}$ and $x_m$.

*(Lipschitz continuity)* Finite convex real functions on $\mathbb{R}$ are continuous, and everywhere left and right differentiable. Denote the left (or right) derivative of $f_n(.,\xi^{n-1})$ by $f'_n(.,\xi^{n-1})$. Since $f_n(.,\xi^{n-1})$ is convex, $f'_n(.,\xi^{n-1})$ is nondecreasing. In order to show that $f_n(.,\xi^{n-1})$ is Lipschitz continuous it is (necessary and) sufficient that the slopes are bounded. Indeed,

$$-\infty < -\Sigma_{j=n}^{N+1}\tilde{\gamma}_j = \bar{g}'_n(-\infty,\xi^{n-1}) \leqq f'_n(-\infty,\xi^{n-1}) \leqq$$

$$f'_n(+\infty,\xi^{n-1}) \leqq \bar{g}'_n(+\infty,\xi^{n-1}) = \Sigma_{j=n}^{N+1}\tilde{\gamma}_j < +\infty.$$

Consequently, each $f_n(.,\xi^{n-1})$ is Lipschitz continuous, and so is each $g_n(.,\xi^{n-1})$.

*(infima attained)* Finally we show that $X_n^*(s_{n-1},\xi^{n-1}) \neq \emptyset$. For all $(s_{n-1},\xi^{n-1}) \in \mathbb{R} \times \Xi^{n-1}$ and for all $n = 1,\ldots,N$ the function

$$x_n \to k_n.x_n + g_n(s_{n-1} + x_n,\xi^{n-1})$$

is continuous, and bounded below by

$$k_n.x_n + g_n(s_{n-1} + x_n, \xi^{n-1}) = \delta_n.x_n + g_n(s_{n-1}, \xi^{n-1}).$$

Since $SP_o$ is *strictly* proper, we know that $\delta_n > 0$, so that

$$\lim_{x_n \to \infty} \{k_n.x_n + g_n(s_{n-1} + x_n, \xi^{n-1})\} = +\infty.$$

Consequently, the infimum over $x_n \in [0,\infty)$ is attained, that is $X_n^*(s_{n-1}, \xi^{n-1}) \neq \emptyset$. □

COROLLARY 8.2. *Assume* (8.22). *Then* $SP_o$ *has optimal policies. Moreover,* $SP_o$ *is solved by the dynamic programming recursion* (8.21). *By this we mean that*

(i) $f_0 = \inf SP_o$.

(ii) *A policy* $x^*$ *is optimal for* $SP_o$ *iff* $\lambda^{n-1}$*-almost surely* $x_n^*(\xi^{n-1}) \in X_n^*(s_{n-1}^*(\xi^{n-1}), \xi^{n-1})$ *for* $n = 1,\dots,N$, *where* $s_{n-1}^*(\xi^{n-1})$ *is the inventory at the end of period* $n-1$ *resulting from* $x^*$ *and* $\xi^N$*; that is,*

$$s_0^*(\xi^0) := -\xi_0,$$

$$s_n^*(\xi^n) := s_{n-1}^*(\xi^{n-1}) + x_n^*(\xi^{n-1}) - \xi_n \text{ for } n = 1,\dots,N.$$

*For any optimal policy*

$$E\, x_n^* < \infty,$$

$$E|s_n^*| < \infty, \; n = 0,1,\dots,N+1.$$

PROOF. Recall that $X_n^*(s_{n-1}, \xi^{n-1}) \neq \emptyset$ for all $(s_{n-1}, \xi^{n-1}) \in \mathbb{R} \times \Xi^{n-1}$ and all $n = 1,\dots,N$, and that $g_n$ is lower semianalytic. From the Jankov-von Neumann selection theorem for lower semianalytic functions [1 Propositions 7.49 and 7.50] it follows that universally measurable functions $\tilde{x}_n: \mathbb{R} \times \Xi^{n-1} \to \mathbb{R}_+$ exist with $\tilde{x}_n(s_{n-1}, \xi^{n-1}) \in X_n^*(s_{n-1}, \xi^{n-1})$ for all $(s_{n-1}, \xi^{n-1}) \in \mathbb{R} \times \Xi^{n-1}$, $n = 1,\dots,N$. Define

$$\hat{x}_n(\xi^{n-1}) := \tilde{x}_n(\tilde{s}_{n-1}(\xi^{n-1}), \xi^{n-1}), \; \xi^{n-1} \in \Xi^{n-1}, \; n = 1,2,\dots,N,$$

where $\tilde{s}_0(\xi_0) := -\xi_0$, and iteratively for $n = 1,\ldots,N$

$$\tilde{s}_n(\xi^n) := \tilde{s}_{n-1}(\xi^{n-1}) + \tilde{x}_n(\tilde{s}_{n-1}(\xi^{n-1}),\xi^{n-1}) - \xi_n.$$

These $\hat{x}_n$ are universally measurable, so that *Borel* measurable functions $x_n^*: \Xi^{n-1} \to \mathbb{R}_+$ exist which coincide with $\hat{x}_n$ almost everywhere with respect to $\lambda^{n-1}$, $n = 1,2,\ldots,N$. Hence $x^* := (x_1^*,\ldots,x_N^*)$ is a policy for $SP_o$, and it is easily verified that its expected total cost is equal to $f_0$. Hence $\inf SP_o \leq f_0$. On the other hand, a standard backward inductive proof shows, that $f_0 \leq \inf SP_o$. Therefore, $f_0 = \inf SP_o = \min SP_o$. Inspection of the inductive proof of $f_0 \leq \inf SP_o$ with the additional knowledge that the equality holds, reveals that (ii) is true. The details are left to the reader as the proof is similar to that of Proposition 4.22.

Finally, recall that for any policy x the expected total cost is bounded below by

$$\Sigma_{n=1}^{N+1} h_n \cdot (-\Sigma_{i=0}^{n} E\,\xi_i) + \Sigma_{n=1}^{N} \delta_n \cdot E\,x_n(\xi^{n-1}).$$

Therefore, any production $x_n$ with $E\,x_n(\xi^{n-1}) = +\infty$ will result in a unbounded high expected total cost since each $\delta_n$, $n = 1,\ldots,N$, is strictly positive. Optimal productions $x_n^*$ satisfy $E\,x_n^* < \infty$, necessarily, and therefore also the optimal inventory levels $s_n^*$ have finite expectations. □

Just as in the deterministic case the optimal policies for $SP_o$ can be characterized completely by optimal inventory levels. We call $y := (y_1,\ldots,y_N)$ with $y_n: \Xi^{n-1} \to [-\infty,\infty)$ a Borel measurable function, $n = 1,2,\ldots,N$, a *level policy* for $SP_o$. The number $y_n(\xi^{n-1})$ is to be interpreted as the target value of the inventory after the production in period n. With this interpretation, *each level policy* y *determines a production policy* x in the following way:

$$x_n(\xi^{n-1}) := (y_n(\xi^{n-1}) - s_{n-1}(\xi^{n-1}))^+, \quad \xi^{n-1} \in \Xi^{n-1}, \quad n = 1,2,\ldots,N,$$

where $s_{n-1}(\xi^{n-1})$ denotes the inventory at the end of period n-1.

Related to the dynamic programming recursion (8.21) we define

$$(8.23) \qquad Y_n^*(\xi^{n-1}) := \arg\min_{y_n \in [-\infty,\infty)} \{k_n \cdot y_n + g_n(y_n,\xi^{n-1})\},$$
$$\xi^{n-1} \in \Xi^{n-1}, \; n = 1,\dots,N.$$

Since for each $\xi^{n-1} \in \Xi^{n-1}$ the function $v_n(y_n) := k_n \cdot y_n + g_n(y_n,\xi^{n-1})$ is convex, continuous and strictly increasing for large $y_n$, we know that $Y_n^*(\xi^{n-1})$ is a nonempty closed convex subset of $\mathbb{R}$. That is $Y_n^*(\xi^{n-1}) = [a,b]$ for certain $a = a_n(\xi^{n-1})$, $b = b_n(\xi^{n-1})$ satisfying $-\infty \leqq a \leqq b < \infty$. Moreover, since the function $v_n$ is even linear on $(-\infty,0]$, which follows from the nonnegativity of the demands and of $\xi_{N+1}$, there are only three possibilities: $a = b = -\infty$, $a = -\infty$ and $b \in [0,\infty)$, and $[a,b] \subset [0,\infty)$, respectively.

A level policy $y^* = (y_1^*,y_2^*,\dots,y_N^*)$ is called *optimal* if $y_n^*(\xi^{n-1}) \in Y_n^*(\xi^{n-1})$ for $\lambda^{n-1}$ - almost every $\xi^{n-1}$, $n = 1,\dots,N$, and the sets $Y_n^*(\xi^{n-1})$ are called the sets of *optimal inventory levels*. These are natural definitions, since from the convexity of $g_n(.,\xi^{n-1})$ it follows that

$$(8.24) \qquad X_n^*(s_{n-1},\xi^{n-1}) = \{(y_n - s_{n-1})^+ : y_n \in Y_n^*(\xi^{n-1})\},$$
$$(s_{n-1},\xi^{n-1}) \in \mathbb{R} \times \Xi^{n-1}, \; n = 1,\dots,N,$$

so that we have the next corollary.

<u>COROLLARY 8.3</u>. *Assume* (8.22). *A production policy* $x^* \in X$ *is optimal for* $SP_o$ *iff it is determined by an optimal level policy* $y^*$.

We note that an optimal policy $y^* = (y_1^*,\dots,y_N^*)$ satisfies $E\,y_n^* < \infty$, but $E\,y_n^* = -\infty$ is not excluded.

<u>REMARK 8.4</u>. In (8.21) we used nice regular conditional distributions $\mu_n(d\xi_n|\xi^{n-1})$. If just arbitrary regular conditional distributions are used instead, then every conclusion on $f_n(.,\xi^{n-1})$ (, and $g_n(.,\xi^{n-1})$, $X_n^*(.,\xi^{n-1})$, $Y_n^*(\xi^{n-1})$) holds $\lambda^{n-1}$ - almost surely instead of everywhere. As said before, two different choices for the regular conditional distributions give rise to two different functions $f_n(.,\xi^{n-1})$ but the

difference is restricted to a set of $\lambda^{n-1}$-measure 0. In fact, both are (universally measurable) versions of the conditional expectation of the minimal costs from period n onward, given $s_{n-1}$ and $\xi^{n-1}$, as might be expected [10 Theorem 2.2]. However, if one uses just an arbitrary version of the conditional expectation given $\xi^{n-1}$ then one may get essentially different functions $f_n$ in (8.21), as pointed out by Olsen [10].

REMARK 8.5. If $\xi_0,\xi_1,\ldots,\xi_N$ are *independent* random variables, then the dynamic programming recursion (8.21) can be simplified. In that case $g_n$, $f_n$, $X_n^*$ and $Y_n^*$ do not depend on $\xi^{n-1}$, $n = N,N-1,\ldots,1$. Then the conclusions on (8.21) and its relation to $SP_o$ are a direct consequence of Corollary 6.12, see Example 6.13.

REMARK 8.6. If all cost coefficients $k_n$, $h_n$, $q_n$ and $-r$ are *nonnegative* then the conclusions regarding (8.21) and its relation to $SP_o$ are a direct corollary of Proposition 4.28, which is a result of Bertsekas and Shreve [1 Proposition 8.2]. This proposition is then applicable, since assumption (4.38 part a) holds in this case. The assumption of strict properness is weaker, since it allows for an arbitrary low expected cost of a period n. However, such an arbitrary low cost would be more than nullified by resulting high costs in future periods. Our approach to (8.21) can be interpreted as a simple adjustment of that of Bertsekas and Shreve. This adjustment is necessary since we have weaker boundedness assumptions about the cost functions and it is possible since we have a simple state transformation function $s_n = s_{n-1} + x_n - \xi_n$.

8.3.4. The linear program and its dual. Just as in the deterministic case, also $SP_o$ can be written as a linear programming problem of the type

$$\text{minimize}\{E\, ez(\xi):\ Az(\xi) = \xi \text{ a.s.},\ z(\xi) \geqq 0 \text{ a.s.}\}$$

where, as before, A is the incidence matrix of the graph G without the last (redundant) row and e is again the row vector of unit costs. In order to describe the decision functions $z(\xi)$ more precisely, we use again the partition $z := (z_0,z_1,\ldots,z_N)^T$ where $z_n := (u_n,v_n,x_{n+1})^T$

for $n = 0,1,\ldots,N-1$ and $z_N := (u_N, v_N, x_{N+1}, w)$. The corresponding partition of e is $(e_0, e_1, \ldots, e_N)$ with $e_n := (h_n, q_n, k_{n+1})$ for $n = 0,1,\ldots,N-1$ and $e_N := (h_N, q_N, k_{N+1}, -r)$. The reason for the partition of the decision functions $z(\xi)$ is, of course, that in order to be consistent with $SP_o$ we have to require that the decisions should be *non-anticipative*: i.e. $z_n$ must be a Borel measurable function of $\xi^n$ rather than of $\xi^N$, $n = 0,1,\ldots,N$. In fact, the problem is a special case of the multistage stochastic linear programming problem described in Section 3.3: (the end of) period n, $n = 0,1,\ldots,N$, corresponds to stage n, and at stage n the realization of the vector $\xi^n$ is known; $z_n$ represents the decisions at stage n. Notice that the matrix A, partitioned as in the previous section, has the lower block triangular structure of the constraint matrix in (3.33).

It is convenient (especially for the formulation of the dual problem) to restrict the attention to decision functions which belong to a topological vector space. Since we already know from Corollary 8.2 that optimal policies for $SP_o$ give integrable productions $x_n^*$ and integrable inventories $s_n^*$ it is natural to ask for *integrable* decision functions $z_n$. Therefore we define

$$Z := \Pi_{n=0}^{N} Z_n, \; Z_n := L_1(\Xi^n, \mathcal{B}^n, \lambda^n; \mathbb{R}^{m_n})$$

with $m_n := 3$ if $n < N$ and $m_N := 4$. That is, $Z_n$ is the Banach space of equivalence classes of Borel functions $z_n : \Xi^n \to \mathbb{R}^{m_n}$, interpreted as column vectors, which satisfy

$$\|z_n\| := E\, \Sigma_{k=1}^{m_n} |(z_n)_k| := \int \Sigma_{k=1}^{m_n} |(z_n)_k(\xi^n)| \lambda^n(d\xi^n) < +\infty.$$

Here equivalence means equality almost surely with respect to $\lambda^n$ (or, what amounts to the same, with respect to $\lambda^N$). As a consequence, by requiring that $z \in Z$ we do not distinguish between decision functions which are equal almost surely. This is no drawback, since equivalent decision functions give the same value for the expected total cost. So we are led to the following linear program in function spaces, which is the stochastic analogue of problem P in the previous section:

$$\text{(SP)}\qquad \text{minimize}_{z\in Z}\left\{E\,\Sigma_{n=0}^{N} e_n z_n(\xi^n):\ \begin{array}{l}\Sigma_{n=j-1}^{j} A_{jn} z_n(\xi^n) = b_j(\xi^j)\ \text{a.s.},\\ \qquad\qquad j = 0,1,\ldots,N,\\ z_n(\xi^n) \geqq 0\ \text{a.s.},\ n = 0,1,\ldots,N\end{array}\right\}$$

where $A_{0,-1} := 0$, $b_j(\xi^j) := \xi_j$ for $j < N$ and $b_N(\xi^N) := (\xi_N, \xi_{N+1})^T$. Notice that the special structure of the matrix A and the properness assumptions $q_n + h_n \geqq 0$, $n = 0,1,\ldots,N+1$, imply that without loss of generality one may restrict oneself to solutions of SP for which almost surely $u_n(\xi^n).v_n(\xi^n) = 0$ and $x_{N+1}(\xi^N).w(\xi^N) = 0$. Recall that $u_n$ and $v_n$ are the first two components of $z_n$ and that $x_{N+1}$ and w are the last two components of $z_N$. This establishes the equivalence of $SP_o$ and SP.

The main reason to formulate problem SP is that we want to study its dual problem. As described in Chapter 2 there exist in general a lot of dual problems, each based on a particular convex perturbation of the primal problem. We intend to study the sensitivity of the linear program SP with respect to its right-hand side $\xi$, giving rise to the usual form of the dual problem. It is not surprising that the approach to linear programming duality in "abstract" spaces, developed in Chapter 4, is of help. To begin with, notice that an appropriate definition of the constraint space of SP is given by

$$H := \Pi_{j=0}^{N} H_j,\ H_j := L_1(\Xi^j, B^j, \lambda^j;\ \mathbb{R}^{1_j})$$

with $1_j := 1$, $j < N$, and $1_N := 2$. Indeed, the equality constraints in SP can be written as $G_1 z = b$ with $b := (b_0, b_1, \ldots, b_N)^T \in H$ given by

$$(G_1 z)_j(\xi^j) := \Sigma_{n=j-1}^{j} A_{jn} z_n(\xi^n),\ \xi^j \in \Xi^j,\ j = 0,1,\ldots,N.$$

In order to come to a dual pair of linear programs in the sense of Definition 4.1, we introduce - not very surprisingly - the spaces of row vectors of almost surely bounded functions

$$\Pi := \Pi_{j=0}^{N} \Pi_j,\ \Pi_j := L_\infty(\Xi^j, B^j, \lambda^j; \mathbb{R}^{1_j}),$$

$$V := \Pi_{n=0}^{N} V_n,\ V_n := L_\infty(\Xi^n, B^n, \lambda^n; \mathbb{R}^{m_n}).$$

Then <V,Z> and <Π,H> are dualities determined by the bilinear forms

$$<\pi,\eta> := E\ \pi\eta = \Sigma_{j=0}^{N}\int_{\Xi^j}\pi_j(\xi^j)\eta_j(\xi^j)\lambda^j(d\xi^j),\ \pi \in \Pi,\ \eta \in H,$$

$$<v,z> := E\ vz = \Sigma_{n=0}^{N}\int_{\Xi^n}v_n(\xi^n)z_n(\xi^n)\lambda^n(d\xi^n),\ v \in V,\ z \in Z.$$

Moreover, let $G_2$: Π → V be defined by

$$(G_2\pi)_n(\xi^n) := E^n\Sigma_{j=n}^{n+1}\pi_j(\xi^j)A_{jn}$$

$$:= \int_{\Xi^N}(\Sigma_{j=n}^{n+1}\pi_j(\xi^j)A_{jn})\mu_n^N(d\xi^N|\xi^n),\ \xi^n \in \Xi^n,$$

$$n = 0,1,\dots,N,$$

with $A_{N+1,N} := 0$, where $\mu_n^N(d\xi^N|\xi^n)$ denotes the regular conditional distribution of $\xi^N$ given $\xi^n$ determined by $\mu_n * \mu_{n+1} * \dots * \mu_N$. For the definition of this composition we refer to Subsection 8.3.2. Recall that $\mu_1(d\xi_1|\xi^0),\dots,\mu_N(d\xi_N|\xi^{N-1})$ are arbitrary nice regular conditional distributions; without loss of generality we will assume that they are the same as those used in the dynamic programming approach to $SP_o$. Since $E^n$ gives a version of the conditional expectation given the value of $\xi^n$, it is easy to verify that the linear maps $G_1$ and $G_2$ are adjoint. Indeed, for all z ∈ Z and π ∈ Π we have

$$<G_2\pi,z> = E[\Sigma_{n=0}^{N}E^n[\Sigma_{j=n}^{n+1}\pi_j(\xi^j)A_{jn}]z_n(\xi^n)]$$

$$= E[\Sigma_{n=0}^{N}E^n[\Sigma_{j=n}^{n+1}\pi_j(\xi^j)A_{jn}z_n(\xi^n)]]$$

$$= E[\Sigma_{j=0}^{N}\pi_j(\xi^j)\Sigma_{n=j-1}^{j}A_{jn}z_n(\xi^n)]$$

$$= <\pi,G_1z>.$$

It is now an elementary exercise to verify that we get a dual pair of linear programs in the sense of Definition 4.1 if together with SP we consider the linear program

(SD) $\qquad \text{maximize}_{\pi\in\Pi}\{<\pi,b>:\ G_2\pi \leqq e\}.$

Here we take the sets of nonnegative functions in V and Z as the positive cones $V^+$ and $Z^+$ in V and Z, respectively, whereas $\Pi^+ := \Pi$ and $H^+ := \{0\}$. In more detail the dual problem can be written as

$$\text{maximize}_{\pi\in\Pi} \; E\, \Sigma_{n=0}^{N+1}\pi_n(\xi^n).\xi_n$$

subject to

$$\text{(SD)}\quad \begin{aligned} -h_n \leqq \pi_n(\xi^n) - E^n\pi_{n+1}(\xi^{n+1}) &\leqq q_n, \text{ a.s., } n = 0,1,\dots,N,\\ E^n\pi_{n+1}(\xi^{n+1}) &\leqq k_{n+1}, \text{ a.s., } n = 0,1,\dots,N,\\ r \leqq \pi_{N+1}(\xi^{N+1}), &\text{ a.s.} \end{aligned}$$

For the sake of convenience we use here (and in the sequel) a slightly different notation: in order to have only *scalar* functions as variables in SD, $\pi_N(\xi^N) \in \Pi_N := L_\infty(\Xi^N,B^N,\lambda^N;\mathbb{R}^2)$ is replaced by $(\pi_N(\xi^N),\pi_{N+1}(\xi^{N+1})) \in \Pi_N \times \Pi_{N+1} := L_\infty(\Xi^N,B^N,\lambda^N;\mathbb{R}) \times L_\infty(\Xi^N,B^N,\lambda^N;\mathbb{R})$. Since $\xi_{N+1}$ is a deterministic constant we may identify $\pi_{N+1}(\xi^N)$ and $\pi_{N+1}(\xi^{N+1})$, which change is notationally very convenient. The ambiguity in the meaning of $\Pi$ is not disturbing.

As might be expected, SD reduces to the deterministic dual problem D if the probability distribution is degenerated.

Applying Theorem 4.5 we will prove a strong duality theorem for the dual pair (SP,SD). Since $L_\infty$ is the norm dual of $L_1$ we know that the norm topology on H is compatible with the duality $\langle\Pi,H\rangle$. This makes the application of part a of Theorem 4.5 straightforward.

<u>THEOREM 8.7</u>. *Assume* (8.22). *Then*

$$(8.25)\qquad +\infty > \min SP = \min SP_o = \max SD > -\infty.$$

*Moreover,* $z^* = (z_0^*,z_1^*,\dots,z_N^*)$, $z_n^* \in Z_n$, *with* $z_n^* = (u_n^*,v_n^*,x_{n+1}^*)$ *for* $n < N$ *and* $z_N^* = (u_N^*,v_N^*,x_{N+1}^*,w^*)$, *and* $\pi^* = (\pi_0^*,\pi_1^*,\dots,\pi_{N+1}^*)$, $\pi_n^* \in \Pi_n$, *are optimal for* SP *and* SD, *respectively, if and only if they are feasible and the complementary slackness conditions hold almost surely:*

$$u_n^*(\xi^n).(h_n + \pi_n^*(\xi^n) - E^n\pi_{n+1}^*(\xi^{n+1})) = 0, \ n = 0,1,\dots,N,$$

(8.26)
$$v_n^*(\xi^n).(q_n - \pi_n^*(\xi^n) + E^n\pi_{n+1}^*(\xi^{n+1})) = 0, \ n = 0,1,\dots,N,$$

$$x_{n+1}^*(\xi^n).(k_{n+1} - E^n\pi_{n+1}^*(\xi^{n+1})) = 0, \qquad n = 0,1,\dots,N,$$

$$w^*(\xi^N).(\pi_{N+1}^*(\xi^N) - r) = 0.$$

PROOF. From Corollary 8.2 we know already that inf $SP_o$ = min $SP_o$ is finite. The equivalence of SP and $SP_o$ gives min SP = min $SP_o$. The other statements in Theorem 8.7 follow as direct applications of Theorems 4.2 and 4.5 part a. Since (SP,SD) is a dual pair we only have to verify the conditions of Theorem 4.5. Indeed, both dualities $\langle V,Z\rangle$ and $\langle \Pi,H\rangle$ are separated. For example, if for any $\bar{v} \in V$ it holds that $\langle \bar{v},z\rangle = 0$ $\forall z \in Z$, then it follows that $E\|\bar{v}\| = 0$ so that $\bar{v} = 0$ a.s.; just take $z := \bar{z} = (\bar{z}_0,\dots,\bar{z}_N)$ defined by

$$\bar{z}_n(\xi^n) := \text{sgn}\ \bar{v}_n(\xi^n),\ \xi^n \in \Xi^n,\ n = 0,1,\dots,N;$$

$\bar{z} \in Z$ since each $\lambda^n$ is a finite measure. A similar proof shows that the positive cone $V^+$ is induced by the positive cone $Z^+$ and the bilinear form $\langle v,z\rangle$: if $\langle \bar{v},z\rangle \geqq 0$ $\forall z \in Z^+$ then by taking $\bar{z} := \text{sgn}\ (\bar{v})^-$ one gets $E(\bar{v})^- \leqq 0$ so that $\bar{v} \geqq 0$ a.s. Trivially $\Pi^+ = \Pi$ is induced by $H^+ = \{0\}$. It remains to show that the optimal value function of SP is bounded above in a neighbourhood of $0 \in H$. That is, if we replace the right-hand side b in SP by $b-\eta$ with $\eta \in H$, then the optimal value $\Theta(b-\eta)$ of SP must satisfy $\Theta(b-\eta) \leqq M$ if $\|\eta\| := \Sigma_{n=0}^{N+1} E|\eta_n| < \varepsilon$. That this is true follows directly by considering the feasible solution of SP corresponding to the policy "never to produce anything". From the beginning of this section we know that

$$\Theta(b-\eta) \leqq c_0.\|b-\eta\| \leqq c_0.\|b\| + c_0.\|\eta\|.$$

This completes the proof. □

As we have seen it is possible to get duality results for SP by a straightforward application of Theorem 4.5. A direct derivation of SD as the conjugate dual is also not difficult, see [9]. For an analogous treatment of duality of rather general multistage stochastic linear programs we refer to Eisner and Olsen [5]. They work with the primal problem in $L_p$, $1 < p < \infty$, and get a dual in $L_q$, $1/p + 1/q = 1$. Rockafellar and Wets [15,16] derive similar results for multistage stochastic convex problems for $p = \infty$ and $q = 1$. In [17] they give nice duality theorems for convex Bolza-type problems in discrete time, which generalize our results considerably. On the other hand, because of its simplicity our model SP allows for an explicit solution of the dual problem, as we will show in the sequel.

From the dynamic programming approach to $SP_o$ we know already that the optimal solutions of SP are characterized completely by optimal inventory levels $y_n^*(\xi^{n-1})$. With Theorem 8.7 this leads to a complete characterization of the solution of SD:

<u>COROLLARY 8.8</u>. *Assume* (8.22). *The level policy* $y^* = (y_1^*,\ldots,y_N^*)$ *is optimal for* SP *and* $\pi^* = (\pi_0^*,\pi_1^*,\ldots,\pi_{N+1}^*)$, $\pi_n^* \in \Pi_n$, *is an optimal solution for* SD *if and only if the following relations hold a.s.:*

$$\text{(8.27a)} \qquad \pi_{N+1}^*(\xi^N) \begin{cases} = r & \text{if } s_N^*(\xi^N) > \xi_{N+1}, \\ \in [r, k_{N+1}] & \text{if } s_N^*(\xi^N) = \xi_{N+1}, \\ = k_{N+1} & \text{if } s_N^*(\xi^N) < \xi_{N+1}, \end{cases}$$

*and for* $n = 0,1,\ldots,N$

$$\text{(8.27b)} \qquad \pi_n^*(\xi^n) - E^n\pi_{n+1}^*(\xi^{n+1}) \begin{cases} = -h_n & \text{if } s_n^*(\xi^n) > 0, \\ \in [-h_n, q_n] & \text{if } s_n^*(\xi^n) = 0, \\ = q_n & \text{if } s_n^*(\xi^n) < 0, \end{cases}$$

*and for* $n = 1,2,\ldots,N$

$$\text{(8.27c)} \qquad E^{n-1}\pi_n^*(\xi^n) \begin{cases} \leqq k_n & \text{if } s_{n-1}^*(\xi^{n-1}) \geqq y_n^*(\xi^{n-1}), \\ = k_n & \text{if } s_{n-1}^*(\xi^{n-1}) < y_n^*(\xi^{n-1}). \end{cases}$$

*Here* $s_n^*(\xi^n)$ *is the inventory at the end of period* n *resulting from* $y^*$

*and* $\xi^n$.

It is interesting to note that (8.27a,b) has the structure of a backward recursion. In fact, given any optimal level policy $y^*$ this recursion determines the optimal dual solution $\pi^*$ completely if the events $\{s^*_{N+1}(\xi^N) = \xi_{N+1}\}$ and $\{s^*_n(\xi^n) = 0\}$, $n = N,N-1,\dots,1$, have zero probability. This is the case, for instance, if the distribution of $\xi^N$ has a density. Two questions come up.

1. Is it possible to solve (8.27a,b)? That is, is it possible to give explicit *formulae for* $\pi^*$ *as functions of* $y^*$?
2. Is it possible to present a *dual dynamic programming method*? That is, is it possible to modify the recursion (8.27a,b) in such a way that $\pi^*$ and $y^*$ are determined simultaneously?

The answer to both questions is the affirmative, as we will show. Just as in the deterministic case a direct relation between the dual solution and the derivatives of the primal cost-to-go functions will appear to exist. Therefore we will first study these derivatives. It will be shown, that they too satisfy a backward recursion. This recursion will appear to be a suitable instrument for answering both questions.

8.3.5. The differentiable form of the (primal) dynamic programming recursion. From now on we make the assumption

(8.28) $\mu_n(.|\xi^{n-1})$ has a *density* $\kappa_n(.|\xi^{n-1})$ for all $\xi^{n-1} \in \Xi^{n-1}$, $n = 1,\dots,N$.

As a consequence $\kappa(\xi) := \Pi^N_{n=1}\kappa_n(\xi_n|\xi^{n-1})$ is a density for $\mu^N_1(.|\xi_0)$ for all $\xi_0 \in \Xi_0$. The assumption (8.28) is equivalent to the seemingly weaker condition that $\mu^N_1(.|\xi_0)$ has a density for $\lambda_0$-almost every $\xi_0$. Indeed, if $\kappa(\xi) = \kappa(\xi_1,\xi_2,\dots,\xi_N|\xi_0)$ is a density for $\mu^N_1(d\xi^N_1|\xi_0)$ for almost all $\xi_0$, then by redefinition on sets with probability zero we may equivalently assume that

$$\kappa(\xi) \geqq 0 \text{ for all } \xi \in \Xi^N,$$

$$\int\kappa(\xi^N_1|\xi_0)d\xi^N_1 = 1 \text{ for all } \xi_0 \in \Xi_0,$$

$$\int|\xi_k|\kappa(\xi^N_1|\xi_0)d\xi^N_1 < \infty \text{ for all } \xi_0 \in \Xi_0,\ k = 0,1,\dots,N.$$

Then

$$\kappa_n(\xi_n|\xi^{n-1}) := \int_{\Xi^N_{n+1}} \kappa(\xi^n_1,\xi^N_{n+1}|\xi_0)d\xi^N_{n+1}$$

is a density for $\mu_n(d\xi_n|\xi^{n-1})$ for all $\xi^{n-1}$, at least after a modification on subsets with $\lambda^{n-1}$-measure 0. It is easy to see that these modifications of $\mu_n$ can be made such that it remains a nice regular conditional distribution.

<u>THEOREM 8.9</u>. *Assume* (8.22) *and* (8.28). *Then the cost-to-go functions* $g_n(.,\xi^{n-1})$ *and* $f_n(.,\xi^{n-1})$ *defined in* (8.21b) *are differentiable for all* $\xi^{n-1} \in \Xi^{n-1}$, *and the derivatives satisfy*

$$f'_{N+1}(s_N,\xi^N) = f'_{N+1}(s_N) = -k_{N+1} + \gamma_{N+1}\cdot\chi_{(\xi_{N+1},\infty)}(s_N), \tag{8.29a}$$

*and for* $n = N,N-1,\dots,1$, $y_n \in \mathbb{R}$, $s_{n-1} \in \mathbb{R}$ *and* $\xi^{n-1} \in \Xi^{n-1}$

$$\begin{aligned} g'_n(y_n,\xi^{n-1}) &= \int\{-q_n + \gamma_n\cdot\chi_{(-\infty,y_n)}(\xi_n) + \\ &\quad + f'_{n+1}(y_n - \xi_n,\xi^{n-1},\xi_n)\}\mu_n(d\xi_n|\xi^{n-1}), \\ f'_n(s_{n-1},\xi^{n-1}) &= \max\{-k_n, g'_n(s_{n-1},\xi^{n-1})\} \\ &= g'_n(\max\{y^*_n,s_{n-1}\},\xi^{n-1}) \text{ for any } y^*_n \in Y^*_n(\xi^{n-1}), \end{aligned} \tag{8.29b}$$

*where for* $n = 1$ *we read*

$$\begin{aligned} f'_1(s_0) &= \max\{-k_1, g'_1(s_0,-s_0)\} \\ &= g'_1(\max\{y^*_1,s_0\},-s_0) \text{ for any } y^*_1 \in Y^*_1(-s_0); \end{aligned}$$

*here* $\chi_A(s) := 1$ *if* $s \in A$ *and* $\chi_A(s) := 0$ *otherwise. Moreover, for all* $\xi^{n-1} \in \Xi^{n-1}$

$$\begin{aligned} f'_n(-\infty,\xi^{n-1}) &= -\pi^{(n)}_n, & f'_n(+\infty,\xi^{n-1}) &= \Sigma^{N+1}_{j=n}h_j, \quad 1 \leq n \leq N+1, \\ g'_n(-\infty,\xi^{n-1}) &= -(q_n + \pi^{(n+1)}_{n+1}), & g'_n(+\infty,\xi^{n-1}) &= \Sigma^{N+1}_{j=n}h_j, \quad 1 \leq n \leq N, \end{aligned} \tag{8.30}$$

*for* $\pi_n^{(n)}$ *defined in* (8.9).

PROOF. From Theorem 8.1 we know that both $g_n(.,\xi^{n-1})$ and $f_n(.,\xi^{n-1})$ are finite convex functions, so that the left as well as the right derivatives exist. It is easy to show that (8.29a) and (8.29b) are valid if $f_n'$ and $g_n'$ are interpreted as left derivatives: the formula for $g_n'$ follows from (8.21b) by use of the Lebesgue dominated convergence theorem (which is applicable since $f_{n+1}(.,\xi^n)$ is Lipschitz continuous), and the other formulae are shown by elementary analysis. An identical proof shows that (8.29a,b) is also true for $f_n'$ and $g_n'$ being interpreted as right derivatives if $\chi_{(\xi_{N+1},\infty)}(s_N)$ and $\chi_{(\xi_n,\infty)}(y_n)$ are replaced by $\chi_{[\xi_{N+1},\infty)}(s_N)$ and $\chi_{[\xi_n,\infty)}(y_n)$, respectively. Therefore, we have for all $y_N \in \mathbb{R}$

$$g_N'(y_N + 0,\xi^{N-1}) - g_N'(y_N - 0,\xi^{N-1}) = \gamma_N\cdot\mu_N(\{y_N\}|\xi^{N-1}) + \delta_{N+1}\cdot\mu_N(\{\xi_{N+1}\}|\xi^{N-1})$$

and this difference vanishes since $\mu_N(d\xi_N|\xi^{N-1})$ has a density. Consequently, $g_N(.|\xi^{N-1})$ is differentiable, and therefore so is $f_{N-1}(.|\xi^{N-1})$. By backward recursion to n it follows in the same way that the left and the right derivatives of $g_n$ (and those of $f_n$) coincide. In other words, each $f_n(.,\xi^{n-1})$ and $g_n(.,\xi^{n-1})$ is differentiable, and the derivatives satisfy (8.29b).

By taking limits in (8.29a,b) we get, applying the Lebesgue dominated convergence theorem again,

$$\begin{aligned} f_{N+1}'(-\infty,\xi^{n-1}) &= -k_{N+1}, \\ g_n'(-\infty,\xi^{n-1}) &= -q_n + E^{n-1}f_{n+1}'(-\infty,\xi^n),\ n = N,N-1,\dots,1, \\ f_n'(-\infty,\xi^{n-1}) &= \max\{-k_n, g_n'(-\infty,\xi^{n-1})\},\ n = N,N-1,\dots,1. \end{aligned}$$

Since bij definition, see (8.9), $\pi_{N+1}^{(N+1)} := k_{N+1}$, $\pi_n^{(n)} := \min\{k_n, q_n + \pi_{n+1}^{(n+1)}\}$, n = N,N-1,...,1, it follows by induction that the left part of (8.30) is true. The other part of (8.30) follows similarly. □

Obviously, (8.29a,b) can be interpreted as a backward recursion. We call it the *differentiable form* of the dynamic programming recursion (8.21). A similar recursion can be formulated if the functions $f_n(.,\xi^{n-1})$ and $g_n(.,\xi^{n-1})$ are not differentiable, but only subdifferentiable. We refer to [6] for the precise formulation of a subdifferentiable form of a generalization of (8.21).

Not surprisingly, the optimal inventory levels can be expressed in terms of the derivatives of cost-to-go functions:

$$\text{(8.29c)} \qquad Y_n^*(\xi^{n-1}) = \{y \in \mathbb{R}: g_n'(y,\xi^{n-1}) = -k_n\} \cup \{-\infty: g_n'(-\infty,\xi^{n-1}) \geqq -k_n\},$$

as follows from (8.23) and the convexity of $g_n(.|\xi^{n-1})$. Notice that for each $y_n \in Y_n^*(\xi^{n-1})$

$$\text{(8.31)} \qquad g_n'(y_n,\xi^{n-1}) = \max\{-k_n,-q_n - \pi_{n+1}^{(n+1)}\} = -\pi_n^{(n)}.$$

Indeed, if $y_n > -\infty$ then $-k_n = g_n'(y_n,\xi^{n-1}) \geqq g_n'(-\infty,\xi^{n-1}) = -q_n - \pi_{n+1}^{(n+1)}$ because of (8.23), the equality being true since $g_n(.,\xi^{n-1})$ is convex. If $y_n = -\infty$ (8.31) follows directly from (8.30) and (8.29c).

Optimal solutions of the dual problem are closely related to the derivatives of the primal cost-to-go functions as can be seen from the next theorem.

<u>THEOREM 8.10</u>. *Assume* (8.22) *and* (8.28). *Then* $\pi^* = (\pi_0^*,\pi_1^*,\ldots,\pi_{N+1}^*)$, $\pi_n^* \in \Pi_n$, *is an optimal solution for* SD *if and only if almost surely*

$$\text{(8.32a)} \qquad \pi_{N+1}^*(\xi^N) = -f_{N+1}'(s_N^*(\xi^N)),$$

$$\text{(8.32b)} \qquad \pi_n^*(\xi^n) = q_n - \gamma_n \cdot \chi_{\{s_n^*(\xi^n)>0\}}(\xi^n) - f_{n+1}'(s_n^*(\xi^n),\xi^n),$$

$$n = N,N-1,\ldots,1,$$

$$\text{(8.32c)} \qquad \pi_0^*(\xi^0) = -f_1'(-\xi_0),$$

*where* $s_n^*(\xi^n)$ *is the inventory at the end of period* n, *resulting from*

$\xi^n$ *and an arbitrary optimal level policy* $y^*$. *Moreover, for each such* $\pi^*$ *and* $y^*$

(8.33) $$E^{n-1}\pi_n^*(\xi^n) = -f_n'(s_{n-1}^*(\xi^{n-1}),\xi^{n-1}) \ a.s., \ n = N,N-1,\dots,1.$$

PROOF. (Only if) Suppose that $\pi^*$ is optimal for SD. Choose any optimal $y^*$. Then (8.27a) and (8.29a) imply (8.32a). From (8.27b) it follows then that a.s.

$$\pi_N^*(\xi^N) = q_N - \gamma_N\cdot\chi_{\{s_N^*(\xi^N)>0\}}(\xi^N) - f_{N+1}'(s_N^*(\xi^N)).$$

For convenience set $y := \max\{s_{N-1}^*(\xi^{N-1}), y_N^*(\xi^{N-1})\}$ so that $s_N^*(\xi^N) = y - \xi_N$. Then we have a.s., because of (8.29b)

$$-E^{N-1}\pi_N^*(\xi^N) = \int\{-q_N + \gamma_N\cdot\chi_{(-\infty,y)}(\xi_N) + f_{N+1}'(y-\xi_N)\mu_N(d\xi_N|\xi^{N-1})$$

$$= g_N'(y,\xi^{N-1})$$

$$= f_N'(s_{N-1}^*(\xi^{N-1}),\xi^{N-1})$$

the last equality being true because of the definition of y. So we proved (8.32b) and (8.33) for n = N. For n < N we proceed by backward induction to n, in an almost identical way.

(If) Suppose that $\pi_n^* \in \Pi_n$, $n = 0,1,\dots,N+1$, and that (8.32) holds almost surely for an arbitrary optimal $y^*$. Then we will show that $\pi^*$ is optimal for SD. By integrating $\pi_n^*(\xi^n)$ with respect to $\mu_n(d\xi_n|\xi^{n-1})$ one derives (8.33) from (8.32). Because of (8.32) and (8.33) $\pi^*$ satisfies all conditions in (8.27) so that Corollary 8.8 guarantees its optimality. □

It is surprising that *nonrecursive* formulae for $f_n'$ and $g_n'$ can be derived. In these formulae optimal inventory levels of periods $\geqq n$ occur, as can be seen in the next theorem.

THEOREM 8.11. *Assume* (8.22) *and* (8.28). *Then the following formulae hold for* $\lambda^{n-1}$*-almost every* $\xi^{n-1}$.

$$f_n'(s,\xi^{n-1}) = -\pi_n^{(n)} + \Sigma_{j=n}^{N+1}\gamma_j.\mu_n^N(D_{nj}(s)\,|\,\xi^{n-1}), \qquad s \in \mathbb{R},$$

(8.34)

$$g_n'(y,\xi^{n-1}) = -q_n - \pi_{n+1}^{(n+1)} + \Sigma_{j=n}^{N+1}\gamma_j.\mu_n^N(C_{nj}(y)\,|\,\xi^{n-1}), \; y \in \mathbb{R},$$

*where* $D_{nj}(s)$, $s \in \mathbb{R}$, *and* $C_{nj}(y)$, $y \in \mathbb{R}$, *both with* $n = 1,\dots,N$, $j = n,n+1,\dots,N+1$, *are subsets of* $\Xi^N$ *defined by*

$$D_{nj}(s) := \{\xi^N \in \Xi^N\colon s > \xi_n + \xi_{n+1} + \dots + \xi_j \geqq y_n^*(\xi^{n-1}),$$

(8.35)

$$\xi_k + \xi_{k+1} + \dots + \xi_j \geqq y_k^*(\xi^{k-1}),\; k \in K_{nj}\},$$

$$C_{nj}(y) := \{\xi^N \in \Xi^N\colon \xi_n + \xi_{n+1} + \dots \xi_j < y,$$

(8.36)

$$\xi_k + \xi_{k+1} + \dots \xi_j \geqq y_k^*(\xi^{k-1}),\; k \in K_{nj}\}.$$

*Here* $(y_1^*,\dots,y_N^*)$ *is an arbitrary optimal level policy, and* $K_{nj} := \{n+1,\dots,\min\{j,N\}\}$. *The formula for* $f_n'(s,\xi^{n-1})$ *is also true for* $n = N+1$ *provided* $s \neq \xi_{N+1}$ *if the following definitions are made:*

$$D_{N+1,N+1}(s) := \{\xi^N \in \Xi^N\colon s > \xi_{N+1}\},$$

$$\mu_{N+1}^N(D\,|\,\xi^N) := \chi_D(\xi^N).$$

PROOF. First we will show that (8.34) is true for *all* $\xi^{n-1} \in \Xi^{n-1}$ if in (8.34) $D_{nj}(s)$ and $C_{nj}(y)$ are replaced by slightly different sets. Recall that since $SP_o$ is strictly proper universally measurable functions $\tilde{y}_k$ exist with $\tilde{y}_k(\xi^{k-1}) \in Y_k^*(\xi^{k-1})$ for all $\xi^{k-1} \in \Xi^{k-1}$, $k = 1,\dots,N$. For any such $(\tilde{y}_1,\dots,\tilde{y}_N)$ define $\tilde{D}_{nj}(s)$ and $\tilde{C}_{nj}(y)$ as $D_{nj}(s)$ and $C_{nj}(y)$, respectively, but with the modification that in (8.35) and (8.36) $y_k^*(\xi^{k-1})$ is replaced by $\tilde{y}_k(\xi^{k-1})$, $k = n,n+1,\dots,N$. These are universally measurable sets, so that $\mu_n^N(\tilde{D}_{nj}(s)\,|\,\xi^{n-1})$ and $\mu_n^N(\tilde{C}_{nj}(y)\,|\,\xi^{n-1})$ are well defined for all relevant values of $n$, $j$, $\xi^{n-1}$, $s$ and $y$. We will now show that (8.34), with $D_{nj}(s)$ and $C_{nj}(y)$ replaced by $\tilde{D}_{nj}(s)$ and $\tilde{C}_{nj}(y)$, is true for all $\xi^{n-1} \in \Xi^{n-1}$. The proof is based on (8.29b), and uses backward induction with respect to n. Clearly,

$$f'_{N+1}(s) = -k_{N+1} + (k_{N+1} - r).\chi_{(\xi_{N+1},\infty)}(s) =$$
$$-\pi^{(N+1)}_{N+1} + \gamma_{N+1}.\mu^N_{N+1}(\widetilde{D}_{N+1,N+1}(s)\,|\,\xi^N),$$

at least if $s \neq \xi_{N+1}$. Assume that we already proved that for some $m \in \{N,N-1,\ldots,1\}$

$$f'_{m+1}(s,\xi^m) = -\pi^{(m+1)}_{m+1} + \Sigma^{N+1}_{j=m+1}\gamma_j.\mu^N_{m+1}(\widetilde{D}_{m+1,j}(s)\,|\,\xi^m).$$

First we will prove the formula for $g'_m(y,\xi^{m-1})$. Equation (8.29b) shows that $g'_m(y,\xi^{m-1})$ equals

$$\int\{-q_m + \gamma_m.\chi_{(-\infty,y_m)}(\xi_m) - \pi^{(m+1)}_{m+1} +$$
$$+ \Sigma^{N+1}_{j=m+1}\gamma_j.\mu^N_{m+1}(\widetilde{D}_{m+1,j}(y - \xi_m)\,|\,\xi^m)\}\mu_m(d\xi_m\,|\,\xi^{m-1}).$$

The desired result follows immediately from the definitions of $\mu^N_m(d\xi^N\,|\,\xi^{m-1})$ and $\widetilde{C}_{mj}(y)$, $j \geqq m$:

$$\int\chi_{(-\infty,y)}(\xi_m)\mu_m(d\xi_m\,|\,\xi^{m-1}) = \mu^N_m(\widetilde{C}_{mm}(y)\,|\,\xi^{m-1}), \text{ and}$$

$$\int\mu^N_{m+1}(\widetilde{D}_{m+1,j}(y - \xi_m)\,|\,\xi^m)\mu_m(d\xi_m\,|\,\xi^{m-1}) = \mu^N_m(\widetilde{C}_{mj}(y)\,|\,\xi^{m-1}), \text{ for } j \geqq m+1,$$

so that, indeed,

$$g'_m(y,\xi^{m-1}) = -q_m - \pi^{(m+1)}_{m+1} + \Sigma^{N+1}_{j=m}\gamma_j.\mu^N_m(\widetilde{C}_{mj}(y)\,|\,\xi^{m-1}).$$

Next we will prove the formula for $f'_m(s,\xi^{m-1})$. By substitution of the formula for $g'_m(y,\xi^{m-1})$ in (8.29b) with $n := m$, we get

$$f'_m(s,\xi^{m-1}) = \max\{-k_m, -q_m - \pi^{(m+1)}_{m+1} + \Sigma^{N+1}_{j=m}\gamma_j.\mu^N_m(\widetilde{C}_{mj}(s)\,|\,\xi^{m-1})\}.$$

Recall that $-\pi^{(m)}_m = \max\{-k_m, -q_m - \pi^{(m+1)}_{m+1}\}$. Therefore, since $\max\{x,y+z\} = \max\{x,y\} + (z - (-x + y)^-)^+$ for arbitrary real numbers $x$, $y$, $z$ with $z \geqq 0$, it follows that

$$(8.37) \qquad f'_m(s,\xi^{m-1}) = -\pi_m^{(m)} + (\Sigma_{j=m}^{N+1}\gamma_j \cdot \mu_m^N(\tilde{C}_{mj}(s)\,|\,\xi^{m-1}) + - (k_m - q_m - \pi_{m+1}^{(m+1)})^-)^+.$$

A comparison of (8.31) with n := m and the formula for $g'_m(y,\xi^{m-1})$ proved a moment ago, both with $y := \tilde{y}_m(\xi^{m-1})$, reveals that

$$(8.38) \qquad \Sigma_{j=m}^{N+1}\gamma_j \cdot \mu_m^N(\tilde{C}_{mj}(\tilde{y}_m(\xi^{m-1})\,|\,\xi^{m-1}) = (k_m - q_m - \pi_{m+1}^{(m+1)})^-.$$

Since $\tilde{C}_{mj}(s)$ is increasing with s, we find by substitution of (8.38) in (8.37) that

$$f'_m(s,\xi^{m-1}) = -\pi_m^{(m)} + \Sigma_{j=m}^{N+1}\gamma_j \cdot \mu_m^N(\tilde{C}_{mj}(s) \setminus \tilde{C}_{mj}(\tilde{y}_m(\xi^{m-1}))\,|\,\xi^{m-1})$$

and this is nothing other than the desired formula for $f'_m$:

$$f'_m(s,\xi^{m-1}) = -\pi_m^{(m)} + \Sigma_{j=m}^{N+1}\gamma_j \cdot \mu_m^N(\tilde{D}_{mj}(s)\,|\,\xi^m).$$

This completes the inductive proof. It remains to show, however, that (8.34) also holds without the modifications. Fix an arbitrary optimal level policy $(y_1^*,\ldots,y_N^*)$. By definition, each $y_n^*$ is Borel measurable, and $y_n^*(\xi^{n-1}) \in Y_n^*(\xi^{n-1})$ for all $\xi^{n-1} \in B^{n-1}$ for $B^{n-1}$ a Borel set in $\Xi^{n-1}$ with $\lambda^{n-1}(B^{n-1}) = 1$. Let for $n = 1,\ldots,N$ $\hat{y}_n$ be any universally measurable selection of $Y_n^*$, then so is $\tilde{y}_n$ defined by

$$\tilde{y}_n(\xi^{n-1}) := y_n^*(\xi^{n-1}) \text{ if } \xi^{n-1} \in B^{n-1},$$

$$\tilde{y}_n(\xi^{n-1}) := \hat{y}_n(\xi^{n-1}) \text{ if } \xi^{n-1} \notin B^{n-1}.$$

If the $y_{n+1}^*,\ldots,y_j^*$ are used in the definition of $D_{nj}(s)$ and $C_{nj}(y)$ and $\tilde{y}_{n+1},\ldots,\tilde{y}_j$ in the definition of $\tilde{D}_{nj}(s)$ and $\tilde{C}_{nj}(y)$, then obviously

$$\lambda^N(\tilde{D}_{nj}(s)) = \lambda^N(D_{nj}(s)),\ \lambda^N(\tilde{C}_{nj}(y)) = \lambda^N(C_{nj}(y)),$$

and consequently, for $\lambda^{n-1}$-almost every $\xi^{n-1}$, the corresponding conditional probabilities coincide too:

$$\mu_n^N(\widetilde{D}_{nj}(s)\,|\,\xi^{n-1}) = \mu_n^N(D_{nj}(s)\,|\,\xi^{n-1}),$$

$$\mu_n^N(\widetilde{C}_{nj}(y)\,|\,\xi^{n-1}) = \mu_n^N(C_{nj}(y)\,|\,\xi^{n-1}).$$

This completes the proof of Theorem 8.11. □

REMARK 8.12. The events $D_{nj}(s)$ and $C_{nj}(y)$ describe conditions on $\xi^N$ which relate the inventories in the periods n and j, under the assumption that the level policy $y^*$ is employed. With

| | |
|---|---|
| $s_{n-1} := s_{n-1}^*(\xi^{n-1})$, | the inventory at the beginning of period n, |
| $t_n := s_{n-1}^*(\xi^{n-1}) + x_n^*(\xi^{n-1})$, | the inventory immediately after the production in period n, |
| $s_j := s_j^*(\xi^j)$, | the inventory at the end of period j, |

we have

$$\xi^N \in D_{nj}(s) \Leftrightarrow \begin{cases} \text{if } s_{n-1} = s, & \text{then } s_j > 0, \\ \text{if } s_{n-1} = (y_n^*(\xi^{n-1}))^+ & \text{then } s_j \leqq 0, \end{cases}$$

$$\xi^N \in C_{nj}(y) \Leftrightarrow \begin{cases} \text{if } t_n = y & \text{then } s_j > 0, \\ \text{if } t_n = \xi_n + (y_{n+1}^*(\xi^n))^+ & \text{then } s_j \leqq 0. \end{cases}$$

The proof is based on

$$s_j = \max\{s_{n-1} - \Sigma_{i=n}^j \xi_i, \\ \max_{k=n,\dots,\min\{j,N\}}\{y_k^*(\xi^{k-1}) - \Sigma_{i=k}^j \xi_i\}\}$$

and

$$s_j = \max\{t_n - \Sigma_{i=n}^j \xi_i, \\ \max_{k=n+1,\dots,\min\{j,N\}}\{y_k^*(\xi^{k-1}) - \Sigma_{i=k}^j \xi_i\}\},$$

respectively, and makes use of the nonnegativity of the demands.

REMARK 8.13. For the *one-period* problem $(N := 1)$ with a fixed initial stock $-\xi_0$ the formula for $g_1'(y) = g_1'(y,\xi_0)$ is

$$g_1'(y) = -q_1 + (q_1 + h_1).F_1(y) - k_{N+1} + (k_{N+1} - r).F_1(y - \xi_{N+1}),$$

where $F_1$ is the distribution function of $\xi_1$. Since the optimal inventory level $y_1^*$ (if $> -\infty$) is determined by $g_1'(y_1^*) = -k_1$, we get the well-known *quantile formula*

$$y_1^* \in F_1^{-1}\left(\frac{q_1 - k_1}{q_1 + h_1}\right) \text{ if } -h_1 < k_1 < q_1,$$

if the terminal costs are neglected $(k_{N+1} := r := 0)$. From (8.34) we see that this result can be generalized to multi-period problems, in the sense that optimal inventory levels are characterized as quantiles of "a *weighted average of certain conditional distribution functions of future demands*". Of course, such a characterization is only possible as the costs of production, holding and shortage are all linear and deterministic.

REMARK 8.14. In general, formulae (8.34) do not hold if $\xi^N$ is *deterministic*, not even in points where the derivative exists. For example, consider $f_n'(s)$ for $\xi_n < s < \xi_{n+1}$, then from Subsection 8.2.3 we know that $f_n'(s) = \rho_{n,n+1}$. Formula (8.34) would give

$$f_n'(s) = -\pi_n^{(n)} + \gamma_n.d_{nn}$$

where $d_{nn} := 0$ if $\xi_n < y_n^*$ and $d_{nn} := 1$ if $\xi_n \geq y_n^*$. A direct verification shows that this equality is correct if $y_n^* \neq \xi_n$. If, however, $y_n^* = \xi_n$, which is the case if $-h_n < k_n - \pi_{n+1}^{(n+1)} < q_n$, this equality is only true if we define $d_{nn}$ as

$$d_{nn} := \frac{k_n + h_n - \pi_{n+1}^{(n+1)}}{h_n + q_n} \in (0,1).$$

From this exercise we may conclude that the definition of $D_{nn}(s)$ is too rough for the deterministic case: one needs a specific convex combination of two sets, namely $D_{nn}(s) := \{s > \xi_n \geq y_n^*\}$ and $D_{nn}^>(s) := \{s > \xi_n > y_n^*\}$. A major cause for the difficulty can be pointed out by means of (8.31), which was used in the proof of Theorem 8.11. If $g_n$ is not differentiable at $y_n^* \in Y_n^*$, which is the normal case if the demands are deterministic, then not every element of $\partial g_n(y_n^*)$ will make (8.31) true, because only $-k_n$, which indeed belongs to this set, will do. Similar complications arise if $\xi^N$ is random but with a *discrete distribution*. As a consequence, one should not expect that Theorem 8.11 will still hold if the assumption regarding the continuity of the distributions is dropped.

REMARK 8.15. We now want to indicate an alternative proof of (8.34), under the somewhat restrictive assumption that the demands are independent random variables. The alternative approach is based on formulae for derivatives which are important on their own because of their natural interpretation. We will restrict ourselves to $f_n'$ for $n = 1$, and even for this case we only give the more interesting part of the proof. For a complete proof we refer to [8].

Let $y_n \in [-\infty,\infty)$, $n = 1,\dots,N$, be arbitrary fixed numbers and denote the corresponding constant level policy by $y := (y_1,y_2,\dots,y_N)$. Denote the deterministic initial stock by s, and the random demands by $\xi = (\xi_1,\dots,\xi_N)$. For the moment they are supposed to be independent. Recall that in that case an optimal level policy $y^*$ exists which is indeed constant. With $T_{y,\xi}(s)$ := the total cost of the production-inventory system as a function of the initial stock s, where y is the policy used and $\xi$ is the demand vector, with $t_y(s) := E\, T_{y,\xi}(s)$, and with $y^*$ a constant optimal level policy we have

$$f_1(s) = t_{y^*}(s) = E\, T_{y^*,\xi}(s), \quad s \in \mathbb{R}.$$

We want to find $f_1'(s)$ by differentiation under the integral. For that reason we are interested in a formula for the derivative of

$$\tau(s) := T_{y,\xi}(s), \quad s \in \mathbb{R},$$

for fixed $(y,\xi)$. Such a formula follows from

$$T_{y,\xi}(s) = \Sigma_{j=1}^{N} k_j.x_j + \Sigma_{j=1}^{N+1}\{h_j.(s_j)^+ + q_j.(s_j)^-\}$$

where, with $y_{N+1} := -\infty$,

$$s_j = s_j(s) = \max\{s - \Sigma_{i=1}^{j}\xi_i, \max\{y_k - \Sigma_{i=k}^{j}\xi_k: k = 1,2,\dots,j\}\},$$
$$j = 1,2,\dots,N+1,$$
$$x_j = x_j(s) = (y_j - s_{j-1}(s))^+, \ j = 1,2,\dots,N.$$

These expressions can be simplified by means of the definitions

$$\eta_0 := 0, \ \eta_j := \Sigma_{i=1}^{j}\xi_i, \ j = 1,2,\dots,N+1,$$
$$\beta_0 := -\infty, \ \beta_j := \max\{y_k + \eta_{k-1}: k = 1,2,\dots,j\}, \ j = 1,2,\dots,N+1,$$

which imply that

$$0 = \eta_0 \leqq \eta_1 \leqq \eta_2 \leqq \dots \leqq \eta_N \leqq \eta_{N+1} < +\infty,$$
$$-\infty = \beta_0 \leqq y_1 = \beta_1 \leqq \beta_2 \leqq \dots \leqq \beta_N = \beta_{N+1} < +\infty,$$
$$\beta_{j+1} = \max\{\beta_j, y_{j+1} + \eta_j\}, \ j = 0,1,\dots,N.$$

Of course, $\eta_j$ depends on $\xi$ and $\beta_j$ depends on $(y,\xi)$. By substitution one derives

$$s_j(s) = -\eta_j + \max\{s,\beta_j\}, \ j = 0,1,\dots,N+1,$$
$$x_{j+1}(s) = (\beta_{j+1} - \max\{s,\beta_j\})^+, \ j = 0,1,\dots,N-1.$$

The number $\beta_j$ is the smallest value for the initial stock s for which there is no production in any of the periods 1,2,...,j. That is

$$s \geqq \beta_j \Leftrightarrow x_1 = x_2 = \dots = x_j = 0, \ j = 1,2,\dots,N.$$

The verification is straightforward, and is left to the reader. As a consequence,

$$-\infty < s < \beta_1 \Leftrightarrow x_1 > 0,$$

$$\beta_j \leqq s < \beta_{j+1} \quad \Leftrightarrow \quad x_1 = x_2 = \ldots = x_j = 0,\ x_{j+1} > 0, \qquad j = 1,\ldots,N-1,$$

$$\beta_N \leqq s < +\infty \quad \Leftrightarrow \quad x_1 = x_2 = \ldots = x_N = 0.$$

The period of the first strictly positive production, which is period j+1 if $\beta_j \leqq s < \beta_{j+1}$, is important for the dependence of $x_i$ and $s_i$ of s. Obviously, if the initial inventory is increased by a marginal amount, then $s_i$, i = 1,2,...,j, is increased by the same amount, and $x_{j+1}$ is decreased by the same amount; all other $s_i$ and $x_i$ remain unchanged. This explains the formulae for the derivatives

$$s_j'(s) = \begin{cases} 0 & \text{if } s < \beta_j, \\ 1 & \text{if } s > \beta_j, \end{cases} \qquad j = 1,\ldots,N+1,$$

$$x_{j+1}'(s) = \begin{cases} -1 & \text{if } \beta_j < s < \beta_{j+1}, \\ 0 & \text{if } s < \beta_j \text{ or } s > \beta_{j+1}, \end{cases} \qquad j = 0,1,\ldots,N-1.$$

These expressions show that $\tau$ is differentiable in all $s \in \mathbb{R}$ except possibly in $s = \beta_j$ and $s = \eta_j$, j = 0,1,...,N+1; in the remaining points its derivative is given by

$$\tau'(s) = -k_1 \qquad \text{if } -\infty < s < \beta_1,$$

$$\tau'(s) = -k_{j+1} + \Sigma_{i=1}^{j}(h_i \cdot \chi_{(\eta_i,\infty)}(s) - q_i \cdot \chi_{(-\infty,\eta_i)}(s))$$

$$\text{if } \beta_j < s < \beta_{j+1} \text{ for a } j \in \{1,2,\ldots,N-1\},$$

$$\tau'(s) = \Sigma_{i=1}^{N+1}(h_i \cdot \chi_{(\eta_i,\infty)}(s) - q_i \cdot \chi_{(-\infty,\eta_i)}(s)) \text{ if } \beta_N < s.$$

Since we assume that $\xi$ has a continuous distribution, we know that

$$\lambda^N(\{\eta_i = s\}) = 0,\ i = 1,2,\ldots,N+1, \text{ for all } s \in \mathbb{R},$$

$$\lambda^N(\{\beta_i = s\}) = 0,\ i = 1,2,\ldots,N+1, \text{ for all } s \neq y_1,$$

so that $T_{y,\xi}$ is differentiable almost everywhere except maybe in $s = y_1$.

Even so, we have for all s

$$t_y'(s) = -\Sigma_{j=0}^{N-1} k_{j+1} \cdot \lambda^N(\{\beta_j < s \leqq \beta_{j+1}\}) +$$

$$(8.39) \qquad + \Sigma_{j=1}^{N+1} h_j \cdot \lambda^N(\{\beta_j < s,\ \eta_j < s\}) +$$

$$- \Sigma_{j=1}^{N+1} q_j \cdot \lambda^N(\{\beta_j < s,\ \eta_j > s\})$$

where the $\leqq$ sign is crucial for $s = y_1$.

Formula (8.39) has some resemblance with (8.34). Not only its form is similar, but we also have that

$$t_y'(-\infty) = -k_1 \leqq -\pi_1^{(1)} = f_1'(-\infty),$$

$$t_y'(+\infty) = +\Sigma_{j=1}^{N+1} h_j \qquad = f_1'(+\infty).$$

On the other hand, it is not true that $t_y$ is convex for an arbitrary constant level policy y as it is for an optimal one $y^*$. Nevertheless, it is possible to derive (8.34) from (8.39) by exploiting the optimality of $y^*$. We will not work out the tedious details, for which we refer to [8]. There we used explicitly the assumption that the demands are independent. By the way, it is not to be expected that a similar proof is possible in the general case. A major reason is, that for $y = (y_1, y_2, \ldots, y_N)$ with $y_n = y_n(\xi^{n-1})$, $n = 1,2,\ldots,N$, it is not true any more that

$$\lambda^N(\{\beta_i = s\}) = 0 \text{ for all } s \in \mathbb{R},\ i = 2,3,\ldots,N.$$

This concludes Remark 8.15.

Let us return to the dual problem SD. By combining Theorems 8.10 and 8.11 we get explicit expressions for the optimal solution of SD in terms of the optimal inventory levels.

<u>THEOREM 8.16</u>. *Assume* (8.22) *and* (8.28). *Then* $\pi^* = (\pi_0^*, \pi_1^*, \ldots, \pi_{N+1}^*)$, $\pi_n^* \in \Pi_n$, *is an optimal solution for* SD *if and only if almost surely*

$$\pi^*_{N+1}(\xi^N) = k_{N+1} - \gamma_{N+1}\cdot\chi_{\{\xi_{N+1}<s^*_N(\xi^N)\}}(\xi^N),$$

$$(8.40)\qquad \begin{aligned} \pi^*_n(\xi^n) &= \pi^{(n+1)}_{n+1} + q_n - \Sigma^{N+1}_{j=n}\gamma_j\cdot\mu^N_{n+1}(D_{n+1,j}(s^*_n(\xi^n))\,|\,\xi^n), \\ &\qquad\qquad n = 1,2,\ldots,N, \\ \pi^*_0(\xi_0) &= \pi^{(1)}_1 - \Sigma^{N+1}_{j=1}\gamma_j\cdot\mu^N_1(D_{1j}(-\xi_0)\,|\,\xi_0), \end{aligned}$$

*where* $s^*_n(\xi^n)$ *is the inventory at the end of period* $n$, *resulting from* $\xi^n$ *and an arbitrary optimal level policy* $y^*$. *Moreover, for each such* $\pi^*$ *and* $y^*$

$$(8.41)\qquad E^{n-1}\pi^*_n(\xi^n) = \pi^{(n)}_n - \Sigma^{N+1}_{j=n}\gamma_j\cdot\mu^N_n(D_{nj}(s^*_{n-1}(\xi^{n-1}))\,|\,\xi^{n-1}) \ \textit{a.s.}, \qquad n = 1,\ldots,N.$$

*In* (8.40) *and* (8.41) *the sets* $D_{nj}(s)$, $1 \leqq n \leqq j \leqq N+1$, *are defined as in Theorem* 8.11, *whereas for* $n = 1,2,\ldots,N$

$$D_{n+1,n}(s) := \Xi^N \ \textit{if } s > 0,$$
$$D_{n+1,n}(s) := \emptyset \ \textit{otherwise}.$$

PROOF. A straightforward substitution of (8.34) into (8.32) and (8.33). □

In Section 8.4 we will make some remarks on the economic interpretation of (8.34) and (8.40).

8.3.6. A dual dynamic programming method. After having established the relationship between the optimal solution of the dual problem SD and the derivatives of the cost-to-go functions of the dynamic programming algorithm, it is not difficult to present a backward recursion, which solves SD (and SP as well). Assume again that each $\mu_n(d\xi_n\,|\,\xi^{n-1})$ has a density. Inspired by (8.29), (8.32) and (8.33) we propose the following iterative scheme

$$(8.42a)\qquad \pi^{(N+1)}(s_N,\xi^N) := \pi^{(N+1)}(s_N) := \begin{cases} k_{N+1} & \text{if } s_N \leqq \xi_{N+1}, \\ r & \text{if } s_N > \xi_{N+1}, \end{cases}$$

and for $n = N, N-1, \ldots, 1$, $\xi^{n-1} \in \Xi^{n-1}$,

$$\pi^{(n)}(s_{n-1}, \xi^{n-1}) := \min\{k_n,\ q_n - \gamma_n \cdot \mu_n((-\infty, s_{n-1}) | \xi^{n-1}) +$$

(8.42b) $$+ \int \pi^{(n+1)}(s_{n-1} - \xi_n, \xi^{n-1}, \xi_n) \mu_n(d\xi_n | \xi^{n-1})\},$$

$$s_{n-1} \in \mathbb{R},$$

and

$$Y_n^*(\xi^{n-1}) := \{s \in \mathbb{R}:\ k_n = q_n - \gamma_n \cdot \mu_n((-\infty, s) | \xi^{n-1}) +$$

(8.42c) $$+ \int \pi^{(n+1)}(s - \xi_n, \xi^{n-1}, \xi_n) \mu_n(d\xi_n | \xi^{n-1})\}$$

$$\cup \{-\infty:\ k_n \geqq q_n + \pi_{n+1}^{(n+1)}\},$$

and call it again a *dual dynamic programming* method. It solves SP since the optimal level sets $Y_n^*(\xi^{n-1})$ are found, and it solves SD simultaneously since $E^n \pi_{n+1}^*(\xi^{n+1}) = \pi^{(n+1)}(s_n^*(\xi^n), \xi^n)$ a.s. which determines $\pi_n^*$ uniquely a.s. by (8.27b). Both claims follow from the observation that $\pi^{(n)} = -f_n'$, $n = N+1, \ldots, 1$, as a comparison of (8.29) and (8.42) shows.

We notice that $\pi^{(n)}(., \xi^{n-1})$ is nonincreasing, reflecting the convexity of $f_n(., \xi^{n-1})$; its extreme values are

$$\pi^{(n)}(-\infty, \xi^{n-1}) = \pi_n^{(n)},$$

$$\pi^{(n)}(\infty, \xi^{n-1}) = -\Sigma_{j=n}^{N+1} h_j,$$

see (8.30). In fact, $\pi^{(n)}(s, \xi^{n-1}) = \pi_n^{(n)}$ for all $s \leqq \max\{\ y^+:\ y \in Y_n^*(\xi^{n-1})\}$; this follows from (8.29b) and (8.31), together with the fact that each $f_n$ and $g_n$ have a constant slope on $(-\infty, 0]$. This follows from (8.29) as well, since all $\xi_j \geqq 0$, $j \geqq n$.

Let us compare the stochastic dual dynamic programming recursion with the deterministic one (8.9a,b,c). If (8.42a,b) is applied in case $\xi$ is deterministic (no density!), then $\pi^{(n)}(s) \equiv \pi^{(n)}(s, \xi^{n-1})$ is defined for $s \neq \eta_{ni} := \Sigma_{j=n}^{i} \xi_j$, $i = n, n+1, \ldots, N+1$ by (8.42) in the following way

$$(8.43a)\qquad \pi^{(N+1)}(s) = \begin{cases} k_{N+1} & \text{if } s < \xi_{N+1}, \\ r & \text{if } s > \xi_{N+1}, \end{cases}$$

and for $n = N, N-1, \ldots, 1$

$$(8.43b)\qquad \pi^{(n)}(s) = \begin{cases} \min\{k_n, q_n + \pi^{(n+1)}(s - \xi_n)\} & \text{if } s < \xi_n, \\ \min\{k_n, -h_n + \pi^{(n+1)}(s - \xi_n)\} & \text{if } s > \xi_n. \end{cases}$$

These $\pi^{(n)}(s)$ are precisely the same as the $\pi^{(n)}(s)$ defined at the end of Subsection 8.2.5, since in both cases $\pi^{(n)}(s) = -f_n'(s)$, see (8.10). Therefore, the recursion (8.43) is a slight modification (see (8.9e)) of the deterministic dual dynamic programming method (8.9a,b,c). There is a difference, however: the determination of $Y_n^*$ has to be adjusted since $f_n$ is not differentiable everywhere: this explains why in (8.9c) one has to consider convex combinations which is not necessary in (8.42c).

8.3.7. A conjugate dual dynamic programming method. In the deterministic case we discussed a so-called conjugate dual dynamic programming method, which came up as a result of a natural iterative approach to the dual problem. A similar approach to the stochastic dual problem SD is possible. In order to see this, notice that

$$\sup_{(\pi_0,\pi_1,\ldots,\pi_{N+1})\in\Pi} E\ \Sigma_{n=0}^{N+1}\pi_n(\xi^n).\xi_n$$

can be calculated as

$$\sup_{\pi_0} E\ [\pi_0(\xi_0).\xi_0 + \sup_{\pi_1} E^0[\pi_1(\xi^1).\xi_1 + \ldots$$
$$\ldots + \sup_{\pi_N} E^{N-1}[\pi_N(\xi^N).\xi_N + \sup_{\pi_{N+1}} E^N[\pi_{N+1}(\xi^N).\xi_{N+1}]]\ldots]],$$

at least if one takes the interchange

$$(8.44)\qquad \sup_{\pi_n} E^m = E^m \sup_{\pi_n},\ m < n,$$

for granted. This leads to the following recursion:

(8.45a) $$p_N(\pi_N,\xi^N) := p_N(\pi_N) := \sup_{\pi_{N+1}\in\mathbb{R}} \{\pi_{N+1}.\xi_{N+1} : (\pi_N,\pi_{N+1})A_{NN} \leq e_N\}$$
for $(\pi_N,\xi^N) \in \mathbb{R}\times\Xi^N$

and for $n = N-1,\ldots,0$ and $(\pi_n,\xi^n) \in \mathbb{R}\times\Xi^n$,

(8.45b) $$\begin{aligned} p_n(\pi_n,\xi^n) := \sup\{E^n[\pi_{n+1}(\xi_{n+1}).\xi_{n+1} + p_{n+1}(\pi_{n+1}(\xi_{n+1}),\xi^n)]: \\ \pi_n A_{nn} + E^n\pi_{n+1}(\xi_{n+1})A_{n+1,n} \leq e_n, \\ \pi_{n+1}(.) \in L_\infty(\Xi_{n+1},B_{n+1},\lambda_{n+1};\mathbb{R})\}, \end{aligned}$$

where the supremum is taken over $\pi_{n+1}(\xi_{n+1})$, and finally

(8.45c) $$\begin{aligned} \sup SD &= \sup_{\pi_0\in\Pi_0} E\,[\pi_0(\xi_0).\xi_0 + p_0(\pi_0(\xi_0),\xi^0)] \\ &= E \sup_{\pi_0\in\mathbb{R}} \{\pi_0.\xi_0 + p_0(\pi_0,\xi_0)\}. \end{aligned}$$

This is just the standard backward recursion for SD, if we interpret this problem as the (N+2)-stage stochastic dynamic programming problem with the following characteristics. At a typical stage n+1 (where n = 0,1,...,N-1) the following definitions are appropriate.

state : $(\pi_n,\xi^n) \in \mathbb{R}\times\Xi^n$,

action : $\pi_{n+1}(.) \in L_\infty(\Xi_{n+1},B_{n+1},\lambda_{n+1};\mathbb{R})$.

constraint set : $\{\pi_{n+1}(.) \in L_\infty: E^n\pi_{n+1}(\xi_{n+1})A_{n+1,n} \leq e_n - \pi_n A_{nn}\}$; it depends on both state variables.

disturbance : $\xi_{n+1} \in \Xi_{n+1}$; its distribution $\mu_{n+1}(.|\xi^n)$, which is also used in the constraint, depends on the second state variable only.

reward function: $\pi_{n+1}(\xi_{n+1}).\xi_{n+1}$; does not depend on the state.

system function: $(\pi_{n+1},\xi^{n+1}) = F_{n+1}((\pi_n,\xi^n),\pi_{n+1}(.),\xi_{n+1})$
$:= ((\pi_n,\pi_{n+1}(\xi_{n+1})),(\xi^n,\xi_{n+1}))$.

For the stages 0 and N the definitions have to be adjusted slightly. With these definitions the general theory on stochastic dynamic programming (e.g. [1 Proposition 8.2]) can be applied to prove that (8.45) is well defined and that its terminal value is equal to sup SD, indeed. Taking this proof for granted too it is then obvious that (8.45) solves SD: denoting the sets of minimizers in (8.45) by $P^*_{n+1}(\pi_n,\xi^n)$, $n = N,\dots,0$ and $P^*_0$ then $(\pi^*_0,\pi^*_1,\dots,\pi^*_{N+1})$ solves SD if and only if almost surely $\pi^*_0 \in P^*_0$, $\pi^*_{n+1}(\xi^n,.) \in P^*_{n+1}(\pi^*_n(\xi^n),\xi^n)$, $n = 0,1,\dots,N$.

Notice that a typical difference between the primal and the conjugate dual dynamic programming recursion is, that in the latter the action spaces are function spaces rather than Euclidean spaces. One might be tempted to get rid of the function spaces, by executing the supremization in (8.45b) pointwisely, that is, by applying (8.44) for m = n. Indeed, it is not unreasonable to expect that

$$\sup_{\pi_{n+1}\in L_\infty} E^n[.] = E^n[\sup_{\pi_{n+1}\in\mathbb{R}} .].$$

Nevertheless this approach fails since the constraints

$$\pi_n A_{nn} + E^n\pi_{n+1}(\xi_{n+1})A_{n+1,n} \leq e_n$$

cannot be separated into conditions for fixed $\xi_{n+1}$. This observation reveals a major difference between SP and SD. Since the constraint matrix A of SP is *lower* block triangular, a natural dynamic programming method exists consisting of *pointwise* minimizations. However, since the constraint matrix $A^T$ of SD is *upper* block triangular, there does not seem to be a natural recursive algorithm for SD which does not include maximizations in function spaces.

## 8.4 ECONOMIC INTERPRETATIONS

In this final section we deal with economic interpretations of the dual problem, both in the deterministic and in the stochastic case. Special attention is paid to the interpretation of the optimal dual solutions $\pi^* = (\pi^*_n, n = 0,\dots,N+1) \in \mathbb{R}^{N+2}$ in the deterministic case and $\pi^* = (\pi^*_n(\xi^n), n = 0,1,\dots,N+1) \in \Pi$ in the stochastic case. First we consider in some detail the interpretation of the optimal dual solution of a "general" convex programming problem, although we use notations coming from the inventory control models discussed in the current chapter. Secondly an attempt is made to find economic interpretations for the specific results of this chapter.

As described in Section 2.3 a dual problem of a (primal) convex programming problem can be seen as derived via a class of perturbations of the primal problem . Then the dual variables are interpreted as "prices" for the perturbations, and their optimal values $\pi^*$ represent an "equilibrium price system" with the property that *if perturbations could be bought at these prices* no perturbation would be profitable. That is, for $\Theta(\eta)$ the optimal value of the primal problem if the perturbation $\eta$ is incorporated, any optimal dual solution $\pi^*$ is characterized by

$$(8.46) \qquad \Theta(0) \leqq \Theta(\eta) + \langle \pi^*, \eta \rangle \text{ for all } \eta \in H,$$

or, equivalently,

$$(8.47) \qquad -\pi^* \in \partial\Theta(0).$$

Here we use the notations corresponding to the stochastic dual pair (SP,SD), but, of course, for the deterministic case and for the general convex programming problem similar characterizations hold. Notice that by definition equilibrium prices are of a *global* nature: no perturbation $\eta$, small or large, positive or negative, is profitable if $\langle \pi^*, \eta \rangle$ has to be paid for it. Negative payments are here and in the sequel interpreted as receipts. Moreover, recall from Theorem 2.1 that for problems with convex perturbation functions $\Theta$ (such as P and SP in the current chapter) equilibrium prices exist if $\Theta$ is bounded above on a neighbourhood of 0.

In spite of the terminology used, equilibrium prices do not necessarily have an *economic* interpretation. An obvious necessary condition is that the primal problem and its class of perturbations must have an economic interpretation: in particular, its objective function should represent the total cost that is to be minimized, and the perturbations should correspond to commodities to which prices can be attached. This condition is satisfied, of course, in the inventory control models P and SP of this chapter. Nevertheless even then the economic interpretation of the optimal dual solutions $\pi^*$ as prices is rather restricted. One of the shortcomings is that they are *fictitious*: they do not tell anything about the market prices for the commodities described by the perturbations; their values are completely endogeneously determined by the possibilities of the shopkeeper, as far as expressed in the models P and SP. In connection to this it is important to realize that if the model changes (even if some parameter gets a different numerical value), the equilibrium prices $\pi^*$ will change as well: either the optimal value function $\Theta$ will change, or, if "only" an actual perturbation $\eta$ would occur, the set of equilibrium prices would change from $-\partial\Theta(0)$ to $-\partial\Theta(\eta)$. Therefore, it would seem that the economic interpretation of the prices $\pi^*$ should be of a local (or *marginal*) nature: they represent unit prices that the shopkeeper is willing to pay for "small" perturbations $\eta$. But even then the economic interpretation is not completely satisfactory, since no distinction is made between "positive" and "negative" perturbations. So we are led to consider the *directional derivatives* $\Theta'(0;\eta)$ at 0 of the optimal value function $\Theta$ for each perturbation $\eta \in H$ separately. For stable convex programs these directional derivatives are completely determined by the subgradient of $\Theta$ at 0 [14 Theorem 17(c)]:

$$(8.48) \qquad \Theta'(0;\eta) = \max\{<-\pi^*,\eta>: -\pi^* \in \partial\Theta(0)\}, \ \eta \in H.$$

These directional derivatives have the obvious interpretation as the (maximum) amount of money the shopkeeper is willing to pay for a marginal perturbation $\eta$. The directional derivatives describe the well-known *shadow prices*. As an illustration, consider in the deterministic inventory control problem P the perturbation $\eta := (0,0,\ldots,0,\eta_n,0,\ldots,0)$ with $\eta_n = \varepsilon > 0$, corresponding to a decrease of the demand $\xi_n$ with $\varepsilon$. Then $\Theta'(0;\eta) = -\varepsilon.\underline{\pi}_n^*$ where

$$\underline{\pi}_n^* := \min\{\pi_n^*: (\pi_0^*,\ldots,\pi_{N+1}^*) \text{ solves D}\}$$

is the shadow price related to a marginal *decrease* of the demand $\xi_n$. Similarly, the shadow price related to a marginal *increase* of the demand $\xi_n$ in the deterministic case is given by

$$\bar{\pi}_n^* := \max\{\pi_n^*: (\pi_0^*,\ldots,\pi_{N+1}^*) \text{ solves D}\}.$$

That is, the opportunity to decrease the demand $\xi_n$ marginally is only advantageous for the shopkeeper if he has to pay for it at a unit price which does not exceed $\underline{\pi}_n^*$, and the opportunity to increase the demand $\xi_n$ marginally is only advantageous if he receives for it at a unit price which is at least $\bar{\pi}_n^*$. We conclude that the shadow prices determined by the optimal dual solution $\pi^*$ represent their economic interpretation.

The inventory control models P and SP discussed in this chapter are not only convex but even *linear* programs. As a consequence the dual problems D and SD, respectively, do not contain the primal variables anymore; they have been eliminated at the cost of additional constraints on the dual variables (see e.g. page 18). In the sequel we give an economic interpretation of the dual *problems* D and SD (inclusive their constraints) in terms of *imputation*. The advantage of working with imputation is that the dual constraints are given a natural meaning. On the other hand it must be noted that, although prices for imputation by definition are fictitious and endogeneous, they do not reflect the marginal meaning of dual variables [12].

Consider first the *deterministic* problem. Suppose that we want to assign the minimal total costs to the demands by way of an (imputed) price $\pi_n$ for the demand $\xi_n$, n = 0,...,N+1. As one way of finding reasonable values for the $\pi_n$ one may consider the outsider's point of view, as it is called by Van de Panne [11]. Suppose a fictitious outsider wants to take over the delivery of the commodity in such a way that in exchange the shopkeeper has to pay the amount of $\pi_n . \xi_n$ for the delivery of $\xi_n$ as soon as this demand arises. The outsider's objective is to maximize $\Sigma_{n=0}^{N+1} \pi_n . \xi_n$. However, in order that the shopkeeper is willing to give up his activities (producing, holding inventory, postponing delivery) the prices $\pi_n$ must satisfy certain conditions. Clearly, $\pi_n \leqq k_n$, $n \geqq 1$; otherwise it is not economic to give up production. Also $\pi_{N+1} \geqq r$, otherwise it would pay for the shopkeeper to order from the outsider large amounts at the end of period N. Moreover, $\pi_n + h_n \geqq \pi_{n+1}$, otherwise it would be preferable for the shopkeeper to buy the amount of $\xi_{n+1}$ already in period n rather than in period n+1. Analogously, $\pi_n \leqq q_n + \pi_{n+1}$. It is easily seen that by the last two constraints it is also not profitable to keep inventory or to keep shortage over a longer time. This completes the interpretation of the dual problem. The optimal solution of the dual problem reflects this interpretation: if all $\xi_n$ are positive, the optimal solution for $\pi_n$ is $\pi_n^*$ which represents the lowest cost associated with satisfying a unit demand in period n. So the solution is only slightly "marginal": it only changes if $\xi_0$ becomes negative.

Consider now the *stochastic* problem. The crucial point is that there is no other information about the future demands except their probability distribution. The dual of the stochastic problem can be reconstructed analogously to the deterministic case. The outsider offers the possibility to deliver the amount of $\xi_n$ of the commodity as soon as this demand is observed. The price $\pi_n$ he asks for this opportunity must be known to the shopkeeper at that time, so $\pi_n = \pi_n(\xi^n)$. In order to make it not economic for the shopkeeper to stay in business, the outsider has to set his prices in such a way that certain constraints are met. In order to discourage production in period n, $\pi_n$ must be competitive compared with $k_n$. Since $\xi_n$ is unknown to the shopkeeper when the decision on the production $x_n$ has to be made, he can base his decision only on his estimate of $\pi_n$, being $\hat{\pi}_n(\xi^{n-1}) := E^{n-1}\pi_n(\xi^n)$. So the appropriate condition is $\hat{\pi}_n(\xi^{n-1}) \leqq k_n$ for all $\xi^{n-1}$. Analogously, to discourage the shopkeeper to hold storage and to postpone delivery, the prices must satisfy $\pi_n(\xi^n) + h_n \geqq \hat{\pi}_{n+1}(\xi^n)$ and $\pi_n(\xi^n) \leqq q_n + \hat{\pi}_{n+1}(\xi^n)$, respectively. Notice that the prices set by the outsider reflect not only the costs of the activities of the shopkeeper, but also the shopkeeper's (incomplete) knowledge about the future demands, forcing him to restrict himself to nonanticipative policies. One would expect that the optimal value of $\hat{\pi}_n(\xi^{n-1})$, which is the relevant number when the shopkeeper considers production in period n, depends on $\xi^{n-1}$ only via the (optimal) inventory at hand and the conditional distribution of $(\xi_n,\xi_{n+1},\ldots,\xi_N)$ given $\xi^{n-1}$. Indeed, we found that $E^{n-1}\pi_n^*(\xi^n) = \pi^{(n)}(s_{n-1}^*(\xi^{n-1}),\xi^{n-1})$ has this property. Furthermore, the behaviour of $\pi^{(n)}(.,\xi^{n-1})$ is compatible with the identification $\hat{\pi}_n(\xi^{n-1}) \equiv E^{n-1}\pi_n^*(\xi^n)$. If s is small enough, e.g. negative, then $\pi^{(n)}(s,\xi^{n-1}) = \pi_n^{(n)}$, the smallest unit cost associated with satisfying the demand in period n by a production in period n or later. We also found $\pi^{(n)}(s,\xi^{n-1}) = k_n$ if $s < y_n^*(\xi^{n-1})$ as might be expected. If $s \to \infty$ the price $\pi^{(n)}(s,\xi^{n-1})$ falls down to its lowest value $-\Sigma_{i=n}^N h_i + r$: this is the scrap value of a unit which is kept in stock from period n onward. We derived these relations from

$$\pi^{(n)}(s,\xi^{n-1}) = -f_n'(s,\xi^{n-1}).$$

This identity is not surprising either, if as above $\pi^{(n)}(s^*_{n-1}(\xi^{n-1}),\xi^{n-1})$ is identified with $\hat{\pi}_n(\xi^{n-1})$. Recall that this is the shopkeeper's estimate of the price for decreasing the demand $\xi_n$, or equivalently for increasing the stock $s^*_{n-1}(\xi^{n-1})$. Since the shopkeeper is supposed to operate optimally, this price is only competing with his own opportunities if for any $\Delta s$

$$f_n(s^*_{n-1}(\xi^{n-1}) + \Delta s,\xi^{n-1}) + \hat{\pi}_n(\xi^{n-1}).\Delta s \leqq f_n(s^*_{n-1}(\xi^{n-1}),\xi^{n-1}),$$

so that $\hat{\pi}_n(\xi^{n-1})$ must be equal to $-f_n'(s^*_{n-1}(\xi^{n-1}))$. We can go further, and indicate an interpretation for

$$\hat{\pi}_n(\xi^{n-1}) = \pi_n^{(n)} - \Sigma_{j=n}^{N+1}\gamma_j.\mu_n^N(D_{nj}(s^*_{n-1}(\xi^{n-1}))|\xi^{n-1}),$$

see (8.41). Recall that the events $D_{nj}(s^*_{n-1}(\xi^{n-1}))$ have something to do with the relation between the current stock $s^*_{n-1}(\xi^{n-1})$ and the sign of the stock in period j, see Remark 8.12. Suppose that $y^*_n(\xi^{n-1}) > -\infty$; then $\pi_n^{(n)} = k_n$. If the current stock is so high that an otherwise certain shortage in period j is avoided, then the price for a marginal increase of the current stock should reflect this. This "explains" the j-th term. The reader who feels that such a far-fetched interpretation does not contribute significantly to her or his insight in the stochastic dynamic production-inventory control problem is right almost surely.

REFERENCES


1. D.P. BERTSEKAS, S.E. SHREVE (1978). *Stochastic Optimal Control; the Discrete Time Case*, Academic Press, New York-San Francisco-London.
2. L. BREIMAN (1968). *Probability*, Addison-Wesley, Reading Ma.
3. W.H. CUNNINGHAM (1976). A network simplex method. *Math. Programming 11*, 105-116.
4. G.B. DANTZIG (1963). *Linear Programming and Extensions*, Princeton University Press, Princeton NJ.
5. M.J. EISNER, P. OLSEN (1975). Duality for stochastic programming

interpreted as L.P. in $L_p$-space. *SIAM J. Appl. Math. 28*, 779-792.

6. W.K. KLEIN HANEVELD (1977). *Markovian Inventory Control Models*, Report Institute of Econometrics OR 7702, University of Groningen, Groningen.
7. W.K. KLEIN HANEVELD (1978). *Markovian Inventory Control Models - Improvements, Extensions*, Report Department of Mathematics, University of Kentucky, Lexington Ky.
8. W.K. KLEIN HANEVELD (1978). *Dual Inventory Control Models*, Report Department of Mathematics, University of Kentucky, Lexington Ky.
9. W.K. KLEIN HANEVELD (1980). A dual of a dynamic inventory control model: the deterministic and stochastic case. P. KALL, A. PREKOPA (eds.). *Recent Results in Stochastic Programming*, Lecture notes in economic and mathematical systems 179, Springer, Berlin-Heidelberg-New York.
10. P. OLSEN (1976). When is a multistage stochastic programming problem well-defined? *SIAM J. Control 14*, 518-527.
11. C. VAN DE PANNE (1976). *Linear Programming and Related Techniques*, second edition, North Holland, Amsterdam.
12. J. PONSTEIN (1983). Applying some modern developments to choosing your own Lagrange multipliers. *SIAM Rev. 25*, 183-199.
13. R.T. ROCKAFELLAR (1970). *Convex Analysis*, Princeton University Press, Princeton NJ.
14. R.T. ROCKAFELLAR (1974). *Conjugate Duality and Optimization*, SIAM, Philadelphia Pa.
15. R.T. ROCKAFELLAR, R.J.-B. WETS (1976). Nonanticipativity and $L_1$-martingales in stochastic optimization problems. *Math. Programming Study 6*, 160-189.
16. R.T. ROCKAFELLAR, R.J.-B. WETS (1978). The optimal recourse problem in discrete time: $L_1$-multipliers for inequality constraints. *SIAM J. Control 16*, 16-36.
17. R.T. ROCKAFELLAR, R.J.-B. WETS (1983). Deterministic and stochastic optimization problems of Bolza type in discrete time. *Stochastics 10*, 273-312.
18. S.E. SHREVE (1977). *Dynamic Programming in Complete Separable Spaces*, Ph.D. dissertation, University of Illinois, Urbana Ill.

# LIST OF OPTIMIZATION PROBLEMS


| Chapter | Problem | Page | Problem | Page | Problem | Page |
|---|---|---|---|---|---|---|
| Chapter 1 | LP | 1 | | | | |
| Chapter 2 | $P_o$ | 9 | D | 12 | LD | 18 |
| | P | 11 | LP | 18 | | |
| Chapter 3 | | | | | | |
| Chapter 4 | $LP_1$ | 52 | $MOB_1$ | 71 | $MA_2$ | 82 |
| | $LP_2$ | 52 | $MOB_2$ | 72 | SDP | 88 |
| | MO | 66 | $NP_1$ | 75 | $SDP_1$ | 91 |
| | $MO_1$ | 67 | $NP_2$ | 75 | $SDP_2$ | 91 |
| | $MO_2$ | 67 | $NP'_2$ | 75 | SDP' | 93 |
| | $\hat{MO}_1$ | 71 | MA | 81 | $\overline{SDP_1}$ | 104 |
| | $\hat{MO}_2$ | 71 | $MA_1$ | 82 | | |
| Chapter 5 | $ICC_1(\beta)$ | 136 | $ICC_2(\alpha)$ | 136 | $ICC_3(\beta)$ | 137 |
| | $L_1(\lambda)$ | 136 | $L_2(\lambda,\alpha)$ | 136 | $L_3(\lambda)$ | 137 |
| Chapter 6 | MIC | 154 | $MIC_n$ | 157 | | |
| Chapter 7 | P | 168 | $D_o$ | 171 | $I\tilde{D}(T)$ | 187 |
| | P' | 168 | $LP_1$ | 178 | $I\tilde{P}(T)$ | 196 |
| | P'' | 168 | $LP_2$ | 178 | $OD(\alpha)$ | 184 |
| | P''' | 168 | $OP(\alpha)$ | 183 | $O\tilde{D}(\alpha)$ | 208 |
| | D | 170 | IP(T) | 183 | $O\tilde{P}(\alpha)$ | 209 |
| | $P_o$ | 171 | ID(T) | 185 | | |
| Chapter 8 | $P_o$ | 222 | D | 230 | SP | 261 |
| | P | 228,229 | $SP_o$ | 244 | SD | 262,263 |

# SUBJECT INDEX


action space 86
adjusted distance 237

backlogging 222
biconjugate function 14
bilinear form 51
bounding function 61
branching 232

Caratheodory map 149
chance constraint (CC) 26
Chebychev approximation 69
circulation 196
compatible topology 13,55
complementary slackness 17,19,53
concave conjugate function 14
conditional expectation 246
conjugate dual DP method 237,282
conjugate dual problem 12,55
conjugate function 13
constraint 10
constraint set 87
continuous increase 135
convex conjugate function 13
convex ordering 193
convex programming 10
cost-to-go function 95,223
criterion function 25,58,166
critical activity 176
critical path 176
criticality index 182

decision rule 87
deterministic equivalent 40
differentiable form DP 226,269
Dijkstra tree 232
distance 235
distribution problem 168
disturbance 141
dual DP method 236,281
dual pair 52
dual problem 12,52
duality 51
duality gap 13
dynamic production model 173
dynamic programming (DP)
algorithm (DPA) 95,155,224,252
dynamic version 40

economic equivalence 36
epigraph 13
epigraph multifunction 148
equilibrium price 13,285
expected loss constraints 38
extremely risky set 121
extremely safe set 117

feasible 10,11
flow 177,194
flow problem 167

generated (by DPA) policy 96
goal programming 31

growth condition 145

imputation 287
induced constraint 41
infcompactness 148
inner problem 165
integrated chance constr. (ICC) 37,116

Karush-Kuhn-Tucker condition 17

Lagrangean function 15
level policy 161,257
linear programming 10,18,52
lower bound condition 143
lower semianalyticity 252
lower semicontinuity 148
L-superadditivity 172

marginal problem 58,80,168
Markovian inventory control 153
Markovian policy 155
mathematical equivalence 35
mathematical programming 9
measurable multifunction 148
minimax approach SLP 165
moment problem 58,66
multi-attribute utility function 34
multiple simple recourse 32
multistage SLP 40

negative dynamic programming 145
Neyman-Pearson lemma 74
nice regular conditional distr. 251
nonanticipativity 41
normal integrand 148
normality 13,55
NW(Corner Rule)distribution 172
NW(Corner Rule)solution 172

objective function 10
optimal inventory level 223,258
optimal solution 10
optimal tree 232
optimal value 10
optimal value function 12
outer problem 165

penalty cost 26
PERT 214
perturbation 11
perturbation function 12
policy 87,154,244,257
positive cone 51
positive quadrant-dependence 175
primal problem 11
production policy 224
project 166,175
project completion time 176
project delay function 180,183
projection 65
proper behaviour DPA 42,141,155
proper inventory model 222

qualitative risk measure 36
quantitative risk measure 36

rearrangement 169
recourse 26
refined penalty cost 29
regular conditional distr. 247
related flows 196
robustness 166

saddle-point 15
saddle-value 15
semi-infinite linear programming 69
separable function 83,169

separated duality 52
shadow price 286
similarity transformation 107
simple recourse 31
Slater's constraint qualification 17
stability 14
stage 86
state space 86
static version 40
stochastic linear prog. (SLP) 22
stochastic dyn. progr. (SDP) 42,58,86
strict increase 124
strict properness 245
subgradient 15,240
system function 141

total flow 177
total variation 60
transition function 63
transition measure 63
transition probability 63,247
transportation problem 171
transshipment problem 228

universal measurability 249
utility function 28

value function 154,223
vector lattice 59

weak duality 13
weighted integrated CC 38
worst-case distribution 204
worst-case marginal 211

Vol. 184: R. E. Burkard and U. Derigs, Assignment and Matching Problems: Solution Methods with FORTRAN-Programs. VIII, 148 pages. 1980.

Vol. 185: C. C. von Weizsäcker, Barriers to Entry. VI, 220 pages. 1980.

Vol. 186: Ch.-L. Hwang and K. Yoon, Multiple Attribute Decision Making – Methods and Applications. A State-of-the-Art-Survey. XI, 259 pages. 1981.

Vol. 187: W. Hock, K. Schittkowski, Test Examples for Nonlinear Programming Codes. V. 178 pages. 1981.

Vol. 188: D. Bös, Economic Theory of Public Enterprise. VII, 142 pages. 1981.

Vol. 189: A. P. Lüthi, Messung wirtschaftlicher Ungleichheit. IX, 287 pages. 1981.

Vol. 190: J. N. Morse, Organizations: Multiple Agents with Multiple Criteria. Proceedings, 1980. VI, 509 pages. 1981.

Vol. 191: H. R. Sneessens, Theory and Estimation of Macroeconomic Rationing Models. VII, 138 pages. 1981.

Vol. 192: H. J. Bierens: Robust Methods and Asymptotic Theory in Nonlinear Econometrics. IX, 198 pages. 1981.

Vol. 193: J. K. Sengupta, Optimal Decisions under Uncertainty. VII, 156 pages. 1981.

Vol. 194: R. W. Shephard, Cost and Production Functions. XI, 104 pages. 1981.

Vol. 195: H. W. Ursprung, Die elementare Katastrophentheorie. Eine Darstellung aus der Sicht der Ökonomie. VII, 332 pages. 1982.

Vol. 196: M. Nermuth, Information Structures in Economics. VIII, 236 pages. 1982.

Vol. 197: Integer Programming and Related Areas. A Classified Bibliography. 1978 – 1981. Edited by R. von Randow. XIV, 338 pages. 1982.

Vol. 198: P. Zweifel, Ein ökonomisches Modell des Arztverhaltens. XIX, 392 Seiten. 1982.

Vol. 199: Evaluating Mathematical Programming Techniques. Proceedings, 1981. Edited by J.M. Mulvey. XI, 379 pages. 1982.

Vol. 200: The Resource Sector in an Open Economy. Edited by H. Siebert. IX, 161 pages. 1984.

Vol. 201: P. M. C. de Boer, Price Effects in Input-Output-Relations: A Theoretical and Empirical Study for the Netherlands 1949–1967. X, 140 pages. 1982.

Vol. 202: U. Witt, J. Perske, SMS – A Program Package for Simulation and Gaming of Stochastic Market Processes and Learning Behavior. VII, 266 pages. 1982.

Vol. 203: Compilation of Input-Output Tables. Proceedings, 1981. Edited by J. V. Skolka. VII, 307 pages. 1982.

Vol. 204: K.C. Mosler, Entscheidungsregeln bei Risiko: Multivariate stochastische Dominanz. VII, 172 Seiten. 1982.

Vol. 205: R. Ramanathan, Introduction to the Theory of Economic Growth. IX, 347 pages. 1982.

Vol. 206: M.H. Karwan, V. Lotfi, J. Telgen, and S. Zionts, Redundancy in Mathematical Programming. VII, 286 pages. 1983.

Vol. 207: Y. Fujimori, Modern Analysis of Value Theory. X, 165 pages. 1982.

Vol. 208: Econometric Decision Models. Proceedings, 1981. Edited by J. Gruber. VI, 364 pages. 1983.

Vol. 209: Essays and Surveys on Multiple Criteria Decision Making. Proceedings, 1982. Edited by P. Hansen. VII, 441 pages. 1983.

Vol. 210: Technology, Organization and Economic Structure. Edited by R. Sato and M.J. Beckmann. VIII, 195 pages. 1983.

Vol. 211: P. van den Heuvel, The Stability of a Macroeconomic System with Quantity Constraints. VII, 169 pages. 1983.

Vol. 212: R. Sato and T. Nôno, Invariance Principles and the Structure of Technology. V, 94 pages. 1983.

Vol. 213: Aspiration Levels in Bargaining and Economic Decision Making. Proceedings, 1982. Edited by R. Tietz. VIII, 406 pages. 1983.

Vol. 214: M. Faber, H. Niemes und G. Stephan, Entropie, Umweltschutz und Rohstoffverbrauch. IX, 181 Seiten. 1983.

Vol. 215: Semi-Infinite Programming and Applications. Proceedings, 1981. Edited by A.V. Fiacco and K.O. Kortanek. XI, 322 pages. 1983.

Vol. 216: H.H. Müller, Fiscal Policies in a General Equilibrium Model with Persistent Unemployment. VI, 92 pages. 1983.

Vol. 217: Ch. Grootaert, The Relation Between Final Demand and Income Distribution. XIV, 105 pages. 1983.

Vol. 218: P. van Loon, A Dynamic Theory of the Firm: Production, Finance and Investment. VII, 191 pages. 1983.

Vol. 219: E. van Damme, Refinements of the Nash Equilibrium Concept. VI, 151 pages. 1983.

Vol. 220: M. Aoki, Notes on Economic Time Series Analysis: System Theoretic Perspectives. IX, 249 pages. 1983.

Vol. 221: S. Nakamura, An Inter-Industry Translog Model of Prices and Technical Change for the West German Economy. XIV, 290 pages. 1984.

Vol. 222: P. Meier, Energy Systems Analysis for Developing Countries. VI, 344 pages. 1984.

Vol. 223: W. Trockel, Market Demand. VIII, 205 pages. 1984.

Vol. 224: M. Kiy, Ein disaggregiertes Prognosesystem für die Bundesrepublik Deutschland. XVIII, 276 Seiten. 1984.

Vol. 225: T.R. von Ungern-Sternberg, Zur Analyse von Märkten mit unvollständiger Nachfragerinformation. IX, 125 Seiten. 1984

Vol. 226: Selected Topics in Operations Research and Mathematical Economics. Proceedings, 1983. Edited by G. Hammer and D. Pallaschke. IX, 478 pages. 1984.

Vol. 227: Risk and Capital. Proceedings, 1983. Edited by G. Bamberg and K. Spremann. VII, 306 pages. 1984.

Vol. 228: Nonlinear Models of Fluctuating Growth. Proceedings, 1983. Edited by R.M. Goodwin, M. Krüger and A. Vercelli. XVII, 277 pages. 1984.

Vol. 229: Interactive Decision Analysis. Proceedings, 1983. Edited by M. Grauer and A.P. Wierzbicki. VIII, 269 pages. 1984.

Vol. 230: Macro-Economic Planning with Conflicting Goals. Proceedings, 1982. Edited by M. Despontin, P. Nijkamp and J. Spronk. VI, 297 pages. 1984.

Vol. 231: G.F. Newell, The M/M/∞ Service System with Ranked Servers in Heavy Traffic. XI, 126 pages. 1984.

Vol. 232: L. Bauwens, Bayesian Full Information Analysis of Simultaneous Equation Models Using Integration by Monte Carlo. VI, 114 pages. 1984.

Vol. 233: G. Wagenhals, The World Copper Market. XI, 190 pages. 1984.

Vol. 234: B.C. Eaves, A Course in Triangulations for Solving Equations with Deformations. III, 302 pages. 1984.

Vol. 235: Stochastic Models in Reliability Theory. Proceedings, 1984. Edited by S. Osaki and Y. Hatoyama. VII, 212 pages. 1984.

Vol. 236: G. Gandolfo, P.C. Padoan, A Disequilibrium Model of Real and Financial Accumulation in an Open Economy. VI, 172 pages. 1984.

Vol. 237: Misspecification Analysis. Proceedings, 1983. Edited by T.K. Dijkstra. V, 129 pages. 1984.

Vol. 238: W. Domschke, A. Drexl, Location and Layout Planning. IV, 134 pages. 1985.

Vol. 239: Microeconomic Models of Housing Markets. Edited by K. Stahl. VII, 197 pages. 1985.

Vol. 240: Contributions to Operations Research. Proceedings, 1984. Edited by K. Neumann and D. Pallaschke. V, 190 pages. 1985.

Vol. 241: U. Wittmann, Das Konzept rationaler Preiserwartungen. XI, 310 Seiten. 1985.

Vol. 242: Decision Making with Multiple Objectives. Proceedings, 1984. Edited by Y. Y. Haimes and V. Chankong. XI, 571 pages. 1985.

Vol. 243: Integer Programming and Related Areas. A Classified Bibliography 1981–1984. Edited by R. von Randow. XX, 386 pages. 1985.

Vol. 244: Advances in Equilibrium Theory. Proceedings, 1984. Edited by C. D. Aliprantis, O. Burkinshaw and N. J. Rothman. II, 235 pages. 1985.

Vol. 245: J. E. M. Wilhelm, Arbitrage Theory. VII, 114 pages. 1985.

Vol. 246: P. W. Otter, Dynamic Feature Space Modelling, Filtering and Self-Tuning Control of Stochastic Systems. XIV, 177 pages. 1985.

Vol. 247: Optimization and Discrete Choice in Urban Systems. Proceedings, 1983. Edited by B. G. Hutchinson, P. Nijkamp and M. Batty. VI, 371 pages. 1985.

Vol. 248: Plural Rationality and Interactive Decision Processes. Proceedings, 1984. Edited by M. Grauer, M. Thompson and A. P. Wierzbicki. VI, 354 pages. 1985.

Vol. 249: Spatial Price Equilibrium: Advances in Theory, Computation and Application. Proceedings, 1984. Edited by P. T. Harker. VII, 277 pages. 1985.

Vol. 250: M. Roubens, Ph. Vincke, Preference Modelling. VIII, 94 pages. 1985.

Vol. 251: Input-Output Modeling. Proceedings, 1984. Edited by A. Smyshlyaev. VI, 261 pages. 1985.

Vol. 252: A. Birolini, On the Use of Stochastic Processes in Modeling Reliability Problems. VI, 105 pages. 1985.

Vol. 253: C. Withagen, Economic Theory and International Trade in Natural Exhaustible Resources. VI, 172 pages. 1985.

Vol. 254: S. Müller, Arbitrage Pricing of Contingent Claims. VIII, 151 pages. 1985.

Vol. 255: Nondifferentiable Optimization: Motivations and Applications. Proceedings, 1984. Edited by V. F. Demyanov and D. Pallaschke. VI, 350 pages. 1985.

Vol. 256: Convexity and Duality in Optimization. Proceedings, 1984. Edited by J. Ponstein. V, 142 pages. 1985.

Vol. 257: Dynamics of Macrosystems. Proceedings, 1984. Edited by J.-P. Aubin, D. Saari and K. Sigmund. VI, 280 pages. 1985.

Vol. 258: H. Funke, Eine allgemeine Theorie der Polypol- und Oligopolpreisbildung. III, 237 pages. 1985.

Vol. 259: Infinite Programming. Proceedings, 1984. Edited by E. J. Anderson and A. B. Philpott. XIV, 244 pages. 1985.

Vol. 260: H.-J. Kruse, Degeneracy Graphs and the Neighbourhood Problem. VIII, 128 pages. 1986.

Vol. 261: Th. R. Gulledge, Jr., N. K. Womer, The Economics of Made-to-Order Production. VI, 134 pages. 1986.

Vol. 262: H. U. Buhl, A Neo-Classical Theory of Distribution and Wealth. V, 146 pages. 1986.

Vol. 263: M. Schäfer, Resource Extraction and Market Structure. XI, 154 pages. 1986.

Vol. 264: Models of Economic Dynamics. Proceedings, 1983. Edited by H.F. Sonnenschein. VII, 212 pages. 1986.

Vol. 265: Dynamic Games and Applications in Economics. Edited by T. Başar. IX, 288 pages. 1986.

Vol. 266: Multi-Stage Production Planning and Inventory Control. Edited by S. Axsäter, Ch. Schneeweiss and E. Silver. V, 264 pages. 1986.

Vol. 267: R. Bemelmans, The Capacity Aspect of Inventories. IX, 165 pages. 1986.

Vol. 268: V. Firchau, Information Evaluation in Capital Markets. VII, 103 pages. 1986.

Vol. 269: A. Borglin, H. Keiding, Optimality in Infinite Horizon Economies. VI, 180 pages. 1986.

Vol. 270: Technological Change, Employment and Spatial Dynamics. Proceedings 1985. Edited by P. Nijkamp. VII, 466 pages. 1986.

Vol. 271: C. Hildreth, The Cowles Commission in Chicago, 1939–1955. V, 176 pages. 1986.

Vol. 272: G. Clemenz, Credit Markets with Asymmetric Information. VIII, 212 pages. 1986.

Vol. 273: Large-Scale Modelling and Interactive Decision Analysis. Proceedings, 1985. Edited by G. Fandel, M. Grauer, A. Kurzhanski and A. P. Wierzbicki. VII, 363 pages. 1986.

Vol. 274: W.K. Klein Haneveld, Duality in Stochastic Linear and Dynamic Programming. VII, 295 pages. 1986.